KV-013-682
£4.75p

GCSE Biology

QuickCheck Study Guides

Michael Roberts
Formerly Head of Biology,
Marlborough College

KINGSTON COLLEGE OF FURTHER EDUCATION LIBRARY

Nelson

Acc. No. 00002843
Class No. 574
Date Rec. 3.3.89
Order No. T484

Thomas Nelson and Sons Ltd
Nelson House Mayfield Road
Walton-on-Thames Surrey
KT12 5PL UK

51 York Place
Edinburgh
EH1 3JD UK

Thomas Nelson (Hong Kong) Ltd
Toppan Building 10/F
22A Westlands Road
Quarry Bay Hong Kong

Distributed in Australia by

Thomas Nelson Australia
480 La Trobe Street
Melbourne Victoria 3000
and in Sydney, Brisbane, Adelaide and Perth

© Michael Roberts, 1988

First published by Thomas Nelson and Sons Ltd 1988

ISBN 0-17-448153-5

NPN 987654321

Printed in Great Britain by Butler & Tanner Ltd,
Frome and London

All Rights Reserved. This publication is protected in the United Kingdom by the Copyright Act 1956 and in other countries by comparable legislation. No part of it may be reproduced or recorded by any means without the permission of the publisher. This prohibition extends (with certain very limited exceptions) to photocopying and similar processes, and written permission to make a copy or copies must therefore be obtained from the publisher in advance. It is advisable to consult the publisher if there is any doubt regarding the legality of any proposed copying.

Contents

Introduction

How to use this book

This book has been written to help you prepare for your GCSE Biology Examination. It is best to use this book towards the end of your biology course, when you are ready to start revising for the exam.

The book provides all the key ideas needed for your exam, advice on how to study the topics, help in revising and preparing for the exam and specimen questions to give you practice in answering exam questions. In addition, a list of the topics included in the different GCSE syllabuses is included in the form of a QuickCheck Syllabus Grid.

Using the QuickCheck Syllabus Grid

The book covers all GCSE biology syllabuses, so there will be a number of topics in the book which will not be needed for your particular exam. You can check which topics are in your GCSE syllabus by using the QuickCheck Syllabus Grid. Select the topics you need by referring to the symbols in the grid:

● means the whole topic is included
○ means part of the topic is included

Where neither of these symbols appears against a topic, it means that the topic is not included in your syllabus.

The addresses of the examination boards are given after this Introduction. You can write to them for copies of past examination papers and the full syllabus for your course.

Studying the topics

Start by reading the advice on the following pages on preparing and revising for the examination. Then work your way through the book, studying the chapters one at a time.

Within each chapter most of the headings are in the form of questions. You can test your factual knowledge of the chapter by covering the page with a card and lowering it line by line, as shown in figure 1. Answer each question as you uncover it.

Answer the questions at the end of each chapter before you go on to the next chapter. The questions will test your knowledge and understanding of each topic, and will also give you practice in the skills which you will need for the examination.

So that you can use the questions again for revision, it is better not to write the answers in the book. Instead, write them out separately. You can then see how you've done by looking up the answers on pages 142 to 161. Go back to the chapter notes to check on anything you did not understand or remember.

Figure 1 *Testing your factual knowledge.*

The sort of questions you are likely to get in the actual exam are described on pages 136 to 141. Examples are given from specimen GCSE papers. Answering these questions will give you practice in exam technique, and the questions cover all the skills tested in the GCSE exam.

It is best to do these questions after you have gone through all the topics in the book. You will find the answers on pages 162 to 164, with some advice on how to arrive at them.

How to prepare and revise for the GCSE examination

The GCSE examination aims to test what you *know*, *understand* and *can do*. What you can do is tested (assessed) regularly during your GCSE course by your teacher. What you know and understand is tested by the externally assessed written examination at the end of the course. These notes are intended to help you prepare for the written examination.

Knowing and understanding the facts

Knowing the facts means remembering them, and that means learning them. Suppose you have just covered a topic in class and you want to learn it. This is what you should do:

(1) **Read the topic in your notes and/or textbook.** Be sure you *understand* it. Trying to learn facts that you

don't understand is a complete waste of time. Conversely, if you do understand the facts, you will find them much easier to remember.

(2) **Make a summary of the topic.** Here are some useful methods that you can use:

(a) **Lists.** A list of key facts and key words is a useful way of remembering things. For each fact write down an important word or technical term which sums it up and helps you to remember the details. An example is the list of characteristics of life on page 3. The important words to remember in this list are: respond, grow, feed, energy, waste, reproduce, cells.

(b) **Flow charts.** A flow chart is a series of short statements connected by arrows. It is used for summarising a sequence of events. As with lists, include important words and technical terms. An example is figure 1.1 on page 1. Important words to remember in this flow chart are: observation, hypothesis, experiment.

(c) **Diagrams.** Diagrams are a particularly useful aid to learning because most people find it easiest to remember things in pictures. Make your diagrams simple and easy to remember. Label them as fully as necessary. Write the labels well clear of the diagram so that you can cover them up later and test yourself. If necessary write short notes by the labels. A labelled diagram with notes is called an annotated diagram. A lot of useful information can be packed into an annotated diagram. An example is Figure 14.4 on page 67.

Keep all your lists, flow charts, diagrams and summaries so that you can use them again.

(3) **Learn your summaries by heart.** The aim is to put the information contained in your summaries into your long-term memory, so that you'll never forget it — at least not until the exam is over! But do make sure that you understand it first.

NEVER LEARN ANYTHING WITHOUT UNDERSTANDING IT FIRST

Remember that the GCSE examination tests what you know and *understand*.

Basic skills

In addition to testing your knowledge and understanding, the GCSE examination tests certain skills which you should have mastered during your biology course. These skills include:

- observing accurately
- measuring things
- putting forward hypotheses
- using certain pieces of apparatus, including the microscope
- designing and carrying out experiments
- recording information and tabulating data
- drawing graphs (including bar graphs)
- getting information from tables and graphs
- carrying out simple calculations
- drawing conclusions
- communicating information and ideas to other people

For detailed advice on the skills required for GCSE Biology see *Doing Biology, A Handbook of Skills for GCSE*, by Susan Tresman (Nelson, 1987).

Be sure that during your revision programme you get plenty of practice at using these skills. One of the best ways of achieving this is to do the questions at the end of each chapter in this book, and the questions on pages 136 to 141.

How can I learn?

Whether it's a list, flow chart or diagram, the method is the same. Suppose you are trying to memorise a list. This is what you should do:

(1) Place the list in front of you, with a sheet of blank paper beside it.

(2) Copy out the key facts in the list, making a real effort to remember each one. Copying out is an essential part of learning. It makes it an active process, and the very act of writing (or drawing in the case of diagrams) helps to fix the facts in your long-term memory. This is where so many people go wrong; they just stare passively at their notes in the hope that they will sink in. They won't!

DON'T JUST SIT THERE, DO SOMETHING!

(3) Put your copy to one side, take another sheet of paper and copy out the list again, but this time try to do as much of it as you can without looking at the original.

(4) Repeat the process, but this time look at the original even less than before.

(5) Continue repeating the process until you can write out the list without looking at the original at all.

Next day . . .

(6) Try writing out the list again, without looking. Unless you're a genius you'll probably have to look at parts of it now and again to refresh your memory. Don't be discouraged! You can't expect it all to go into your long-term memory straight away. Just repeat the procedure outlined above every day until the information sticks. Then, once a week or so, try writing out the list again to check that you haven't forgotten it and to reinforce it in your memory. This self-testing is another essential part of learning. It also teaches you to recall things quickly. After all, it's no use storing information in your brain if you can't get it out again later!

Some tricks to help you remember

All sorts of tricks can be used to help you remember things. It's best to use a variety of them. Here are some examples:

(1) When learning a list, make the first letters in the list form a word which you can memorise easily. Hopefully, each letter of the word will jog your memory and enable you to quickly recall the list (figure 2).

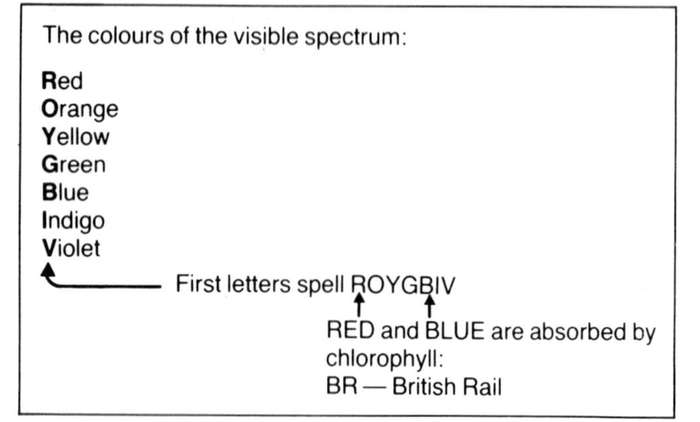

Figure 2 *How to remember a list.*

(2) Make up a silly phrase or sentence from the first letters in the list, then memorise it. This is called a mnemonic (pronounced nemonic) (figure 3). Mnemonics are used a lot by medical students to help them remember the parts of the body. If possible use mnemonics with strong visual images—this makes them easier to remember (figure 4). Don't spend too long thinking up mnemonics—to be useful they should be short and simple and come to mind quickly.

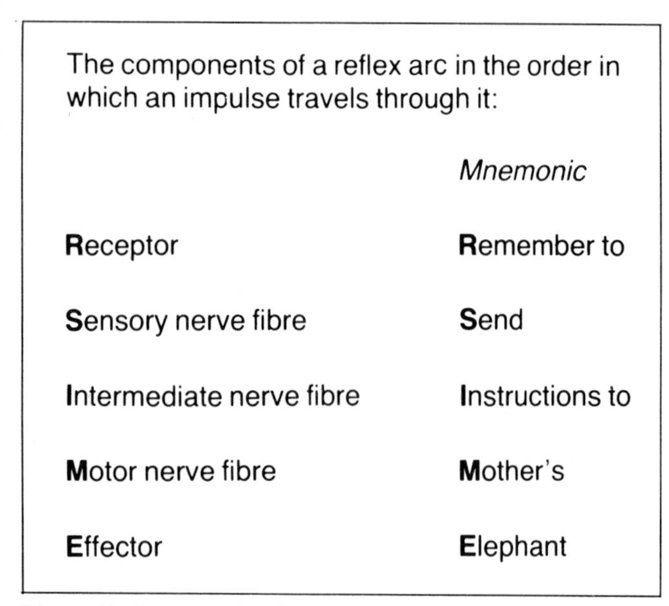

The components of a reflex arc in the order in which an impulse travels through it:

	Mnemonic
Receptor	**R**emember to
Sensory nerve fibre	**S**end
Intermediate nerve fibre	**I**nstructions to
Motor nerve fibre	**M**other's
Effector	**E**lephant

Figure 3 *An example of a mnemonic.*

Figure 4 *A visual picture to go with the mnemonic in figure 3 (from an idea by Pat McGuire).*

(3) In the case of diagrams there are many ways of remembering what goes where. Here's an example. The eye contains two fluids, aqueous humour in front of the lens and vitreous humour behind the lens. How can you remember which is which? Well, aqueous is in front of vitreous just as a is in front of v in the alphabet.

(4) It's often helpful to know the origin of words, particularly technical terms, from Latin or Greek. Take aqueous humour, for example. This fluid is watery. How can you remember this? Well, the word aqueous comes from the Latin word aqua, meaning water. 'But I've never done any Latin', you say. It doesn't matter; just think of other English words with the same meaning, e.g. aquarium, aqualung, aquaduct—they all have something to do with water.

(5) By far the best way of remembering things is to think about what you're learning. Be constantly asking yourself why things are as they are. If you understand the reason for a fact, you'll find it easier to remember that fact. Going back to the eye, why is the eyeball filled with a jelly-like fluid rather than a watery fluid? It's because the function of the fluid is to maintain the shape of the eye and this is best done by a jelly-like fluid. Common sense really, isn't it? A lot of biology is just plain common sense.

Planning and carrying out your revision

Revising for an examination requires organisation and self-discipline. Here are some hints on how to go about it:

(1) Put aside a regular amount of time on certain days for your biology, and stick to it.

(2) Sit in the same place each time, preferably at a desk in a room of your own.

(3) Some people like listening to music while they are working. Okay, but keep it in the background, and listen to tapes rather than the radio to avoid DJs interrupting your train of thought.

(4) Keep your desk tidy at all times; you can't work effectively in a mess.

(5) Don't day-dream. Day-dreaming is most tempting when you're tired so don't go to too many late-night parties!

(6) Avoid being distracted. Don't look out of the window too often, and keep your younger brother out of your room.

(7) Plan how you're going to spend the time; be sure to spend part of the time testing yourself on work covered in previous sessions.

(8) Have a short break at least once an hour: working for too long without a break is as bad as having too many breaks. You will also find that after a short break you can concentrate better and think more clearly.

(9) Keep physically fit. If you don't play a game such as football or tennis, be sure to take the dog for a walk every day.

(10) When you do take a break let it be a complete break: forget all about GCSE!

There is a lot of information in this book, much of it in the form of tables. You don't need to *know* all of it. The QuickCheck Syllabus Grid on the following pages will give you a general idea of what you need to know, but for the detail you must examine your syllabus carefully and look at specimen or past questions.

Use your discretion before you start memorising information.

QuickCheck Syllabus Grid

CHAPTER	TOPIC	LEAG A and B	LEAG C	MEG A	MEG B	NEA	NISEC	SEB	SEG	WJEC
	Exam board Syllabus									
1	Introducing living things	●	●	●	●	●	●	●	●	●
2	The variety of life	○	○	●	○	○	○	●	○	○
3	Habitats: definitions and features	●	●	●	●	●	●	●	●	●
3	Colonisation and succession		●	○	●		○	○	●	○
3	Soil study	A only	●	●	○		●	○	can choose	
3	Humans and environment	●	●	●	●	●	●	●	●	●
4	Energy flow and cycling of matter	●	●	●	●	●	●	●	●	●
5	Estimating and sampling populations	●	●	○	●		○	●	●	
5	Population growth and control	○	●	●	●	●	●	●	●	●
6	Micro-organisms, pests and disease	●	●	●	●	●	●	●	●	●
	(except stopping food going bad)		○				●	●	●	●
7	Cells, tissues and organs	○	○	○	○	○	○	○	○	○
8	Movement in and out of cells	●	●	●	●	●	●	●	●	●
9	The chemistry of life	●	●	●	●	●	●	●	●	●
10	Aerobic respiration	●	●	●	●	●	●	●	●	●
10	ATP		●	●	●					
10	Anaerobic respiration		●	●	●	●	●	●	●	●
11	Nutrition in animals	●	●	●	●	●	●	○	●	●
	(except Minerals and vitamins)	○	○	○	○	○	○		○	○
12	Feeding and digestion	●	●	●	●	●	●	●	●	●
	(except Teeth)	●	●	●	●	○	●	●	○	●
12	Tooth decay	●	●	●	●		●		●	●
12	Herbivores and carnivores		●		●			●		
12	Feeding in other animals		○	○	○			○		
13	Breathing	●	●	●	●	●	●	●	●	●
14	Circulation	●	●	●	●	●	●	●	●	●
	(except Blood groups)					●			●	
14	Named blood vessels	●	●	●	●	●	●	●		
14	Lymph/lymphatic system	○			●		○		○	○
15	Nutrition in plants	○	○	○	○	○	○		○	○

16	Photosynthesis (basic)	●	●	●	●	●	●	●	●	●
16	Photosynthesis (additional)	○	○	○	●	○	○	○		
17	Structure and functions of leaves	●	●	●	●	●	●	●	●	●
17	Stems and roots	○	●	●	○	○	●	○	○	●
18	Homeostasis/feedback		●	●	●	●	○		●	
18	The skin	●	●	●	●	●	●		●	●
18	The liver	○	●	○	○	●	●	●	●	●
18	Excretion and osmo-regulation	●	●	●	●	●	●	●	●	●
19	Nervous coordination	●	●	●	●	●	●	●	●	●
	(except Brain)		○		○		○	●	●	○
19	Taxis				●	●		●		
19	Drugs		●		●				●	
20	Receptors in general	●	●	●	●	●	●	●	●	●
20	The eye	●	●	●	●	●	●			●
20	The ear		●		●			●		
21	Glands and hormones	○	●	●	●	○	●	○	●	●
22	Muscles and movement	○	○	○	○	○	○	○	●	○
23	Producing offspring	●	○	●	○	○	○		●	
24	Sexual reproduction in humans	●	●	●	●	●	●	●	●	●
24	Menstrual cycle	●	●	●		●	●		●	●
24	Contraception			●	●	●			●	
24	Sexually transmitted diseases	○		○	○		○		○	○
25	Reproduction of the flowering plant	●	●	●	●	●	●	●	●	●
26	Measuring growth and growth rates		○	●	○	●	●		○	
26	Phototropism	●	●	●	●	●	●	●	●	●
26	Development in other animals				○	○		○		
27	Chromosomes and genes	●	●	●	●	●	●	●	●	●
	(except Genetic engineering)		●	●	●	●		●	●	
28	Heredity	○	○	○	○	○	○	○	●	○
29	Variation	●	●	●	●	●	●	●	●	●
	(except Evolution/Darwin)		●	●	●	●	●			●

Examination boards

London and East Anglian Group (LEAG)
c/o University of London Schools Examination Board
Stewart House
London WC1B 5DN

Southern Examining Group (SEG)
c/o University of Oxford Delegacy of Local Examinations
Ewert Place
Summertown
Oxford OX2 7BZ

Welsh Joint Education Committee (WJEC)
245 Western Avenue
Cardiff CF5 2YX

Northern Examining Association (NEA)
c/o Joint Matriculation Board
Manchester M15 6EU

Midland Examining Group (MEG)
c/o Oxford and Cambridge Schools Examination Board
Elsfield Way
Oxford OX2 8EP

Northern Ireland Schools Examination Council (NI)
Beechill House
42 Beechill Road
Belfast BT8 4RS

Scottish Examinations Board (SEB)
Ironmills Road
Dalkeith
Midlothian EH22 1BR

Acknowledgements

We are grateful to the following examining bodies for permission to reproduce questions from sample GCSE papers: the London and East Anglian Group (LEAG); the Southern Examining Group (SEG); the Welsh Joint Education Committee (WJEC); the Northern Examining Association (NEA), which is made up of the Associated Lancashire Schools Examining Board, Joint Matriculation Board, North Regional Examinations Board, North West Regional Examinations Board, Yorkshire and Humberside Regional Examinations Board; and the Midland Examining Group (MEG).

Introducing living things

What is biology?

Biology is the study of living organisms. (An **organism** is an object which possesses the characteristics of **life** – see page 3).

The branches of biology

These are some of the main branches of biology

- **Zoology**: the study of animals.
- **Botany**: the study of plants.
- **Microbiology**: the study of very small organisms.
- **Anatomy**: the study of the structure of organisms.
- **Physiology**: the study of how the body works.
- **Nutrition**: the study of food and how organisms feed.
- **Heredity** (genetics): the study of how characteristics are passed from parents to offspring.
- **Ecology**: the study of where organisms live.

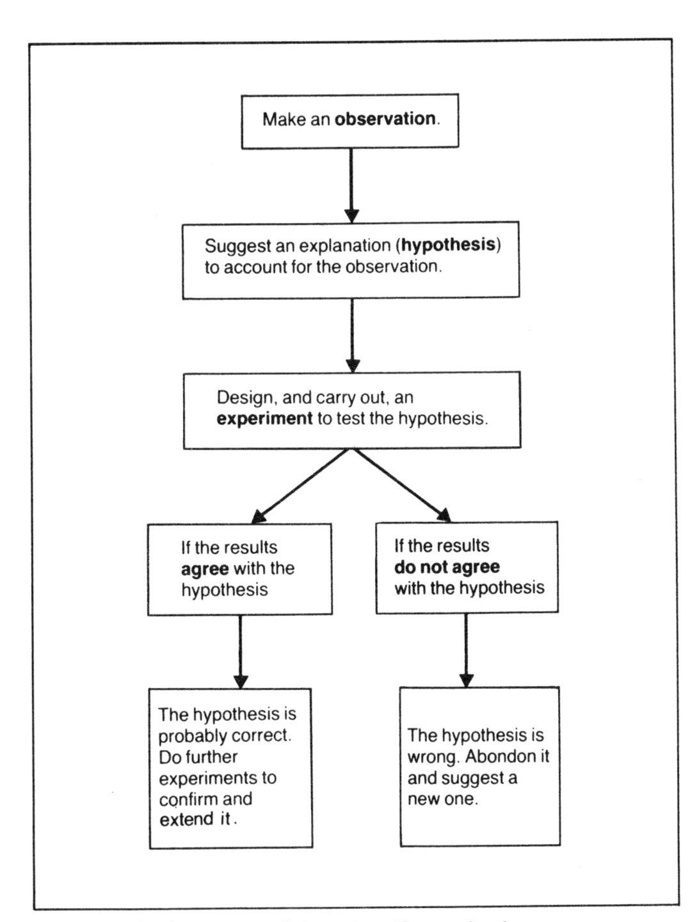

Figure 1.1 *Summary of the scientific method.*

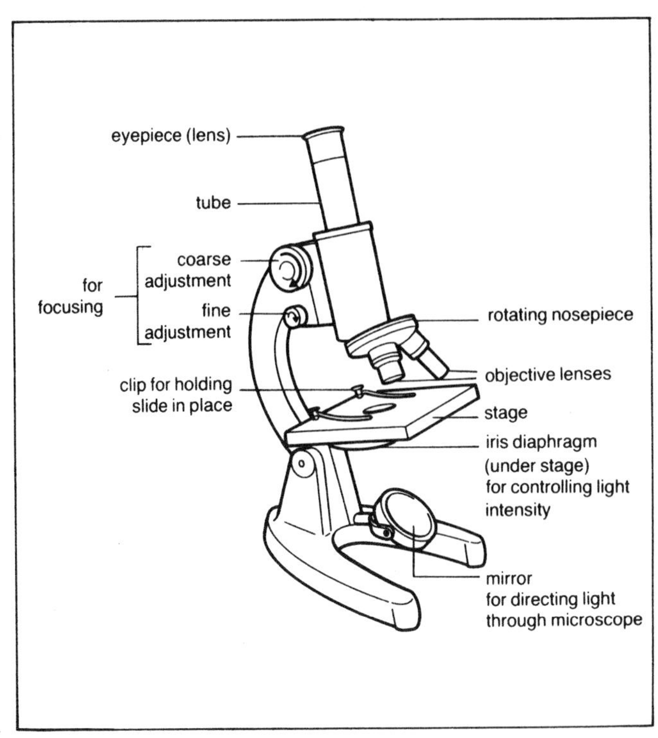

Figure 1.2 *The main parts of a typical light microscope.*

How do biologists work?

As with other scientists, biologists use the **scientific method** (figure 1.1).

The scientific method involves carrying out **experiments**. Every experiment should have a **control**. The control is a standard with which you can compare the result of an experiment.

Observing small objects

Biologists often have to examine small organisms, or parts of organisms. To do this they *magnify* the object by means of a **hand lens** (magnifying glass) or **microscope**.

There are two main types of microscope, the light (optical) microscope and the electron microscope.

The parts of a typical light microscope are shown in figure 1.2.

Magnification

The **magnification** achieved by a lens or microscope is the number of times larger the image of an object is than the object itself.

Resolving power

The maximum effective magnification which a lens or microscope can achieve is determined by its **resolving power**. The resolving power is the *minimum distance apart* which two points have to be in order to be perceived separately.

Table 1.1 *The light and electron microscopes compared. The figures are based on maximum performance of high quality microscopes. A standard school microscope has a maximum magnification of × 400.*

	Type of lenses	Rays transmitted through microscope	Resolving power	Maximum magnification
Light microscope	Glass	Light	0.3 μm	× 1500
Electron microscope	Electro-magnetic	Electrons	0.0002 μm (= 0.2 nm)	× 500 000

Table 1.1 gives the resolving power and other features of the light and electron microscopes.

Sizes and units

It is useful to hold in one's mind the relative sizes of various objects encountered in biology, and to know the **units** in which they are conveniently expressed. This is summarised in table 1.2.

Expressing scales

When you draw a small specimen, larger than it actually is, you should state the **scale**. The scale is the *number of times larger* your drawing is than the specimen itself. A drawing which is twice the size of the specimen is expressed as ×2.

You should also state the scale when you draw a specimen smaller than it actually is. A drawing which is half the size of the specimen is expressed as ×½ (×0.5).

Table 1.2 *The sizes of various biological objects and the units in which they are measured.*

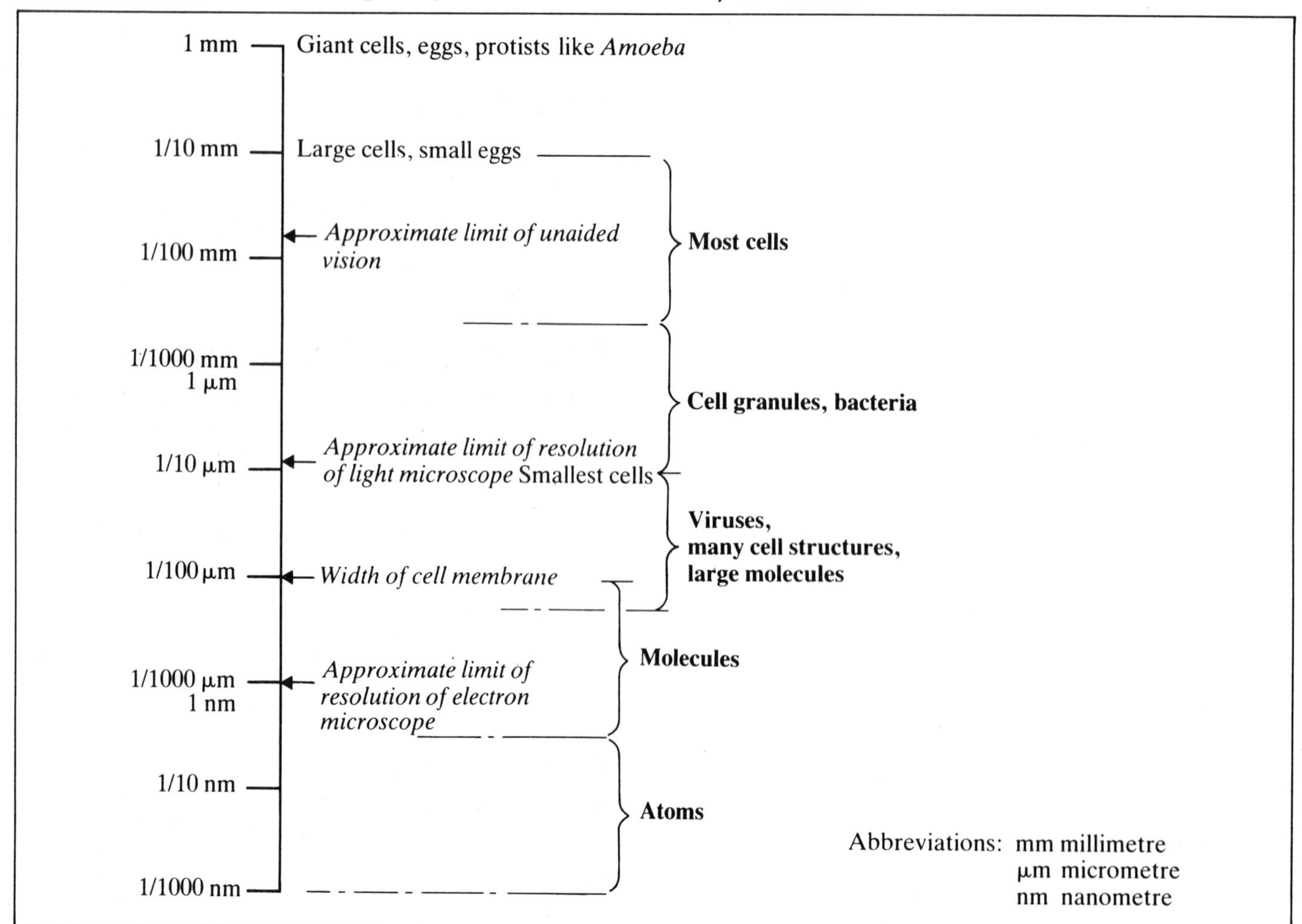

Table 1.3 *The different types of feeding (nutrition) found in organisms.*

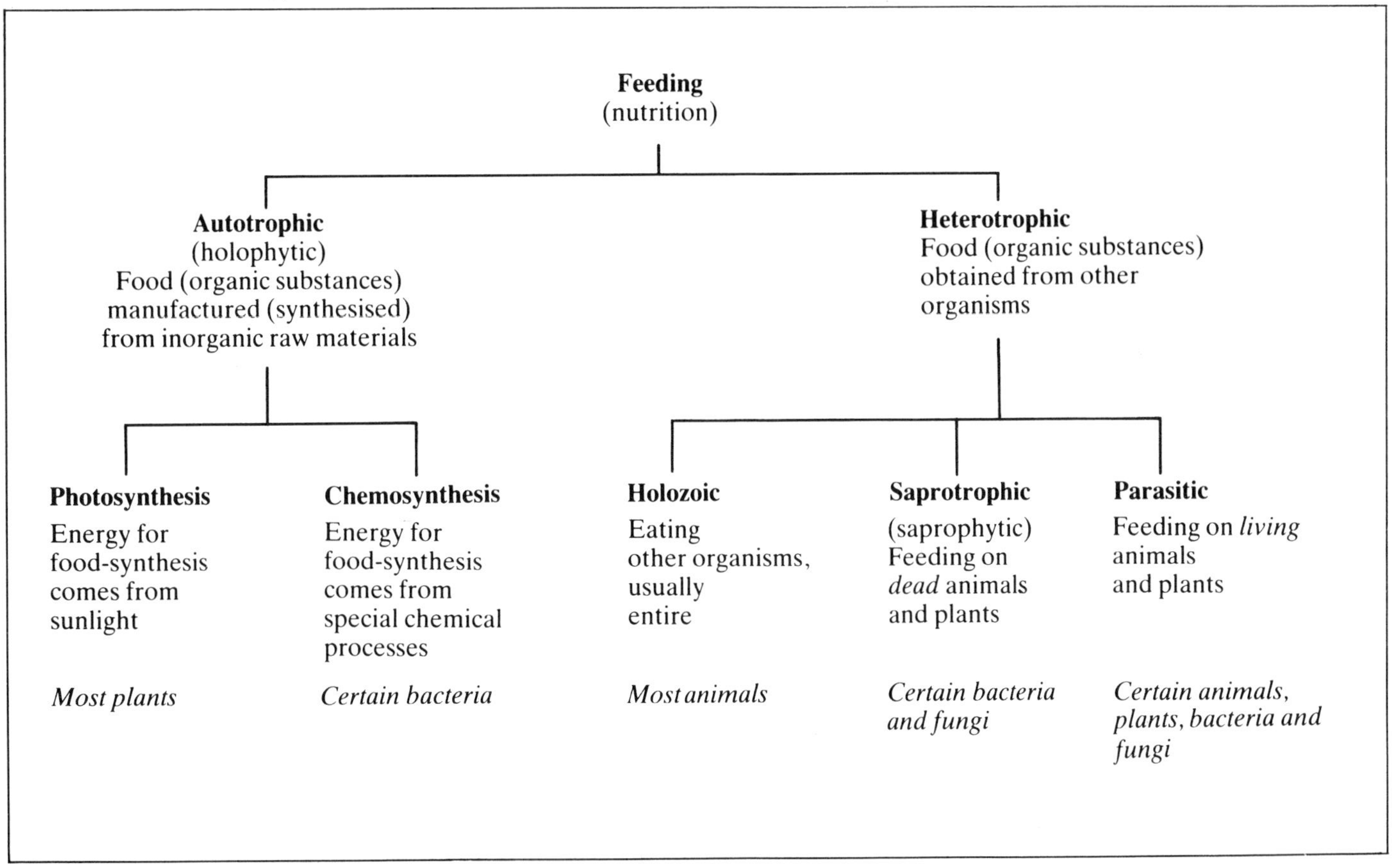

Feeding (nutrition)

- **Autotrophic** (holophytic) Food (organic substances) manufactured (synthesised) from inorganic raw materials
  - **Photosynthesis** Energy for food-synthesis comes from sunlight — *Most plants*
  - **Chemosynthesis** Energy for food-synthesis comes from special chemical processes — *Certain bacteria*
- **Heterotrophic** Food (organic substances) obtained from other organisms
  - **Holozoic** Eating other organisms, usually entire — *Most animals*
  - **Saprotrophic** (saprophytic) Feeding on *dead* animals and plants — *Certain bacteria and fungi*
  - **Parasitic** Feeding on *living* animals and plants — *Certain animals, plants, bacteria and fungi*

The characteristics of living organisms

All living organisms share the following characteristics:

1. They respond to stimuli. Animals generally respond by moving quickly; plants respond by growing (see pages 89 and 121).
2. They grow. Growth takes place by substances being taken into the organism and built up into its structure (**assimilation**) (see page 119).
3. They feed. Feeding (nutrition) takes place in different ways in different organisms (table 1.3) (see pages 55 and 73).
4. They produce energy. The process by which organisms produce energy is **respiration** (see page 46). It involves **gas exchange** (see page 39).
5. They get rid of poisonous waste substances. This process is called **excretion** (see page 85).
6. They reproduce. Reproduction may be **sexual** or **asexual** (see page 103).
7. They are made of cells (see page 31). Most organisms consist of numerous **cells**. Some organisms (known as protists) consist of only one cell.

Animals and plants compared

Table 1.4 summarises the main differences between a typical animal and plant. The main difference is in their methods of feeding. All the other differences stem from this.

Table 1.4 *The differences between a typical animal and plant.*

Typical animal	Typical plant
Feeds on ready-made organic food (heterotrophic feeding)	Makes its own organic food by photosynthesis (autotrophic feeding)
Has feeding structures such as mouth and gut	Lacks feeding structures
Lacks chlorophyll	Has chlorophyll
Lacks leaves	Has leaves
Lacks roots	Has roots
Moves around	Does not move around
Has nerves and muscles	Lacks nerves and muscles
Has receptors such as eyes and ears	Lacks receptors

• Questions

1 (a) Which magnifying aids, if any, would be needed to ascertain the detailed shape of:
(i) a bee's leg;
(ii) a buttercup petal;
(iii) an amoeba;
(iv) a pollen grain;
(v) a virus?

(b) The scale of a biological drawing is given as 0.5. This means that the dimensions of the drawing are:
(i) five times larger than the natural size;
(ii) five times smaller than the natural size;
(iii) half the natural size;
(iv) twice the natural size.
Which statement is correct?

2 The following are types of nutrition found in organisms:
chemosynthesis holozoism parasitism photosynthesis saprotrophism

(a) Which term best describes:
(i) a bird pecking at a worm;
(ii) a fungus spreading through the leaves of a potato plant;
(iii) a geranium plant in a pot;
(iv) a tapeworm in the human gut;
(v) bacteria causing decay of leaves in the soil.

(b) Give *two* examples of organisms which do not fit neatly into one of the categories listed above.

3 A student set up the experiment shown in the illustration below:

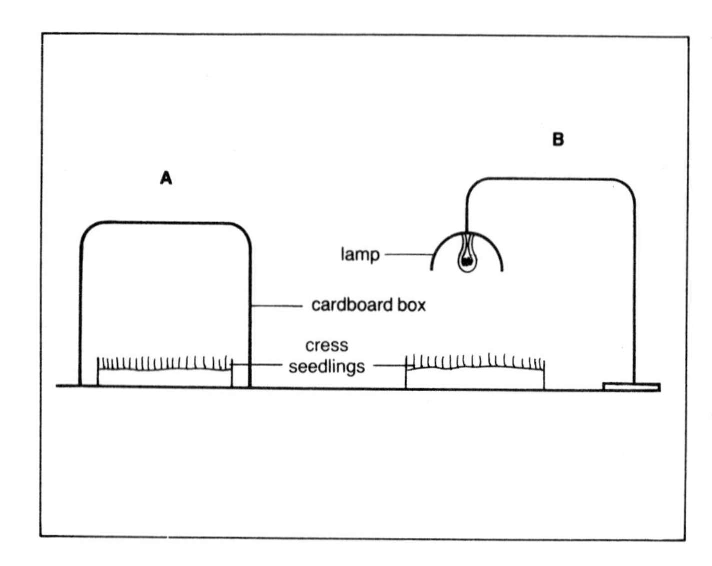

(a) What was the student attempting to investigate?

(b) (i) Which set-up, A or B, is the control?
(ii) Why is the control necessary?

(c) (i) Suggest *two* reasons why this experiment, as carried out by the student, might give inaccurate results.
(ii) Re-design the experiment so that it would give more accurate results.

2

The variety of life

How do we classify organisms?

We classify organisms by splitting them up into groups and sub-groups called, successively, **kingdoms, phyla** (singular **phylum**), **classes, orders, families, genera** (singular **genus**) and **species**. Table 2.1 shows how the human is classified.

The members of a small group such as a species have much more in common with each other than the members of a larger group such as a phylum.

Table 2.1 *This table shows how humans are classified. As we go downwards from top to bottom the number of organisms in each group decreases and the similarities between them increases. Ape-man and primitive human are, of course, extinct and are known only from their fossil remains.*

Name of group	The animals that belong to each group
Kingdom ANIMAL	All animals
Phylum CHORDATA	Animals with a backbone
Class MAMMALIA	Backboned animals, with hair
Order PRIMATE	Mammals with grasping hands and feet
Family HOMINIDAE	Ape-man and primitive human as well as modern human
Genus HOMO	Primitive human and modern human only
Species SAPIENS	Modern man only

How do we name organisms?

Every organism is given a **scientific name** which consists of the name of the genus followed by the name of the species, e.g. *Homo sapiens*. This is known as the **binomial system** (literally 'two name' system) and it was first introduced by Carl Linnaeus, a Swedish naturalist, in the 18th Century. The genus name always starts with a capital letter, the species name with a small letter, and both are written in italics or underlined.

In addition some organisms are given **common names**. For example, 'human' is the common name for *Homo sapiens*.

Why are scientific names better than common names?

Common names can lead to confusion. There are two reasons for this:

- an organism may have more than one common name;
- the same common name may be used for more than one organism.

How can you find the name of an organism?

The best way is to use a **key**. This consists of a list of features, e.g. hard cuticle, six legs etc. Each feature in the list is followed by an instruction which leads you to another feature. Eventually you arrive at the name of the organism.

The only way to learn about keys is to use them. One of the questions at the end of this chapter will give you practice at using simple keys of the kind which occur in examination questions.

How can you make a key?

Suppose you have a collection of organisms and you want to make a key so that other people can identify them. First you split the collection of organisms into two approximately equal sized groups on the basis of a particular feature. Then you split each of these groups into two further groups on the basis of another feature. You continue splitting the organisms up until each one is in a group of its own. This final group is given the name of the organism. You now rearrange the features in the form of a key. For an example, see page 9.

When making a key it is important to avoid using features such as size which vary from one individual to another.

Observing similarities and differences

Using keys and identifying organisms depends on observing accurately, and if necessary recording, similarities and differences between them. This is an important skill; one of the questions at the end of this chapter will give you practice at it.

How can you collect organisms?

Obviously the method used will depend on the types of organisms and where they occur. Here are the main methods:

- pick them up with fingers or forceps, e.g. snails;
- dig for them, e.g. earthworms;
- shake them out of trees or shrubs, e.g. insects;
- attract them with a light trap, e.g. night-flying insects;
- entice them with food (bait), e.g. small mammals;
- provide still water in which they can lay eggs, e.g. mosquitoes;

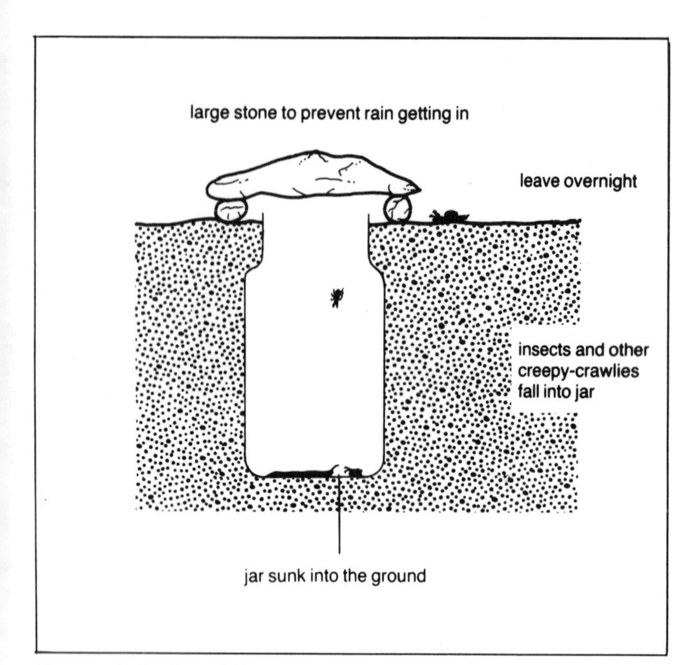

Figure 2.1 *A pitfall trap.*

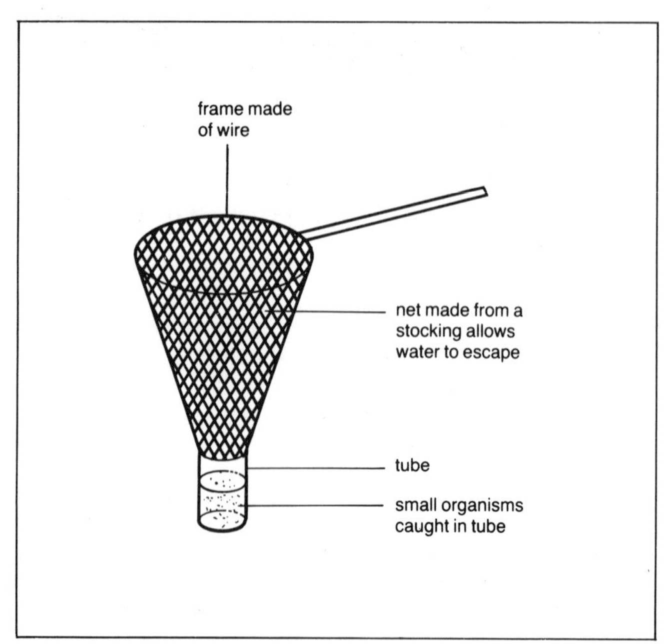

Figure 2.3 *A plankton net.*

- catch them with a net, e.g. butterflies;
- sweep them out of long grass with a net, e.g. insects;
- catch them with a **fishing** or **plankton net** (figure 2.1), e.g. tadpoles;
- capture them in a **pitfall trap** (figure 2.2), e.g. ground beetles;
- suck them into a **pooter** (figure 2.3), e.g. small insects in bark of tree.

Note: For special methods of extracting small organisms from soil, see page 15. In the interests of conservation it is important to kill as few organisms as possible and to return them to their natural habitats after you have observed them.

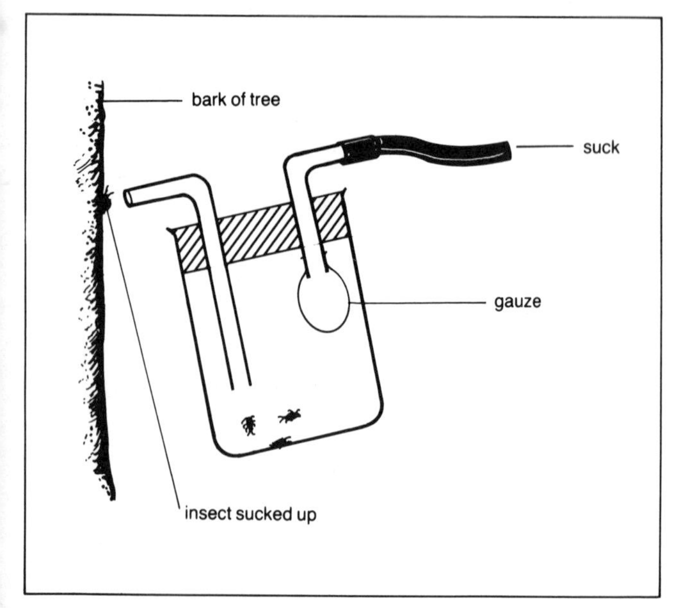

Figure 2.2 *A pooter.*

What do you need to take on a collecting trip?

The answer will depend on the habitat you are intending to visit and the kinds of organisms you are expecting to collect. Here are the main things which should be taken on a general collecting trip:

- scissors or secateurs;
- knife/penknife;
- trowel and/or spoon;
- forceps, spatula and small brush;
- gardening gloves;
- specimen tubes, jars etc;
- plastic and/or polythene bags;
- self-adhesive and/or tie-on labels;
- white dishes or bowls;
- hand lens (×10 magnification);
- notebook;
- pencil;
- apparatus for collecting organisms (see above).

Who's who in the world of living things

Virus kingdom

On the borderline between living and non-living. Can only be seen with the electron microscope. No cell structure. Reproduce inside other organisms and cause diseases.
Note: Because they are not really *living* organisms, viruses are not usually included in classifications.

Examples: viruses causing common cold, influenza, poliomyelitis and AIDS.

Bacteria kingdom

Can only be seen with the high power of the light microscope. Consist of a single cell with a wall; no proper nucleus. Occur in air, water, soil or inside other organisms. Many of them cause diseases.

Examples: decay bacteria (decomposers), nitrifying and nitrogen-fixing bacteria, bacteria causing typhoid, cholera and syphilis.

Protist kingdom

Some can be seen with the low power of the microscope. Consist of a single cell. Some are plant-like and others are animal-like. Live mainly in water or inside other organisms.

Examples: *Euglena*, *Amoeba*, *Paramecium*, malarial parasite.

Fungus kingdom

Consist of a 'network' (mycelium) of fine threads (hyphae). Live in soil or inside other organisms, especially plants. No chlorophyll, feed saprotrophically or parasitically. Reproduce rapidly by spores.

Examples: pin mould (*Mucor*), mushroom, yeast, potato blight fungus.

Lichens
Consist of a fungus and a plant-like protist combined together. Grow on rocks and tree trunks. Very resistant to drying.

Plant kingdom

Many-celled organisms which contain the green substance chlorophyll and make their own food by photosynthesis.

Algae
Simple plants which do not have roots, stems or leaves. Usually green, but sometimes brown or red. Live in water.

Examples: *Spirogyra*, seaweed.

Mosses and liverworts (Bryophytes)
Have simple leaves or leaf-like form. Found mainly in damp places. Reproductive spores are formed in capsule.

Examples: common moss, leafy liverwort.

Ferns (Pteridophytes)
Have proper roots and stems, and leaf-like fronds. Found mainly in damp places. Reproductive spores are formed on the undersides of the fronds.

Examples: common fern, bracken.

Conifers (Gymnosperms)
Large plants with seed-bearing cones for reproduction. Good at surviving in dry or cold climates. Most of them keep their leaves throughout the year.

Examples: pine tree, fir tree.

Flowering plants (Angiosperms)
Wide range of plants with seed-bearing flowers for reproduction. Seeds protected inside fruits. Range from small herbs to massive trees. Divided into dicotyledons and monocotyledons.

Dicotyledons
Seed contains an embryo with two seed-leaves (cotyledons). Broad leaves with branched veins forming a network.

Examples: buttercup, foxglove, oak tree.

Monocotyledons
Seed contains an embryo with one seed-leaf (cotyledon). Narrow leaves with straight, parallel veins.

Examples: grass, iris, palm tree.

Animal kingdom

Organisms that feed on other organisms and usually move around.

Animals without backbones (invertebrates)

Coelenterates
Many-celled animals with tentacles and sting cells. Most of them live in the sea.

Examples: *Hydra*, jellyfish, sea anemone, coral.

Flatworms
Body elongated and flat. Some of them live in ponds and streams, but most are parasites causing diseases.

Examples: fresh-water flatworm, tapeworm, blood fluke.

Roundworms
Body elongated and thread-like, round in cross-section. This group includes some harmful parasites.

Examples: *Ascaris*, hookworm, threadworm.

Annelids
Body divided up by rings into a series of segments.

Examples: earthworm, leech, ragworm, tube worm.

Molluscs
Have a soft body usually protected by a shell. In some the shell is greatly reduced.

Examples: snail, slug, mussel, squid, octopus.

Echinoderms
Have a tough spiny skin. Most of them are star-shaped. They all live in the sea.

Examples: starfish, brittle star, sea urchin.

Arthropods
Segmented animals with a hard cuticle (exoskeleton) and jointed limbs. Split into four groups mainly on the basis of the number of legs.

Crustaceans
Quite a lot of legs. Two pairs of antennae (feelers). Front part of body usually protected by a shield-like cover. Mainly aquatic.
Examples: shrimp, water flea (*Daphnia*), woodlouse, barnacle, lobster, crab.

Myriapods
Lots of legs. One pair of antennae. Body long and clearly segmented. Live on land.
Examples: centipede, millipede.

Arachnids
Four pairs of legs. No antennae. Mouthparts with pincers. Live on land; some are external parasites.
Examples: spider, scorpion, tick, mite.

Insects
Three pairs of legs. One pair of antennae. Body divided into three parts: head, thorax and abdomen. Usually two pairs of wings.

INCOMPLETE METAMORPHOSIS
(egg → nymphs → adult)

Examples: grasshopper, locust, cockroach.

COMPLETE METAMORPHOSIS
(egg → larva → pupa → adult)

Examples: butterfly, moth, housefly.

Animals with backbones (vertebrates)

Fishes
Live in water. Have gills for breathing, scales on their skin, and fins for movement.
Examples: shark, dogfish, ray, minnow, stickleback.

Amphibians
Have moist skin without scales. Live on land but lay eggs in water. Have fish-like tadpole larva which changes into the adult.
Examples: frog, toad, newt.

Reptiles
Have dry waterproof skin with scales. Eggs have a leathery shell and are laid on land.
Examples: lizard, crocodile, tortoise, snake.

Birds
Have feathers. Eggs have hard shells. Wings for flying, and a beak for feeding.
Examples: sparrow, owl, vulture, ostrich.

Mammals
Have hair. The young develop inside the mother and after birth are fed on her milk.

Egg-laying mammals
Young develop inside an egg which is laid by the mother.
Example: duck-billed platypus.

Pouch mammals
Young born at an early stage and finish their development inside a pouch, where they feed on the mother's milk.
Example: kangaroo.

Placental mammals
Young develop inside the mother, attached to a placenta.
Examples: rat, lion, sheep, elephant, bat, whale, monkey, human.

• Questions

1 Give a reason for each of the following.

(a) There are fewer differences between members of a species than between members of a phylum.

(b) It is better for a scientist to refer to an organism by its scientific name than by its common name.

(c) It is better to pick up a very small animal with a paint brush than with a pair of forceps (tweezers).

(d) The net used for catching plankton should have a very fine mesh.

(e) Size is rarely used in keys.

2 The drawings below show six invertebrate animals.

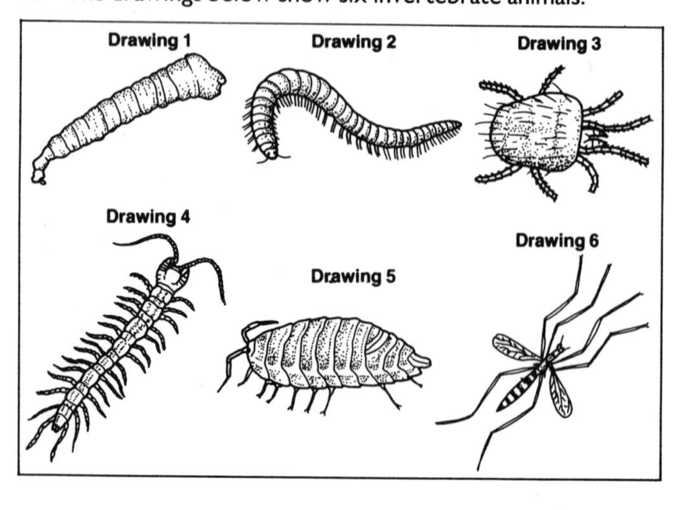

Use the key below to identify each of the animals. In each case write down the letter of the animal represented by the drawing.

Key

①	More than 4 pairs of legs	Go to 2
	4 pairs of legs of less	Go to 4
②	More than 15 body segments	Go to 3
	Less than 15 body segments	Animal A
③	Two pairs of legs on each body segment	Animal B
	One pair of legs on each body segment	Animal C
④	No legs can be seen	Animal D
	Legs can be seen	Go to 5
⑤	4 pairs of legs	Animal E
	3 pairs of legs	Animal F

(NEA)

3 The drawings below show two arthropods.

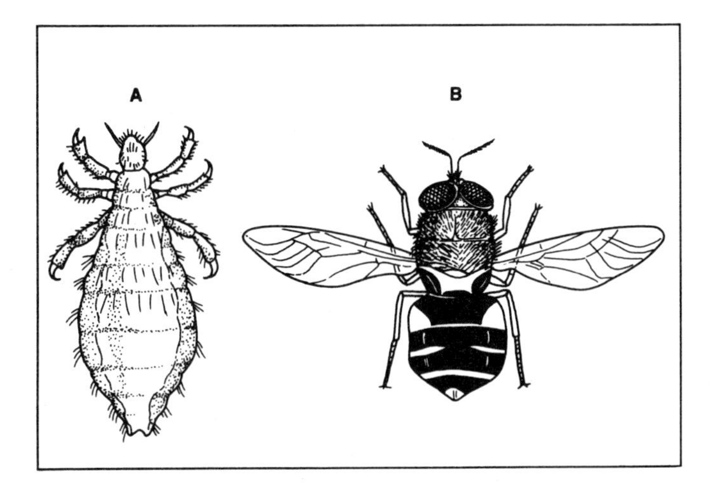

(a) Make a drawing of the hind leg of arthropod **A** which is twice as large as the drawing above (magnification ×2).

(b) Write down *three* differences between the two arthropods.

(c) Give *two* distinguishing characteristics possessed by arthropods.

(d) Arthropod **A** lives on the skin of a mammal. Describe *one* feature shown in the drawing which could help it to live on the skin.

(SEG, modified)

4 (a) Give *one* feature of each of the following groups which would enable it to be distinguished from all other groups.
(i) Angiosperm (ii) Bird
(iii) Fish (iv) Mammal (v) Protist

(b) How could you tell the difference between:
(i) a fungus and an alga;
(ii) an insect and an arachnid;
(iii) a reptile and an amphibian;
(iv) a vertebrate and an invertebrate;
(v) a virus and a bacterium.

5 Which of these plants, A to D, is:

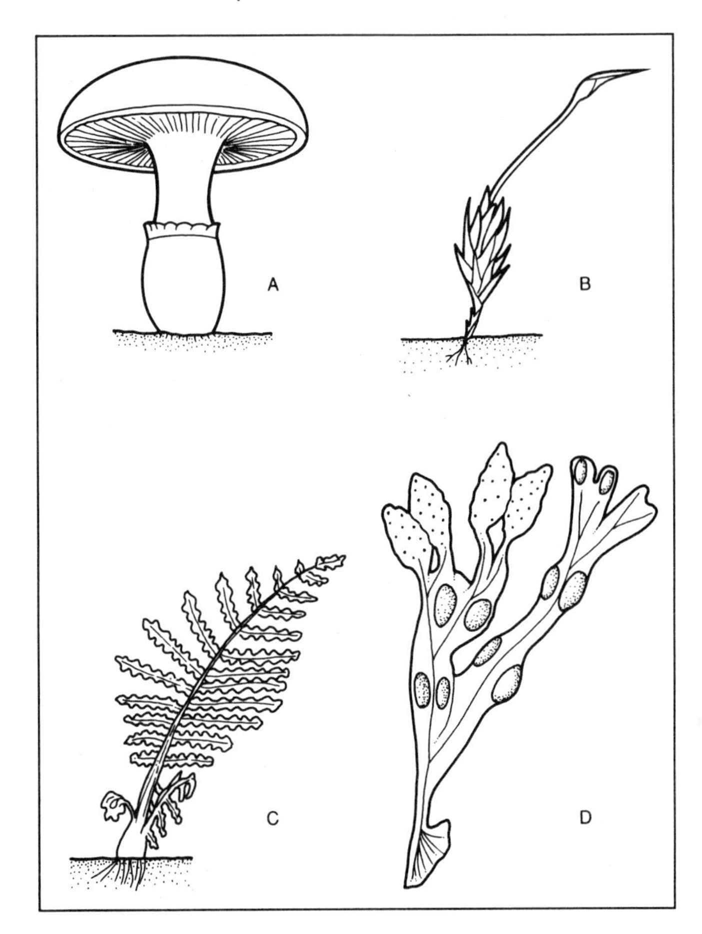

(a) an alga? (b) a moss?
(c) a fern? (d) a fungus?

(NEA)

3 Habitats and environment

What is a habitat?

A **habitat** is the place where an organism lives, e.g. a pond or wood. An organism may live in a particular situation within a habitat, e.g. under a stone or in a crevice in the bark of a tree: these are called **microhabitats**. An organism's habitat or microhabitat provides it with shelter, food and a place to reproduce.

What is the environment?

The **environment** is the conditions which exist in a habitat. An organism's environment is made up of two parts:

- the **physical (abiotic) environment,** i.e. physical features such as temperature, light, etc;
- the **biological (biotic) environment**, i.e. other organisms in the habitat.

Features of the physical environment

Here are ten important features of an organism's physical environment. They apply to both land (terrestrial) and water (aquatic) habitats.

1. Light
2. Temperature
3. Moisture/Humidity
4. Rainfall
5. Wind
6. Waterflow
7. Pressure
8. pH (acidity/alkalinity)
9. Murkiness (visibility)
10. Saltiness (salinity)

Features of the biotic environment

Here are ten important features of an organism's biotic environment. *In this list the organism is referred to as 'you'*.

1. Organisms which are your food.
2. Predators which eat you.
3. Organisms which give you diseases.
4. Organisms which compete with you.
5. Organisms which protect you from predators.
6. Organisms which shelter you from adverse physical conditions.
7. An organism which serves as your host, if you happen to be a parasite.
8. Animals which pollinate you, if you are a plant.
9. Animals which help to disperse your fruits or seeds.
10. Animals or plants to which you are attached.

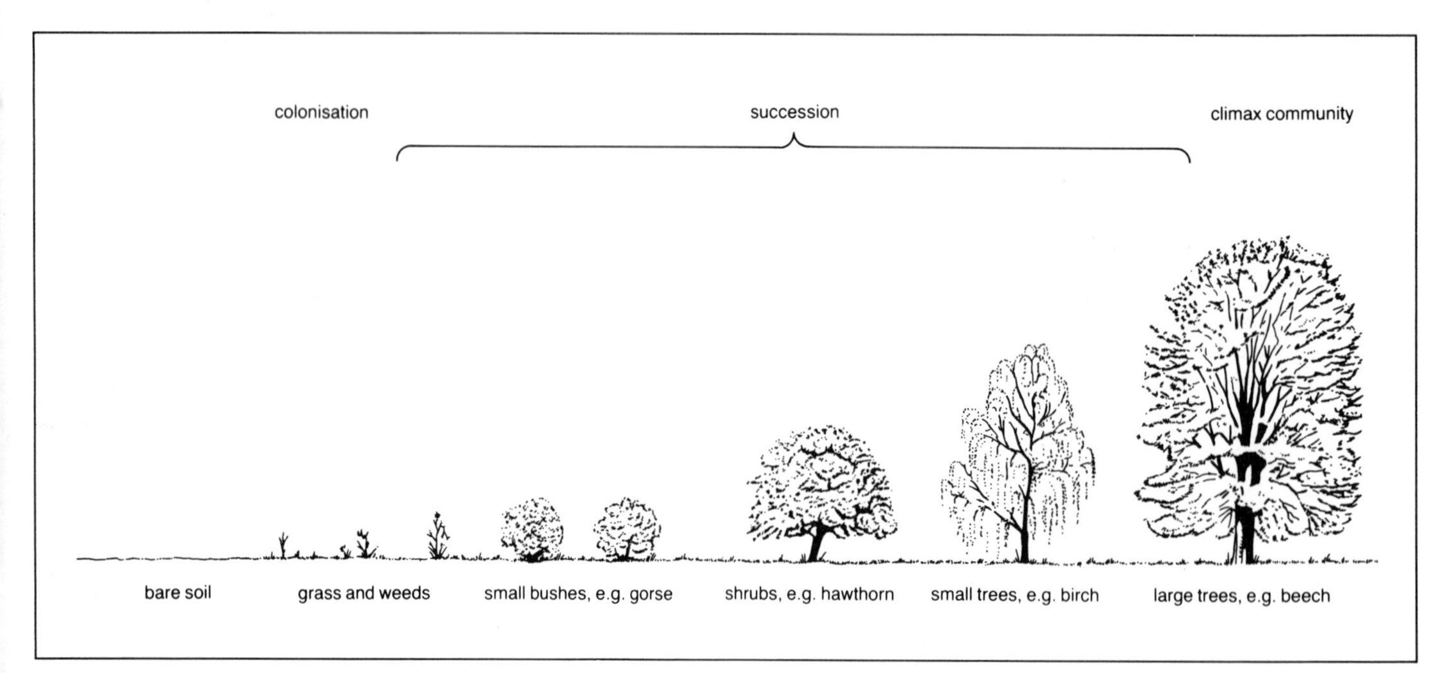

Figure 3.1 *An example of succession.*

What is a community?

A **community** is all the species which live in a particular habitat and belong to the same food web.

Within a community the organisms interact with each other, and with the physical features of the environment, to form an **ecosystem**. The food web is the basis of the ecosystem – its skeleton as it were.

What is colonisation?

Colonisation is the process in which organisms move into, and become established in, an unoccupied area, e.g. bare soil.

What is succession?

Succession follows colonisation. It is the gradual change from one community of organisms to another. Succession results eventually in a **climax community** (figure 3.1).

During succession the arrival of each new species alters the environment in such a way that further species can move in.

Seasonal changes

In temperate regions such as Britain, organisms change in various ways in autumn and spring (table 3.1). These **seasonal changes** enable the organisms to survive the winter.

Some of the changes are brought on by the change in temperature, others by the change in the amount of light received each day, i.e. daylength. The responding of organisms to changes in daylength is called **photoperiodism**.

Table 3.1 *Summary of seasonal changes in a temperate region such as Britain.*

Organism	Autumn	Spring
Flowering plants in general	Fruits and seeds formed.	Seeds germinate and give rise to leafy and flowering shoots.
Herbaceous plants	Die back and leave perennating organs in soil.	Perennating organs give rise to leafy and flowering shoots.
Deciduous trees and shrubs	Leaves change colour and fall off, dormant buds formed.	Buds open and give rise to leaves and flowers.
Birds	Some species migrate.	Migrants return; nesting, courtship and reproduction occur.
Other land vertebrates	Some species hibernate.	Hibernation ends, activity resumed, reproduction occurs.
Insects	Some species form a dormant pupa.	Adult emerges from pupa.
Other organisms	Many species produce dormant spores.	Spores germinate, giving rise to new individuals.

Soil

Soil is the layer of particles and related materials on the surface of the earth's crust (figure 3.2). Soil is a most important component of land habitats, and it affects water habitats too.

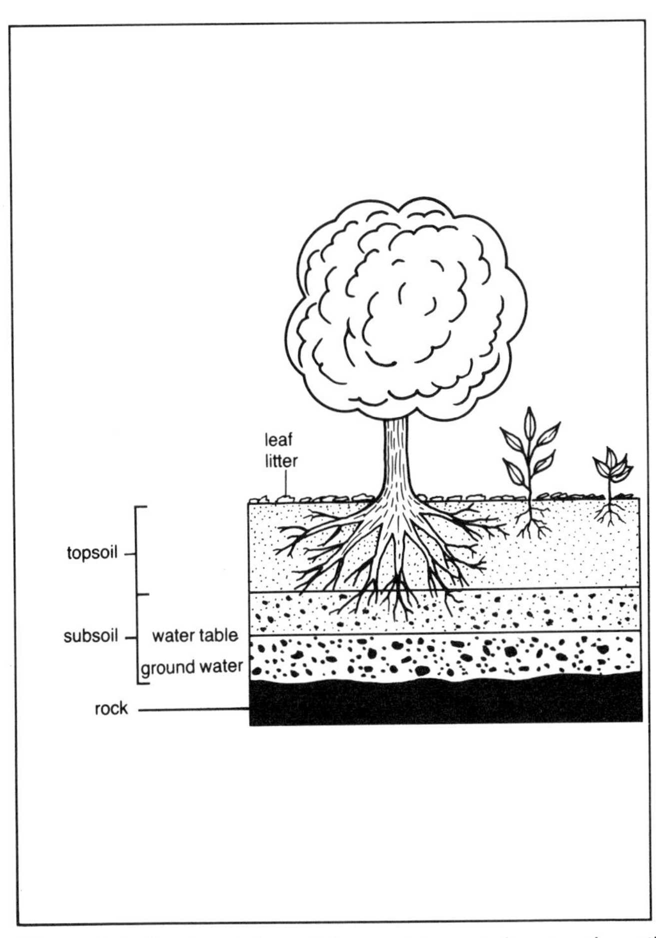

Figure 3.2 *Sectional view of the earth's crust showing the soil and other materials.*

Table 3.2 *The constituents of soil and their importance.*

Constituent	Description	Importance
Soil particles clay silt fine sand coarse sand gravel	 Diameter less than 0.002 mm. Diameter average 0.01 mm. Diameter average 0.1 mm. Diameter average 1.0 mm. Diameter more than 2.0 mm.	 Form the framework of the soil; good soil is a mixture of approximately two thirds sand and one third clay: this is called loam.
Soil water	Film round soil particles.	Provides source of water for plants, absorbed by roots.
Soil air	In spaces between particles.	Contains oxygen which is needed for aerobic respiration of plant roots and soil micro-organisms.
Humus	Sticky black substance formed by decay of dead organisms.	Sticks soil particles together into crumbs which plant roots can grip, provides nutrients for plant growth, holds water and helps to insulate the soil.
Mineral salts (nutrients)	Dissolved in soil water, derived from soil particles and humus.	Source of nitrogen and other elements needed by plants.
Lime (calcium carbonate, chalk)	White clay-like particles mixed in with rest of soil.	Prevents soil being too acidic and provides calcium for plant growth.

Table 3.3 *Problem soils and how to cope with them. Cover the right hand column and try to think of the remedies for yourself.*

Type of soil	Main problems	Remedy
Too much sand	Water drains through it quickly, washing out nutrients (leaching)	Dig in manure and/or compost.
	Dries, and loses heat quickly.	Spread manure or peat over surface (mulching).
Too much clay	Particles tightly packed with little or no air in between, heavy to dig when wet, rock-hard when dry.	Add humus and/or lime. (Lime causes clay particles to clump together into crumbs: this is called flocculation).
Too much water	Air spaces replaced by water (waterlogging), reducing oxygen content and slowing respiration of plant roots and soil micro-organisms.	Drain soil with underground pipes and/or soak-away.
	Decomposers respire anaerobically, producing lactic acid; this prevents decay being complete, so partly decayed material accumulates as peat.	See below.
Too much acid (pH too low)	Caused by too much peat; some plants grow poorly in acidic soil.	Add lime to make soil more alkaline.
Too much alkali (pH too high)	Caused by too much lime (chalk); some plants grow poorly in alkaline soil.	Add peat to make soil more acidic.

What does soil consist of?

The various constituents of soil are summarised in table 3.2. Good soil should contain all these constituents in the right proportions.

How can we measure the properties of soil?

Different types of soil have different properties. It is important for farmers and gardeners that they should know what sort of soil they have got. The main properties of soil are investigated as shown in the following list.

- **Soil particles**. Shake soil sample with water, particles subsequently settle in different layers according to their size.
- **Water content**. Weigh soil sample, heat to evaporate water, then re-weigh dried sample.
- **Humus content**. Weigh dried soil sample, burn off humus, then re-weigh.
- **Air content**. Cover soil with water in measuring cylinder, stir to dislodge air, note fall in level of water.
- **Acidity/alkalinity (pH)**. Shake soil sample with distilled water, dip pH paper into the water, note what colour the paper goes and compare with pH scale (figure 3.3).
- **Presence of lime**. Mix soil sample with concentrated hydrochloric acid, fizzing indicates lime.

Problem soils

Some types of soil give problems to farmers and gardeners. These problem soils are summarised in table 3.3

What is fertile soil?

Fertile soil is soil in which plants grow well. So fertile soil is *productive* soil. To be fertile, soil has to have sufficient nutrients (mineral salts) to support good plant growth (see page 71).

How is soil made fertile?

In natural conditions soil is kept fertile by the cycling of nitrogen and other elements (see page 20). Nutrients are provided by the **humus** in the soil.

In artificial conditions, e.g. a garden or field, soil may be made fertile by applying a **fertiliser** (see page 72).

Which organisms live in soil?

Table 3.4 gives examples of some important soil organisms and explains if they are helpful or harmful. Small organisms can be collected from a sample of soil by means of a **Tullgren funnel** or equivalent apparatus.

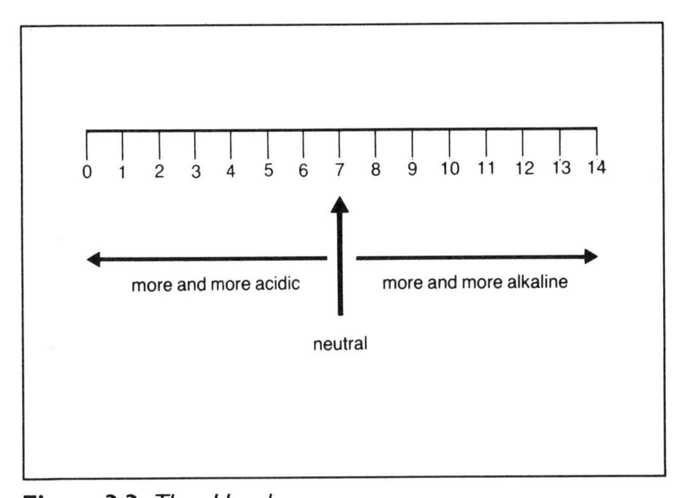

Figure 3.3 *The pH scale.*

Table 3.4 *Some soil organisms and why they are important.*

Organisms	Importance
Helpful	
Grass	Roots help to hold soil particles together; destruction of plants by e.g. over-grazing may result in erosion.
Earthworms	Turn over, drain, aerate, fertilise and refine the soil.
Moles	Drain the soil and eat pests (but also eat earthworms).
Bacteria and fungi	Decomposers bring about decay; nitrifying and nitrogen-fixing bacteria raise nitrate content of soil.
Harmful	
Wireworms (larvae of type of beetle)	Eat plant roots.
Leather jackets (larvae of daddy longlegs)	Eat plants roots.
Millipedes	Some species eat plant roots.
Roundworms (nematodes)	Some species attack plant roots and live inside them as parasites.
Bacteria	De-nitryfying bacteria lower nitrate content of soil.

How do humans affect the environment?

Humans affect the environment in four main ways.

- **Pollution**. Pollution is any process which leads to an increase in a harmful agent in the environment. The harmful agent is called a **pollutant** (table 3.5).
- **Pest control**. Pest control is getting rid of unwanted organisms (see page 26). Although this is necessary for humans, it alters ecosystems and can upset the balance of nature.
- **Modern farming**. Modern farming methods, particularly **monoculture**, alter the appearance of the countryside and can destroy wildlife. (Monoculture is explained on page 72).
- **Urban development**. The building of homes, factories, roads and other amenities inevitably alters the environment and destroys wildlife.

Table 3.5 *Examples of some important pollutants.*

Pollutant	Main components	Why it is harmful
Air pollutants Smoke from industrial plants	Carbon dioxide gas, sulphur dioxide gas, carbon particles. (Smoke plus fog makes smog.)	Reduces visibility, dirties buildings, causes bronchitis, sulphur dioxide harms plants.
Acid rain	Sulphuric acid, formed by sulphur dioxide gas reacting with atmospheric water and oxygen.	Erodes buildings, harms plants.
Motor vehicle exhaust	Carbon monoxide gas, lead (except when lead-free petrol is used). (Action of bright sunlight on exhaust makes photochemical smog.)	Carbon monoxide causes faintness; lead harms brain. Photochemical smog causes eye-irritation and headaches.
Water pollutants Oil	Liquid which floats on water.	Ruins beaches, kills fish and sea birds.
Sewage and fertiliser run-off	Nitrogenous substances.	High nitrogen content makes algae multiply quickly; algae then die, causing decomposers to multiply and use up oxygen so that fish suffocate.
DDT (insecticide)	Complex chemical substance containing chlorine.	Damages animal tissues. (Gets into food chains, becoming more concentrated at each step.)
Land pollutants Mine tips	Lead, cadmium etc.	Harmful to animals and certain plants.
Litter	Natural materials (e.g. apple cores), human-made materials (e.g. plastic)	Unsightly, may attract flies and rats. Natural materials can decay (biodegredable), human-made materials often cannot decay (non-biodegredable).
Noise	Loud sounds.	Discomfort, pain, damage to ears.
Radioactivity from nuclear tests, etc.	Waves given out from radioactive substances.	Damages genes (mutation) causing deformities and/or cancer.

What is conservation?

Conservation means protecting our environment so that animals and plants can maintain their numbers. Conservation can be achieved in the ways shown below.

① Reduce pollution as much as possible (see table 3.5).

② Restore areas which have been devastated by e.g. mining.

③ Do not over-exploit animals which are killed for food and other purposes.

④ Minimise the felling of trees and destruction of hedgerows.

⑤ Do not use natural forests extensively as a source of timber.

⑥ Do not allow one species of animal or plant to flourish at the expense of another.

⑦ Help endangered species to survive so that they do not become extinct.

⑧ Respect all forms of wildlife and observe the countryside code:

Take nothing but photographs
Leave nothing but footprints
Kill nothing but time

Living together

Sometimes two organisms, belonging to different species, live in or on each other. Such associations are called **symbiosis**. The two organisms are part of each other's biotic environment.

Types of symbiosis

- **Parasitism**. An association between two species (parasite and host) in which the parasite gains food and other benefits, and the host is harmed. If the parasite lives on the surface of its host, it is called an **ectoparasite**. If it lives inside its host, it is called an **endoparasite**.
  Examples: flea on dog, head louse on human scalp, tapeworm in human gut, blood fluke in human bloodstream, potato blight fungus in potato plant.
- **Mutualism**. An association between two species in which both benefit. (This used to be called symbiosis but nowadays the term symbiosis is taken to cover all types of association between species).
  Examples: lichen (fungus combined with green protists), nitrogen-fixing bacteria in roots of leguminous plants.

It is sometimes difficult to decide if a symbiotic association is harmful or beneficial, and there are 'in betweens'. In some associations one partner gains and the other neither loses nor gains: this is called **commensalism**.

Epiphytes

An **epiphyte** is a plant which is attached to another plant only for support. The epiphyte does not feed on its host or harm it.

Examples include mosses, ferns and orchids attached to the branches of trees: they root themselves in, and gain nutrients from, humus which collects in crevices where the branches fork.

Social animals

Social animals are animals that live in some kind of organised group (colony or society). Within the group different individuals perform different tasks.

Examples are bees, ants and termites (social insects), baboons and humans.

• Questions

1 Read through the list of features of the physical environment on page 10.

(a) Choose a habitat and explain how you would *measure* each feature in the list.

(b) Choose *one* organism which lives in the habitat and explain how its life is affected by each feature in the list.

2 Read through the list of features of the biotic environment on page 10. Give a specific example of each feature listed. In each case 'you' can be you personally or any *named* organism.

3 Read the list of seasonal changes in table 3.1 (page 11).

(a) These changes apply more to land habitats than water habitats. Why?

(b) Give one specific example of each organism in the list.

(c) Name one organism which does not appear to undergo any changes in autumn. How does it survive the winter?

4 (a) Give the common name of a plant which appears to thrive in each of the following types of soil:
(i) sandy soil; (ii) clayey soil; (iii) acidic soil.

(b) Why is adding manure a good remedy for sandy soil?

(c) Why is waterlogged soil acidic?

5 The diagram below shows a Tullgren funnel which is used for collecting small animals from a sample of soil.

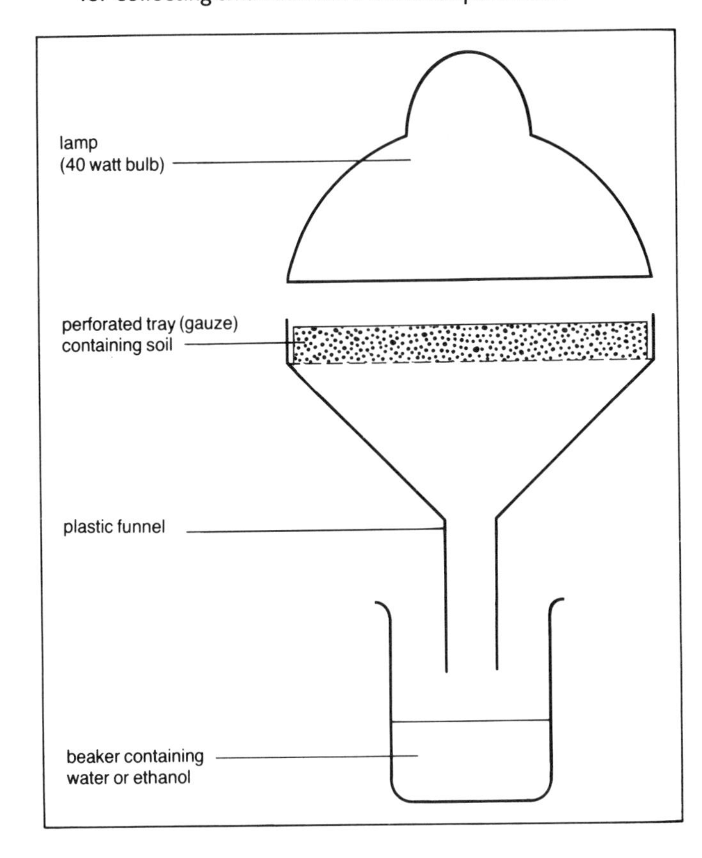

(a) Explain how the apparatus works.

(b) The soil in the perforated tray should not be too deep. Why?

(c) What is the advantage of having water, rather than ethanol, in the beaker?

(d) Name *two* animals which could be collected by this method.

(e) Make a labelled diagram of a similar apparatus which could be used to collect microscopic organisms which live in soil water.

6 Which of the conservation methods listed on page 14 would help each of the following organisms to survive?
(a) river trout; (b) redwood trees;
(c) whales; (d) bluebells; (e) herrings.

7 The graph below shows the relationship between respiratory tract diseases and air pollution along a straight line from a rural area into a city centre.

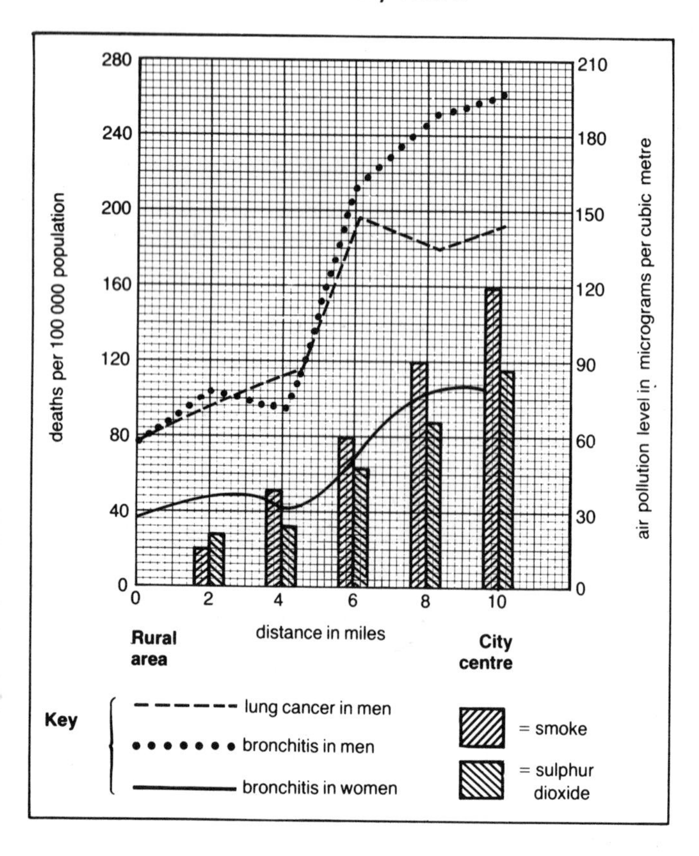

(a) (i) Suggest one factor which contributes to the change in smoke concentration from the rural area to the city centre.
(ii) Suggest a different factor which contributes to the change in sulphur dioxide from the rural area to the city centre.

(b) Suggest a cause for the difference in the death rates from bronchitis between men and women.

(c) The graph below shows the changes in the lichen cover on walls from the rural area to the same city centre. Lichens are simple organisms.

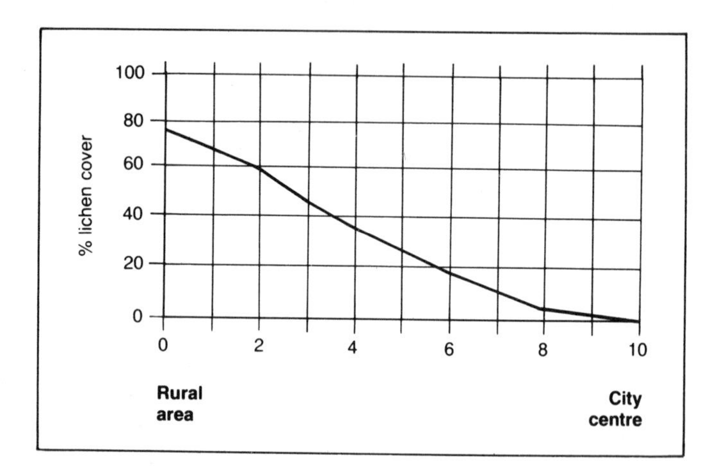

Using the information from the *two* graphs, suggest a relationship between air pollution and lichen cover.

(d) What changes might enable lichens to recolonise the city centre *without* their being artificially introduced?

(NEA)

4

Energy flow and the cycling of matter

What is a food chain?

A **food chain** is a sequence of organisms in which each organism provides food for the next organism. Here is an example of a food chain:

duckweed → tadpole → great diving beetle → pike

What do the arrows mean?

The arrows show the direction in which energy passes along the food chain.

The duckweed uses light energy from the sun to build up organic substances. When the duckweed is eaten by a tadpole, energy is transferred from the duckweed to the body of the tadpole; and when the tadpole is eaten by a water beetle energy is transferred from the tadpole to the beetle – and so on. This step by step transfer of energy through a food chain is called **energy flow**.

Terms used to describe food chains

The steps in a food chain are called **trophic levels**. Trophic means feeding. The terms used in describing food chains are summarised in figure 4.1.

Producers make, i.e. *produce*, organic substances by photosynthesis. Producers include plants and green protists (single-celled organisms).

Consumers feed on, i.e. *consume*, other organisms. Consumers include animals, fungi, non-green protists and most bacteria.

How are animal consumers classified?

They are classified into:

- **Herbivores**; animals which eat plants, e.g. cow, sheep.
- **Carnivores** (predators); animals which eat other animals, e.g. lion.
- **Omnivores**; animals which eat plants and animals, e.g. pig, human.

The final animal in a food chain is called a **top carnivore,** e.g. the pike in figure 4.1. Note that **parasites** are also consumers (see page 15).

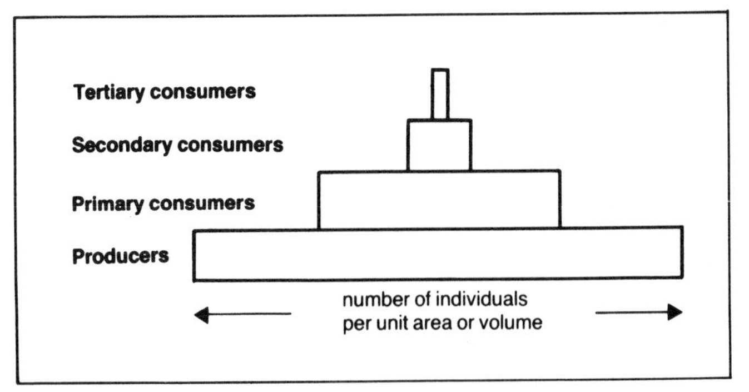

Figure 4.2 *Pyramid of numbers. The number at each trophic level is determined by counting the individuals in a given area or volume.*

What can you say about the number of organisms at each trophic level?

The number of organisms usually decreases at each trophic level. This can be shown as a bar chart; it is called a **pyramid of numbers** (figure 4.2).

What can you say about the mass of organisms at each trophic level?

The total dry mass of organisms (the **biomass**) decreases at each trophic level. A bar chart showing this is called a **pyramid of biomass** (figure 4.3).

Why is there a decrease in numbers and biomass at each trophic level?

When a tadpole feeds on duckweed, only *some* of the energy in the duckweed finishes up in the body of the tadpole. The rest is lost (see below). This loss of energy happens at every step in a food chain.

Figure 4.1 *Terms used to describe a food chain.*

Example	Feeding (trophic) levels	Terms used to describe the organisms at each level	Type of organism at each level
Pike ↑	Fourth trophic level	Third (tertiary) consumer	Carnivore (predator)
Great diving beetle ↑	Third trophic level	Second (secondary) consumer	Carnivore (predator)
Tadpole ↑	Second trophic level	First (primary) consumer	Herbivore
Duckweed	First trophic level	Producer	Plant (photosynthesiser)

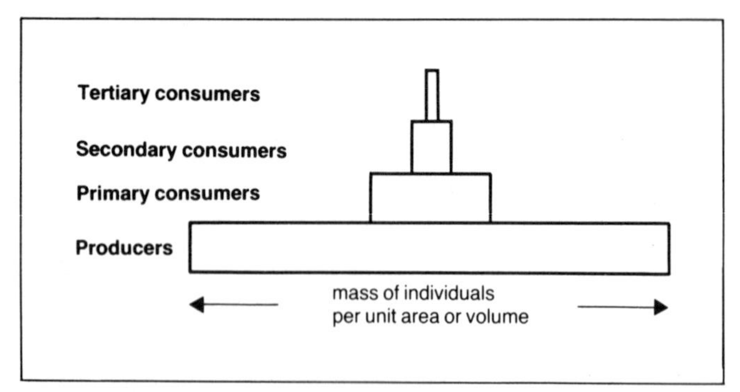

Figure 4.3 *Pyramid of biomass. It is best to measure the dry mass at each trophic level. This is determined by collecting the individuals in a given area or volume, heating them to drive off all traces of water, then weighing the dry residue.*

A pyramid of numbers or of biomass – which is better?

For showing how energy is lost in a food chain, a pyramid of biomass is better than a pyramid of numbers. Numbers can be misleading, giving upside-down pyramids like the ones in figure 4.4. Biomass, if measured correctly, always gives a pyramid.

Food chains important to humans

Here are three important ones:

- grass → cattle → human
- cereal mix → poultry → human
- plant plankton → animal plankton → fish → human

Plankton consists of very small (mainly microscopic) organisms which live in the surface waters of seas and lakes. It provides food for many species of fish and is therefore important to the fishing industry.

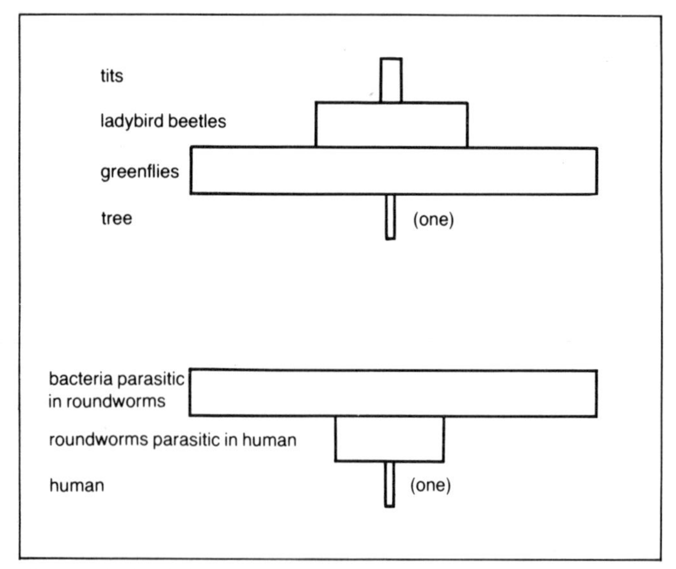

Figure 4.4 *Misleading pyramids of numbers. If these were pyramids of biomass, they would be the normal pyramid shape.*

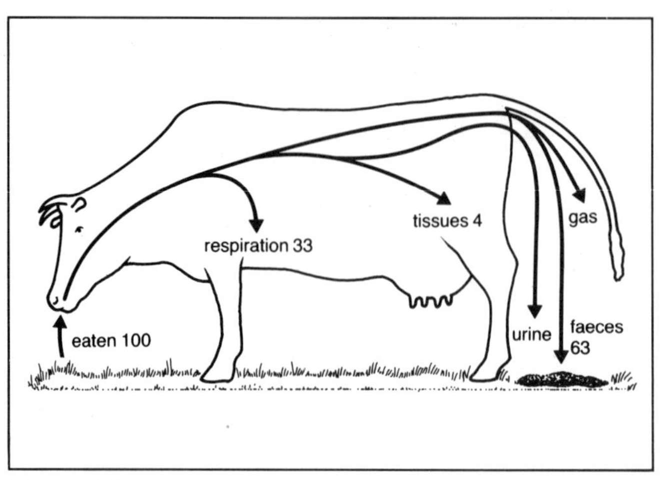

Figure 4.5 *What happens to the energy in the grass eaten by a cow? Suppose 100 units of energy are ingested. Only four units get into the cow's tissues. 63 units are lost in faeces, urine and gas, and 33 units in respiration.*

Why is energy lost when one organism feeds on another?

Consider this step in a food chain:

grass → cow

Figure 4.5 shows what happens to the energy which the cow takes in when it eats the grass. Notice that only four per cent of the energy in the grass gets into the tissues of the cow. The remaining 96 per cent is lost. Similar losses occur when humans eat meat from the cow.

Why is this energy loss important to humans?

More people can be supported by eating producers (e.g. wheat) than by eating consumers (e.g. cattle). So in countries where people are short of food, it is better to grow crops than to raise livestock.

What is a food web?

A **food web** is a series of interconnected food chains (figure 4.6). In a food web most, or all, of the consumers have more than one source of food. This means that if one of the consumer species is destroyed or dies out, the others can still survive.

The various organisms in a food web make up a **community** (see page 11).

An organism's position in a food web is called its **ecological niche**.

What is decay?

Decay is the breakdown of the bodies of dead organisms. It is brought about by **decomposers**. Decomposers are micro-organisms (mainly bacteria and fungi). They feed on the dead material, breaking it down into simple chemical substances.

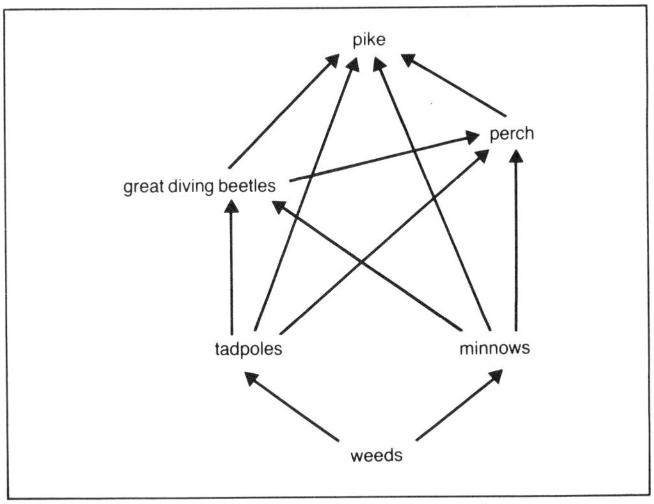

Figure 4.6 *A simple food web. Do not learn this diagram by heart; just understand the principle behind it.*

What conditions are needed for decay to occur?

- **Moisture**. Dry matter, e.g. hay, will not decay.
- **Warmth.** Frozen matter will not decay.
- **Oxygen**. The decomposers need this for respiration.

In addition, no chemicals must be present which might harm the decomposers.

How does lack of oxygen affect decay?

If oxygen is absent the decomposers respire anaerobically (see page 46). Lactic acid builds up. This kills the decomposers, preventing decay from being completed. The result is the formation of, for example, peat or silage.

Why is decay important?

- It prevents dead bodies from piling up.
- It plays an important part in the *cycling* of carbon and other chemical elements.

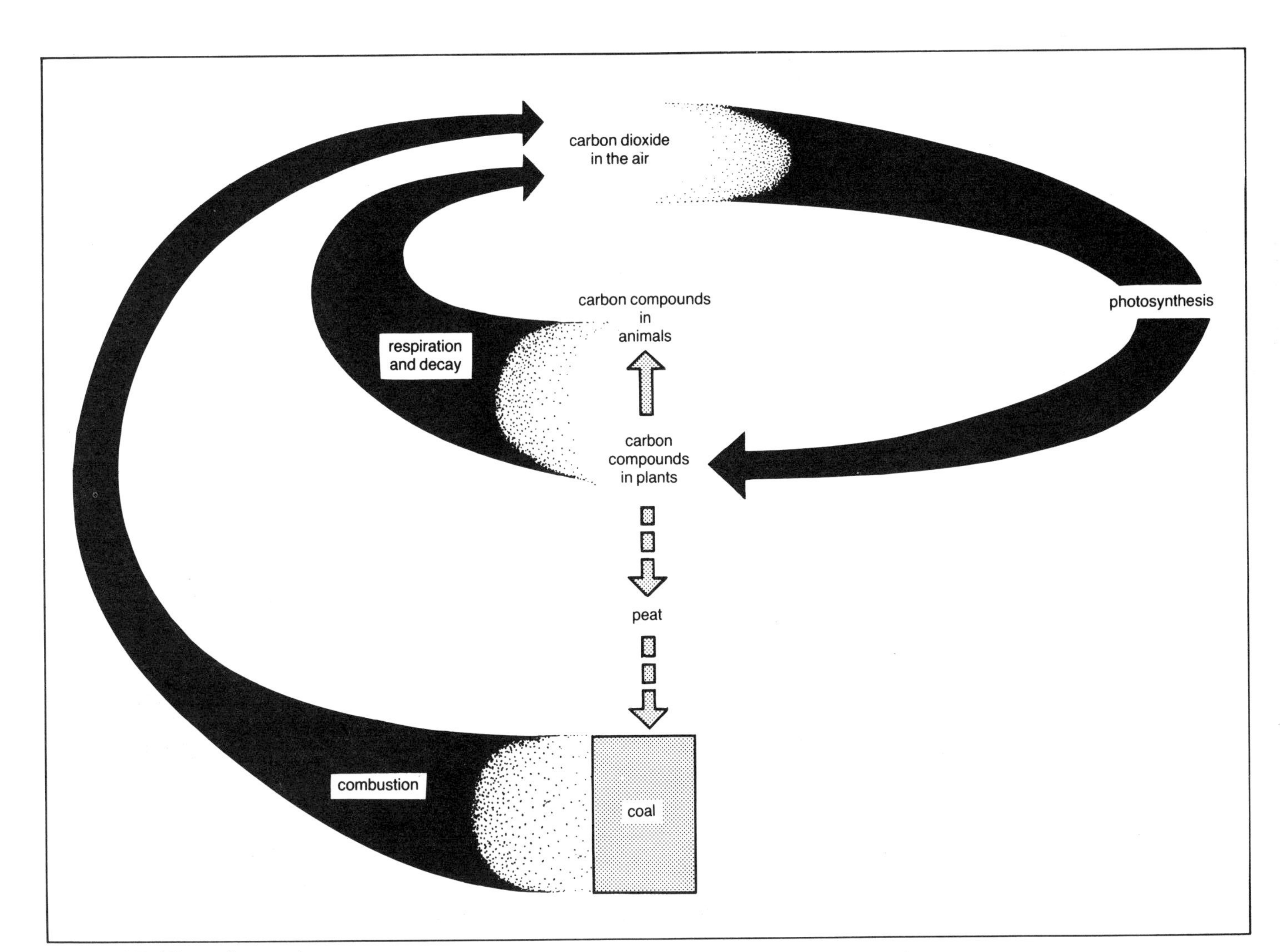

Figure 4.7 *The carbon cycle.*

The carbon cycle

The **carbon cycle** is the way carbon circulates in nature (figure 4.7). Carbon dioxide is taken out of the atmosphere by photosynthesis, and put back into the atmosphere by respiration, decay and combustion (burning).

The oxygen cycle

The **oxygen cycle** is the way oxygen circulates in nature. Oxygen is put into the atmosphere by photosynthesis, and taken out of it by respiration, decay and combustion.

The nitrogen cycle

The **nitrogen cycle** is the way nitrogen circulates in nature (figure 4.8). Plant and animal protein is broken down into ammonia by excretion and decay. The ammonia is then converted by **nitrifying bacteria** into, ultimately, nitrates. The nitrates are absorbed by plants and built up into protein again.

De-nitrifying bacteria decrease the amount of nitrates available to plants by converting them into nitrites and nitrogen. However, **nitrogen-fixing bacteria** in the soil and in the root nodules of leguminous plants (e.g. peas, beans and clover) can convert atmospheric nitrogen into, ultimately, protein.

The water cycle

The **water cycle** is the way water circulates in nature. Water evaporates from seas, lakes and rivers, and from land plants (transpiration). The water vapour condenses in the atmosphere and falls as rain or snow. Water sinks into the soil and is absorbed by plants.

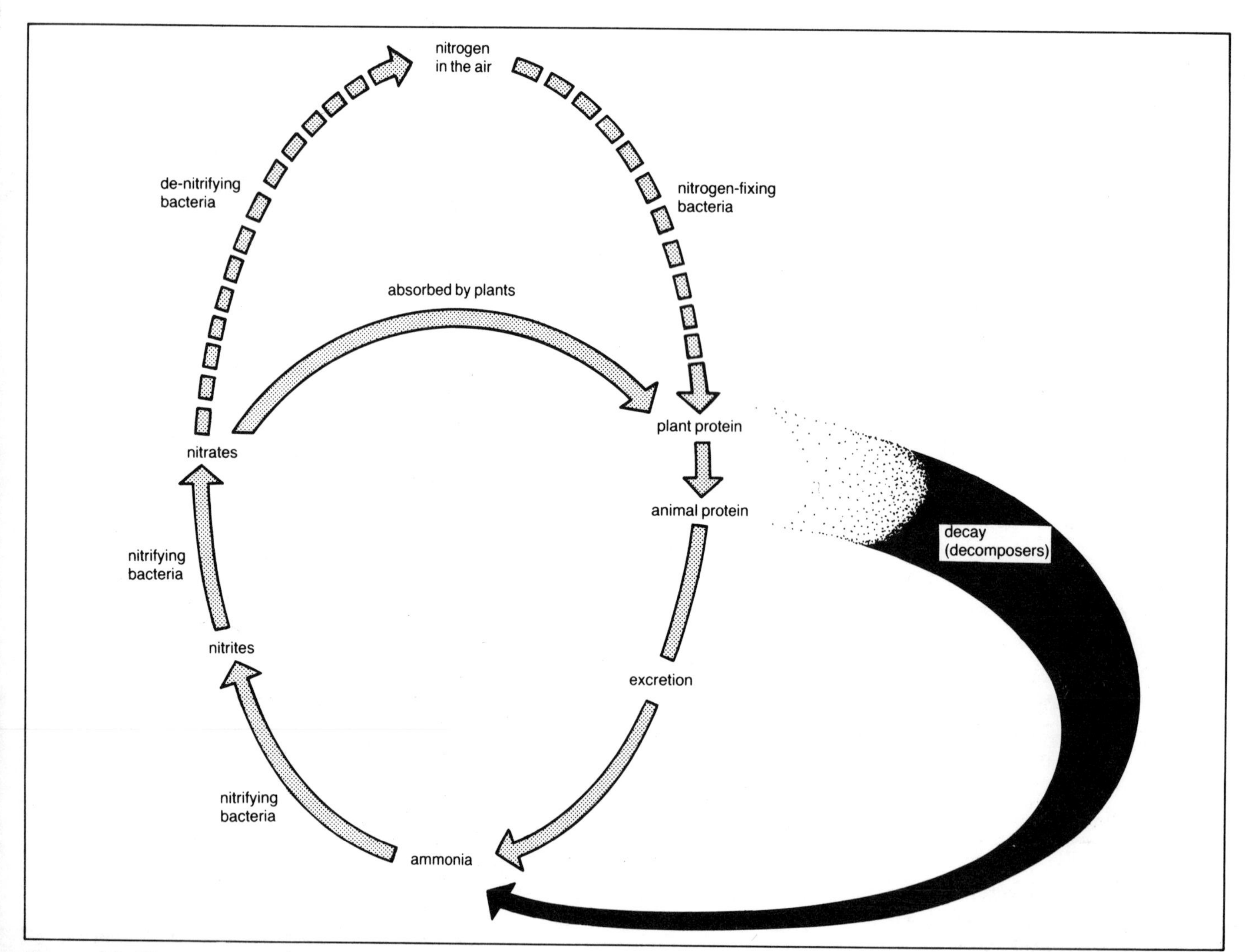

Figure 4.8 *The nitrogen cycle.*

Questions

1 Explain the difference between an organism's habitat, environment and ecological niche.

2 The following diagram shows the feeding relationships of a series of organisms.

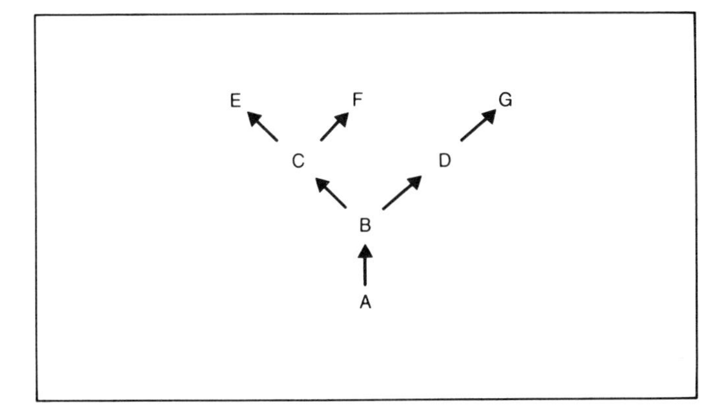

Say whether each of the following changes is likely to cause an increase or a decrease in the population of C. Explain your answers.

(a) An increase in the population of B.
(b) An increase in the population of D.
(c) An increase in the population of E.
(d) A decrease in the population of F.
(e) An increase in the population of G.

3 (a) What is the difference between a food chain and a food web?

(b) A food chain is more easily destroyed than a food web. Why?

(c) The following observations all have the same basic explanation. What is the explanation?

A A given mass of organisms at one trophic level supports a smaller mass of organisms at the next trophic level.

B It is unusual for there to be more than six steps in a food chain.

C As a source of food for humans, crops are more efficient than livestock.

(d) What is the meaning of the word 'efficient' in observation **C** above?

(e) Organisms towards the top of a food chain are usually larger than those lower down.
(i) Why should this be so?
(ii) Give *two* exceptions.

4 The diagram below shows a food web of organisms found in a certain area of soil.

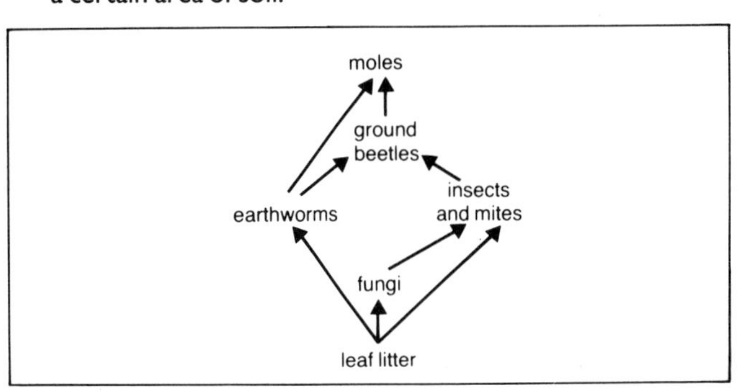

(a) Where does the energy for *all* these organisms come from?

(b) From the diagram, name an organism which is
(i) a primary consumer,
(ii) a secondary consumer,
(iii) a tertiary consumer,
(iv) a decomposer,
(v) a top carnivore.

(c) From the diagram write out a complete food chain consisting of
(i) three different organisms,
(ii) four different organisms,
(iii) five different organisms.

(d) Which organisms in the food web represent producers, and what is their role in the community?

5 Consider this conversion in the carbon cycle:
plant material → peat → coal

(a) How is peat formed?
(b) How is peat converted into coal?
(c) How, and in what form is the carbon in coal returned to the carbon cycle?
(d) Why are the world's supplies of coal running out?

6 (a) In the carbon cycle what natural process carried out by organisms adds carbon dioxide to the atmosphere, and what natural process carried out by organisms removes carbon dioxide from the atmosphere?

(b) In the nitrogen cycle give one conversion which is carried out by each of the following organisms:
(i) decay bacteria (decomposers),
(ii) nitrifying bacteria,
(iii) nitrogen-fixing bacteria.

(c) Certain de-nitrifying bacteria convert nitrates into nitrites.
(i) Are these bacteria helpful or harmful to farmers? Explain your answer.
(ii) Give one other conversion which is brought about by de-nitrifying bacteria.

5

Populations and distribution

What does population mean?

A **population** is all the individual organisms of a particular species in a given locality or habitat.

The **population size** is the number of individuals making up the population.

How do we estimate population size?

- **Descriptive method.** A rough and ready method suitable for stationary organisms which can be seen at a glance. Look at the locality and assess the relative abundance of the different species. Record them on the scale shown in table 5.1.

Table 5.1 *This scale is used for describing the occurrence of plants or animals in a habitat. It is called the DAFOR scale from the initial letters of the words.*

Description	Meaning
Dominant	has the greatest effect
Abundant	hardly ever out of sight
Frequent	constantly found
Occasional	seldom found
Rare	hardly ever found

- **Quadrat method.** An accurate method suitable for stationary organisms which can be counted individually, e.g. thistles in a field. Obtain a square frame (**quadrat**) of known area. Place the quadrat on the ground. Count the individuals inside the quadrat. This gives the number of individuals per unit area and is called the **density**. Repeat in different places, selected at random, within the locality and work out the average density.
- **Grid method.** A modification of the quadrat method for stationary organisms which cannot be counted individually, e.g. grass on waste ground. Obtain a quadrat frame which has been subdivided into 100 squares (**grid**). Place the grid on the ground. Count the squares occupied by the organism. Take into account any squares that are only partially filled. The figure you get is the **percentage cover**. Repeat in different places, selected at random, and work out the average percentage cover.
- **Line transect method.** For seeing how the population of an abundant stationary species varies across a habitat, e.g. mussels between high and low tide marks on the seashore. Obtain a tape which has been marked at regular intervals. Stretch the tape across the habitat. Count the organisms touching or overlapping the tape between one mark and the next. If you wish you can record the exact positions of individual organisms along the tape. In this way you can study the **distribution** of different species in the habitat.
- **Belt transect method.** A modification of the line transect method suitable for less abundant stationary species, e.g. sea anemones between high and low tide marks on the seashore. Stretch two parallel tapes across the habitat and divide the area between them into a row of squares. Estimate the density or percentage cover in each square. (Alternatively lay a quadrat frame or grid in successive positions across the habitat.)
- **Catching/trapping method.** For finding the relative population sizes of moving organisms in different localities, e.g. fish in two different ponds. Select a suitable method of catching or trapping the organisms (see page 5). Catch/trap the organisms in different localities for the same period of time. Count the individuals obtained from each locality.

What is sampling?

Consider the quadrat method of estimating populations described in the previous section. Note that you do not count all the individuals in the entire locality. Instead you count the individuals in several randomly selected places and work out the average. This is called **sampling**.

For greatest accuracy you should sample as many places as possible within the locality.

Why estimate population size?

- Estimating the population size of a species in different localities can tell us about the distribution of the species.
- Estimating the population size of a species at different times can tell us how the population size changes with time.
- Estimating the population size of different species in the same locality may tell us whether or not the species affect each other.

How do populations grow?

When a species is introduced into a new area, its population grows in three stages (figure 5.1):

- **Stage 1**: the numbers increase slowly at first, then gradually more rapidly.
- **Stage 2**: the numbers increase at the maximum rate by **exponential** growth.
- **Stage 3**: the numbers increase more slowly and eventually level off.

What is exponential growth?

Exponential growth is an increase in numbers by multiplication like this:

$$2 \xrightarrow{\times 2} 4 \xrightarrow{\times 2} 8 \xrightarrow{\times 2} 16 \xrightarrow{\times 2} 32 \xrightarrow{\times 2} 64 \xrightarrow{\times 2} \text{etc.}$$

In other words the numbers double at regular intervals of time. The time that it takes for the population to double is called the **doubling time**.

The rate at which the population increases is determined by, amongst other things, the birth rate and the death rate.

What is meant by the birth rate?

The **birth rate** is the rate at which new individuals are born. For humans, the birth rate is expressed as an annual (yearly) percentage: it is the number of births during the year divided by the total number of individuals at the beginning of the year, multiplied by 100.

What is meant by the death rate?

The **death rate** is the rate at which individuals die. For humans, the death rate is expressed as an annual (yearly) percentage: it is the number of deaths during the year divided by the total number of individuals at the beginning of the year, multiplied by 100.

When do populations grow?

Populations grow, i.e. increase, when:

- the birth rate is greater than the death rate, and/or
- more individuals enter the community than leave it (i.e. **immigration** exceeds **emigration**).

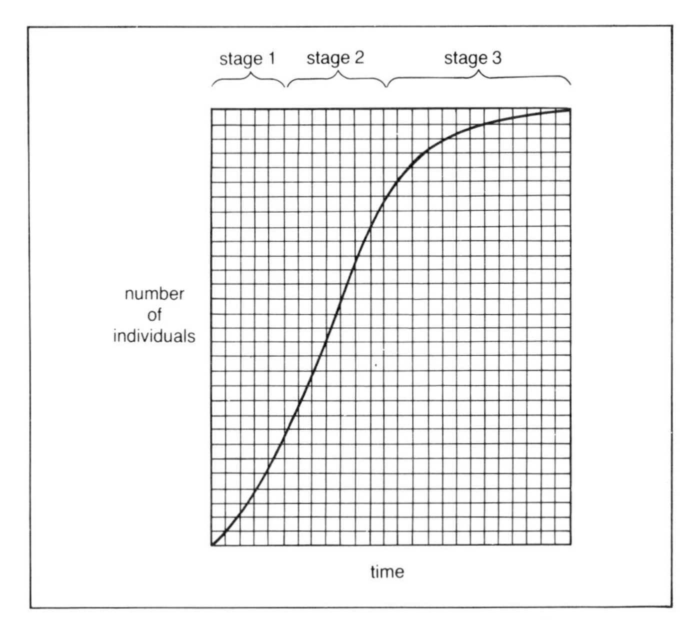

Figure 5.1 *Generalised graph showing growth of a population.*

In general the size of a population depends on the balance between the birth rate, death rate, immigration and emigration.

What stops populations growing?

In general these factors stop populations growing and may cause them to fall:

1. disease.
2. predation.
3. shortage of food.
4. shortage of living space.
5. failure to reproduce successfully.
6. fighting between individuals.
7. emigration.

Some of these factors are associated with competition.

What is competition?

Competition arises when two or more individuals are trying to obtain the same commodity, e.g. food or shelter. There are two types of competition:

1. competition between members of the same species (**intraspecific competition**);
2. competition between members of different species (**interspecific competition**).

How do organisms avoid competing with each other?

- They may move, or be moved, from an overcrowded area to a less crowded area (e.g. dispersal of fruits, seeds and spores).
- They may acquire, and defend, a piece of **territory** (e.g. birds such as the robin).

Human population growth

In the world as a whole the human population is increasing. Moreover, the rate of increase is increasing. Since 1800 the doubling time has decreased from 100 years to only 25 years. At present there is no sign of the world population levelling off.

Why is the human population increasing?

- More babies and young children survive than used to be the case, i.e. **infant mortality** is lower.
- People live longer than they used to, i.e. **life expectancy** is greater.

Both these changes have been brought about by improvements in people's standards of living and by advances in medical science. However, the human population cannot go on increasing for ever.

How can we stop the human population increasing?

By ensuring that the birth rate and death rate are equal. The only humane way of achieving this is through **birth control** (see page 111). Many developed countries have succeeded in limiting their population growth by birth control, but some developing countries have been less successful.

A typical developed country might have a birth rate of 2 per cent and a death rate of 1 per cent, giving an overall population growth of 1 per cent.

A typical developing country might have a birth rate of 5 per cent and a death rate of 2 per cent, giving an overall population growth of 3 per cent. Despite its relatively high death rate, the developing country still has a higher overall population growth rate than the developed country.

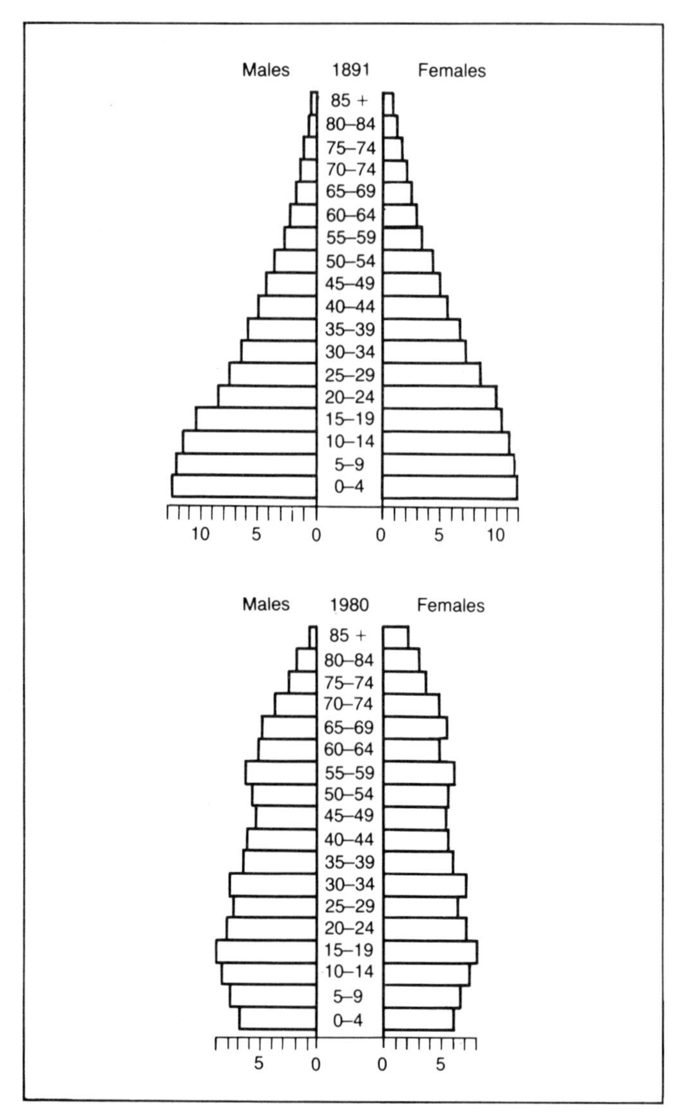

Figure 5.2 *Population pyramids for Great Britain, 1891 and 1980.*

What is population structure?

Population structure is the percentage of people in a country which fall into different age groups and sexes. The population structure is shown as a **population pyramid**. Two population pyramids for Great Britain are shown in figure 5.2.

What conclusions can we draw from these two population pyramids?

The main conclusions are that the proportion of young children in the population is lower, whereas the proportion of elderly people is higher, than in 1891. The decrease in young children has been brought about by birth control, the increase in elderly people by better medical care.

What use are population pyramids?

Population pyramids enable us to forecast the population structure in the future. This enables authorities to plan things like housing, schools and hospitals to meet future needs.

Fluctuating populations

The population size of many animals in the wild goes up and down, i.e. fluctuates. When the population grows too large, various factors cause it to fall (see page 23). When the population has fallen sufficiently, these factors no longer apply, so the population builds up again.

The population sizes of predators and their prey fluctuate together. This is because predators depend on their prey for food. The fluctuations in the predator population usually lag slightly behind those of the prey (figure 5.3).

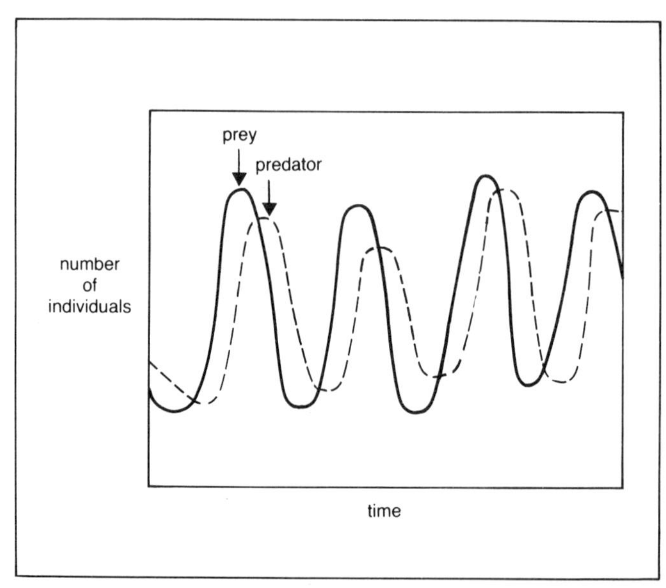

Figure 5.3 *Generalised graph showing fluctuations in the population sizes of predator and prey.*

• Questions

1 (a) What is meant by (i) the density and (ii) the percentage cover of a plant?

(b) You wish to use a one-metre quadrat frame to estimate the density of plantain plants in a lawn.
 (i) How can you ensure that your quadrat is placed randomly on the lawn each time you make a count?
 (ii) What should you do if some of the plants are partly inside and partly outside the frame?

2 The diagram below shows the positions of daisies in a lawn which has a large tree at one end. Each dot represents a daisy plant.

(a) (i) From the diagram, estimate the daisy population at different distances from the tree trunk. Explain your method.
 (ii) Plot your results either as a line graph or as a bar chart, whichever you consider more appropriate.

(b) (i) Suggest one reason why there are relatively few daisies close to the tree trunk.
 (ii) Describe an experiment which you could do to find out if your suggestion is correct.

3 Look at figure 5.1. Suppose the graph shows the increase in the population of rabbits on an island from an original single pair.

(a) Suggest one reason why the population increases slowly in stage 1.

(b) Assuming that there are no predators on the island, suggest four reasons why the rabbit population levels off in stage 3.

4 The table below shows the birth rate and death rate for two different continents:

	Continent A	Continent B
Birth rate	4.63%	1.62%
Death rate	1.98%	1.03%

(a) What is the percentage growth rate of the population in each continent?

(b) Suggest one reason why the birth rate is lower in Continent B than in Continent A.

(c) Suggest two reasons why the death rate is higher in Continent A than in Continent B.

(d) Write down the names of two continents which, in your opinion, are like Continent A, and two which are like Continent B.

5 Look at the population pyramids in figure 5.2.

(a) Draw, in outline, the shape of the population pyramid which you would expect in 50 years time. (There is no need to show all the bars).

(b) Why do you think the pyramid will have the shape which you have shown?

(c) To cope with the population structure shown by your pyramid, what changes would be needed in (i) schools, and (ii) health services between now and the year 2000?

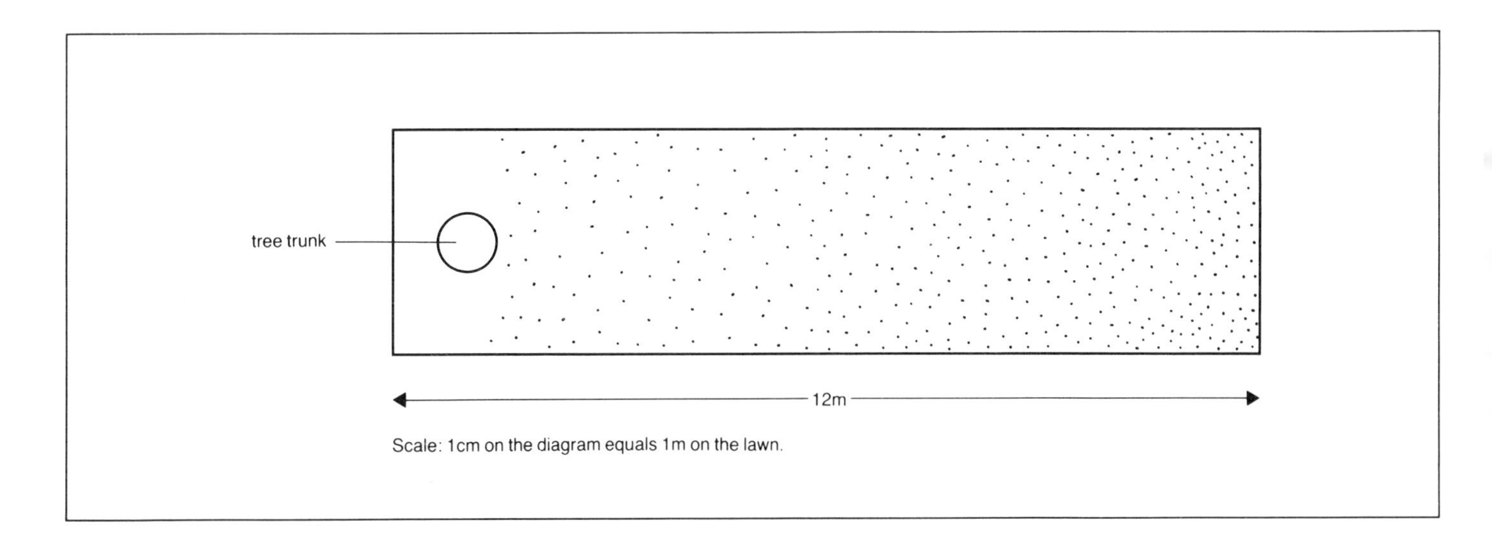

Scale: 1cm on the diagram equals 1m on the lawn.

6

Micro-organisms, pests and disease

What are micro-organisms?

Micro-organisms (microbes) are very small organisms which can only be seen clearly with a microscope. They include viruses, bacteria, protists, fungi and several animal groups. (*Note:* Viruses are included as micro-organisms but they are not really living things.)

Some micro-organisms are useful to humans, others are harmful.

In what ways are micro-organisms useful?

Here are a few important ways in which micro-organisms are useful to humans:

- They help to bring about **decay** (bacteria, fungi and others).
- They are important in the **nitrogen cycle** (bacteria).
- They are used in the **food industry**, e.g. cheese-making (bacteria and fungi), bread-making (yeast).
- They are used in the **brewing industry** (yeast).
- They produce **antibiotics**, e.g. penicillin (fungi).
- They produce **useful enzymes** (bacteria).
- They are used in **genetic engineering** (bacteria).

In what ways are micro-organisms harmful?

Micro-organisms are harmful in two main ways:

- The ones that bring about decay can make food go bad.
- Many of them are parasites and cause disease.

What are pests?

'Pest' is not really a scientific term, but it is used by biologists. We can take it to mean any organism which is harmful to humans. As such, they are a very important part of our biotic environment.

How do pests harm us?

- Some pests are parasites and cause diseases in humans, domestic animals and plants (including crop plants).
- Some pests transmit (i.e. carry) diseases from one human, domestic animal or plant to another.
- Some pests eat useful animals and plants.
- Some pests compete with useful animals and plants.

How can we get rid of pests?

We get rid of pests from the environment by **chemical** and **biological control**.

Table 6.1 *Three important types of pesticide and what they kill. DDT stands for dichlorodiphenyltrichloroethane (don't try to remember that!)*

Type of pesticide	What it kills	Example
Herbicide	Plants e.g. dandelions in a lawn.	Paraquat, hormone weedkiller.
Insecticide	Insects e.g. houseflies, mosquitoes.	DDT
Fungicide	Fungi e.g. potato blight, mildew.	Copper sulphate

What is chemical control?

In chemical control pests are treated with a chemical substance which kills them. The chemical substance is called a **pesticide** (table 6.1). Pesticides are often applied as a spray.

How do pesticides work?

Most animal pesticides either damage the nervous system or inhibit respiratory enzymes (see page 44). Most herbicides either stop photosynthesis or interfere with growth.

Types of herbicide

① **Contact herbicides** kill the parts of the plant which they touch, e.g. paraquat.

② **Transported herbicides** are carried from the part of the plant which they touch to other parts, e.g. hormone weedkiller (see page 122).

Some herbicides act on leaves, others on roots. Some root-acting herbicides remain active in the soil for a long time after they have been applied.

Selective weedkillers

A **selective weedkiller** is a herbicide which, when applied to mixed vegetation, kills some plants but not others. For example, hormone weedkiller, when sprayed in the right concentration onto a lawn, kills dandelions and plantain but not the grass. The reason is that the dandelions and plantain have broad leaves which take up more of the weedkiller than the much narrower-leaved grass.

What are the advantages and disadvantages of pesticides?

Pesticides have the advantage of being quick and efficient. However, they have these disadvantages:

- They may get into food chains and harm humans (see page 14).
- They may kill useful predators, with the result that some other animal becomes a pest.

- The pest may become resistant to the pesticide (see page 134).

Because of these disadvantages, more and more use is being made of biological control.

What is biological control?

In biological control an organism is put in the pests' environment. The organism kills the pests by eating or parasitising them (table 6.2).

Table 6.2 *Four examples of biological control.*

Pest	Harm done by pest	How controlled
Greenfly (aphid)	Sucks plant juices, transmits plant virus diseases.	Ladybird beetle (adult and larva) eats greenflies.
Mosquito	Transmits malaria.	Certain species of fish eat mosquito larvae.
Rabbit	Eats crop plants.	Certain type of virus causes myxomatosis.
Water hyacinth	Clogs up rivers and reservoirs in tropics.	*Tilapia* (fish) eats water hyacinth.

Does biological control have any disadvantages?

When you put a new predator or parasite into a community, it alters the food web which already exists. This may upset the balance of nature.

Nowadays biological and chemical control methods are used together, and are carefully regulated. In this way pests are kept under control without upsetting the balance of nature.

What is disease?

Disease is the condition which arises when something goes wrong with the normal working of the body. A disease is accompanied by various signs, e.g. sore throat, skin rash: these are called **symptoms**.

Sometimes a disease gives rise to a particular pattern of illnesses: this is called a **syndrome**. An example is AIDS (see page 112).

What causes disease?

Diseases may be caused by:

- lack of certain nutrients in the diet (deficiency diseases), e.g. scurvy;
- faults in our genes (inherited diseases), e.g. cystic fibrosis;
- stress, e.g. many types of mental disease;
- harmful micro-organisms.

Diseases caused by micro-organisms

Disease-causing micro-organisms are described as **pathogenic** – 'germs' as they are commonly called (table 6.3).

Table 6.3 *Ten well known diseases of humans and domestic animals and the micro-organisms which cause them.*

Disease	Type of organism which causes it	Main victims
AIDS	Virus	Human
Athlete's foot	Fungus	Human
Common cold	Virus	Human
Foot and mouth disease	Virus	Cattle
German measles (rubella)	Virus	Human
Influenza	Virus	Human
Malaria	Protist	Human
Rabies	Virus	Dog, human
Syphilis	Bacterium	Human
Whooping cough	Bacterium	Human

When germs get into your body, they multiply and spread before you start feeling ill. This is called the **incubation period**.

A person may have germs in his or her body without showing symptoms of the disease. Such a person is called a **carrier**.

How do germs harm us?

- Some germs destroy our cells or stop them working properly.
- Other germs release poisonous substances (**toxins**) into the bloodstream.

What is an infectious disease?

An **infectious disease** can be spread from one person to another.

Infection may cause a disease to spread through a community (**epidemic**) or even through a country, continent or the whole world (**pandemic**).

How are infectious diseases spread?

The micro-organisms which cause infectious diseases are spread in the following ways:

1. By droplets from coughs and sneezes, e.g. common cold, influenza.
2. By dust, e.g. diphtheria, scarlet fever.
3. By touch, e.g. impetigo, athlete's foot.

④ By faeces, e.g. typhoid, cholera.

⑤ By animals such as flies and mosquitoes, e.g. plague, malaria.

⑥ By blood mixing, e.g. viral hepatitis, AIDS.

⑦ By sexual intercourse, e.g. syphilis, AIDS.

An animal such as a fly which carries micro-organisms from one person to another is called a **vector**.

Some vectors carry micro-organisms to humans from other mammals such as rats. The other mammal does not suffer from the disease, but it serves as a **reservoir** of infection.

How are we protected from infectious diseases?

We are protected by our own natural defence mechanisms (e.g. page 66) and by the following artificial ('human-made') methods:

① **Sterilisation**. Sterilisation means freeing an object of germs. It can be achieved by **heat treatment** and/or **disinfectants**.

② **Destroying animal vectors**. Insects such as flies and mosquitoes can be killed by various insecticides.

③ **Isolating infectious individuals**. An infectious individual may be isolated (put in **quarantine**) until he or she is no longer infectious.

④ **Keeping the skin clean**. Regular washing is essential. Broken skin should be treated with a germ-killing **antiseptic** and, if necessary, covered with a **dressing**.

⑤ **Active immunisation (vaccination)**. A small dose of dead or inactivated germs (**vaccine**) of a particular disease is put into your body by injection or scratching the skin (**inoculation**) or by mouth. Once in the bloodstream, the vaccine causes your body to produce antibodies against the disease (see page 66). In this way you are protected from the disease before you ever come into contact with it. For some diseases further doses of vaccine (**boosters**) may be needed to keep up the protoection.

⑥ **Passive immunisation**. If you have already caught a disease and need quick protection, you may be given ready-made antibodies by injection.

⑦ **Antibiotics**. An antibiotic is a chemical substance, produced by fungi and other micro-organisms, which destroy bacteria. The most famous example is **penicillin**: it comes from the fungus *Penicillium* and was discovered by Alexander Fleming in 1928.

⑧ **Drugs**. A drug is a natural or synthetic chemical substance which affects the body in some way. Some drugs are extremely harmful (see page 90). However, many drugs kill or inactivate micro-organisms, or ease the symptoms of diseases. Drugs are particularly useful in treating virus diseases because viruses are not destroyed by antibodies.

How can we stop food going bad?

To stop food going bad we have to kill or inactivate the micro-organisms which would normally make it decay. This is an important aspect of preventing disease because bad food can cause food poisoning, e.g. salmonella infections (*Salmonella* is a type of bacterium).

Food is prevented from going bad by the following methods. These methods also enable us to preserve the food.

① **Heat treatment**. The food is heated then sealed. The heat treatment kills the micro-organisms, e.g. bottling, canning, pasteurisation of milk.

② **Freezing**. This inactivates the micro-organisms and slows down decay.

③ **Drying (dehydration)**. This inactivates the micro-organisms, preventing decay.

④ **Chemical treatment**. The food is treated with chemicals which kill the micro-organisms, e.g. pickling, smoking, salting.

⑤ **Irradiation with radio-active rays**. This kills the micro-organisms.

Human diseases today

Many infectious diseases that were once killers have been brought under control. The main reasons are:

- People have a better diet and are therefore less susceptible to disease.
- Better personal hygiene (cleanliness).
- Better medical care with the use of immunisation and antibiotics.
- Improvements in environmental health, e.g. sewage disposal.

In developed countries the main problems are now heart disease, cancer and the sexually transmitted disease AIDS.

In developing countries tropical diseases such as malaria and bilharzia are still major killers.

Plant diseases

Plant diseases are caused mainly by viruses and parasitic fungi.

An example of a plant virus disease is tobacco mosaic disease of tobacco plants. This virus, the first ever to be seen, was discovered in 1935 .

Some plant fungal diseases are summarised in table 6.4. Each type of fungus is specific to a particular type of host.

How do fungi damage plants?

Fungi consist of branched, thread-like hyphae which grow through the host plant, feeding on the contents of its cells (see page 59).

Dutch elm disease fungus blocks the xylem vessels (water-conducting tubes) in the trunk and branches of the tree.

Table 6.4 *Four important fungal diseases of plants and the ways they affect their hosts.*

Fungus	Host	Effect on host
Potato blight	Potato plant	Dies quickly, tubers rot.
Mildew	Cereals and many garden plants	Weakened, less grain produced.
Rust	Cereals	Leaves die, less grain produced.
Dutch elm disease	Elm tree	Dies slowly.

How are plant fungal diseases spread?

① **By wind and air currents**: they blow the spores from plant to plant, e.g. potato blight, mildew.

② **By the soil**: the spores rest in the soil during the winter and infect new plants the next year.

③ **By insects**: they carry the spores from plant to plant, e.g. Dutch elm disease which is spread by a beetle.

④ **By seeds**: the spores are carried on, or in, the seeds of the host plant; when the seeds germinate, the fungus develops in the seedling.

How are plant fungal diseases controlled?

① By prohibiting the import, and sale, of disease-carrying plants.

② By growing plants in conditions in which the fungus cannot thrive. For example, potato blight fungus needs a moist, humid atmosphere for the spores to germinate, so growing potatoes in a dry place can help to prevent the disease.

③ By growing plants across, rather than down, the path of the prevailing wind. This prevents the spores being blown onto other plants of the same type.

④ By rotating crops. If a particular plant is not grown in a field for several years, there will be no host available for the spores to infect.

⑤ By sterilising the soil, particularly soil which is used in greenhouses.

⑥ By treating plants with a fungicide (see page 26). The fungicide may prevent the spores germinating and/or reduce the severity of the disease.

⑦ By burning all diseased plants. This prevents others becoming infected.

⑧ By developing varieties of plants which are immune (resistant) to fungal diseases.

• Questions

1 (a) Give an example of a pest which:

- (i) eats the leaves of a useful plant;
- (ii) eats the roots of a useful plant;
- (iii) transmits a disease from one human to another;
- (iv) transmits a disease from one plant to another;
- (v) competes with a useful plant.

(b) State two ways, besides those listed above, in which pests can be harmful, and give one example of each.

2 Here are descriptions of four types of herbicide:

Type A: Does not sink into the soil more than 2–3 cm but remains active for over a month.
Type B: Kills broad-leaved, but not narrow-leaved, plants on contact.
Type C: Kills plants quickly but does not remain active in the soil.
Type D: Remains active in all layers of the soil for over a year.

Which of the four herbicides would you use for each of the following jobs?

(a) Clear weeds from soil in which you want to sow seeds next week.

(b) Keep a gravel path free of weeds.

(c) Prevent weeds growing up in a bed of soil which contains large flowering plants.

(d) Remove plantain and dandelions from a grass tennis court.

3 The bar chart below shows the numbers of acute cases of poliomyelitis in Britain between 1957 and 1964. Poliomyelitis is caused by a virus which is spread by consuming contaminated food or water. Widespread vaccination against poliomyelitis started in 1958.

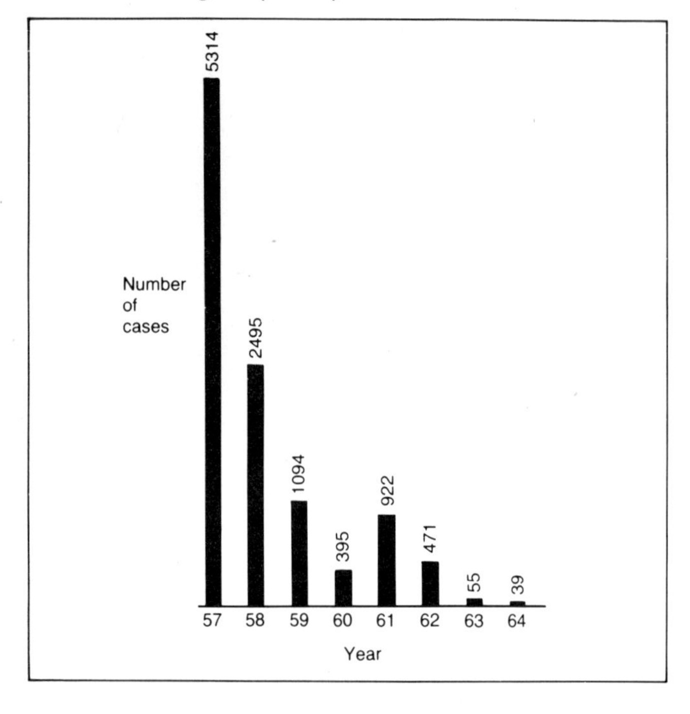

(a) What is meant by the term vaccination?

(b) How does vaccination protect a person against poliomyelitis?

(c) (i) What was the percentage drop in the number of cases between 1957 and 1960?
(ii) Suggest two reasons for this drop in the number of cases.

(d) From the bar chart, when do you think vaccination against poliomyelitis became a routine procedure for everyone? Give a reason for your answer.

4 Rust is a fungus which occurs in, amongst other places, North America. A map of North America is shown below.

(a) In what part of North America are spores of the rust fungus most likely to survive the winter?

(b) Copy out the following passage and write down the missing words.
Spores are ....(i)........ and are easily spread by the ....(ii)........ which carries them in a ....(iii)........ direction. The spreading of the fungus is aided by the practice of growing ....(iv)........ over vast areas in the path of the spreading spores.

(c) Suggest two ways in which rust disease could be controlled.

(d) What is the economic importance of rust disease?

(MEG, modified)

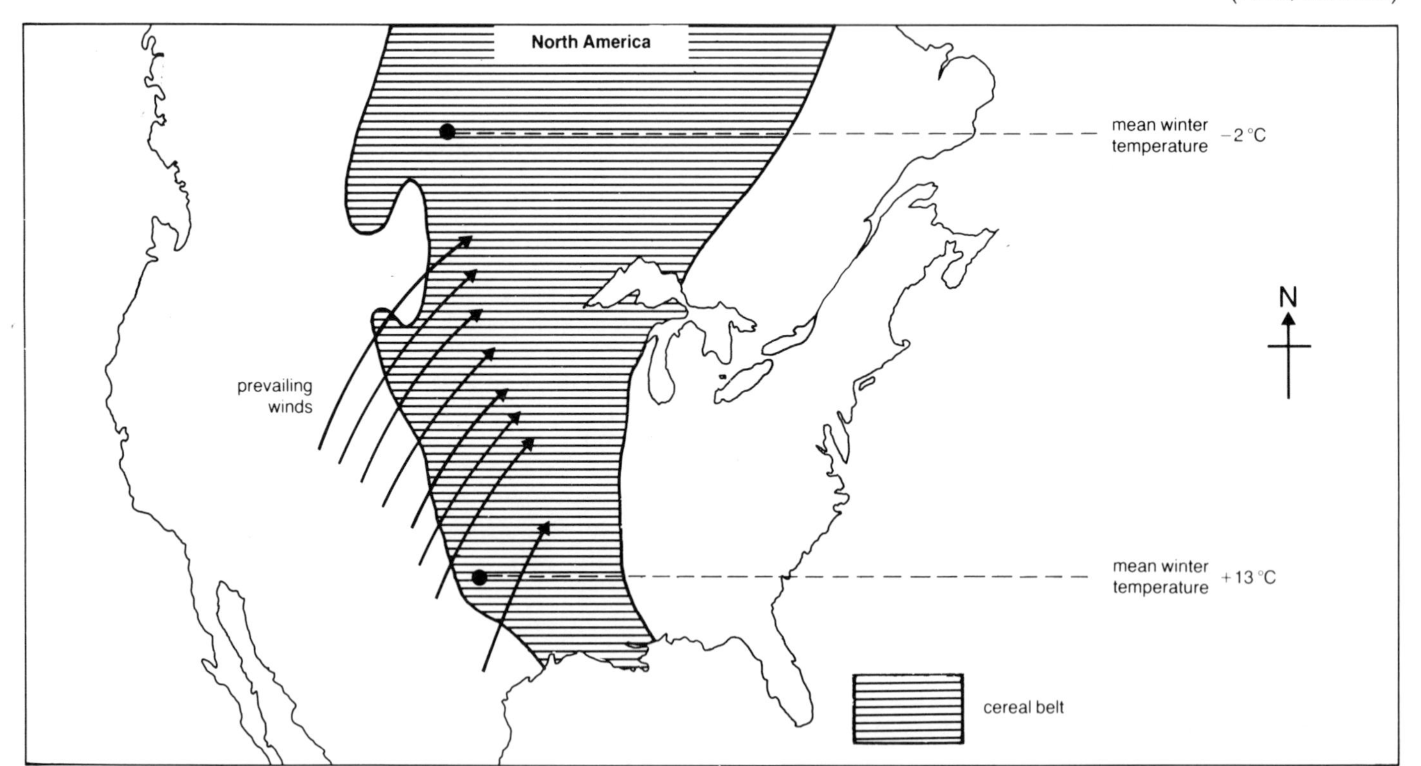

Cells, tissues and organs

What are cells?

The cell is the basic unit of an organism: *it is the smallest part of the organism which shows all the characteristics of life.*

A typical cell is approximately 20 micrometres wide. The symbol for micrometre is μm. There are 1000 micrometres in one millimetre.

All living organisms are made of cells. The human body is composed of about one hundred million million cells, that is 100 000 000 000 000 or 10^{14}. Even a small animal like *Hydra* has over 100 000 cells.

Some organisms consist of only one cell: they are described as **unicellular**. Examples include *Amoeba* and *Euglena*. Organisms which consist of more than one cell are described as **multicellular**.

Structure of an animal cell

Figure 7.1 is a diagram of a typical animal cell. It includes structures which can be seen with a light microscope and with the much more powerful electron microscope.

The cell has two main parts: **nucleus** and **cytoplasm**. The nucleus is a round body in the centre of the cell and the cytoplasm surrounds it. The nucleus and cytoplasm contain important structures which are shown in the diagram.

Structure of a plant cell

Figure 7.2 is a diagram of a typical plant cell. Like the diagram of the animal cell, it includes structures which can be seen with the electron microscope as well as the light microscope.

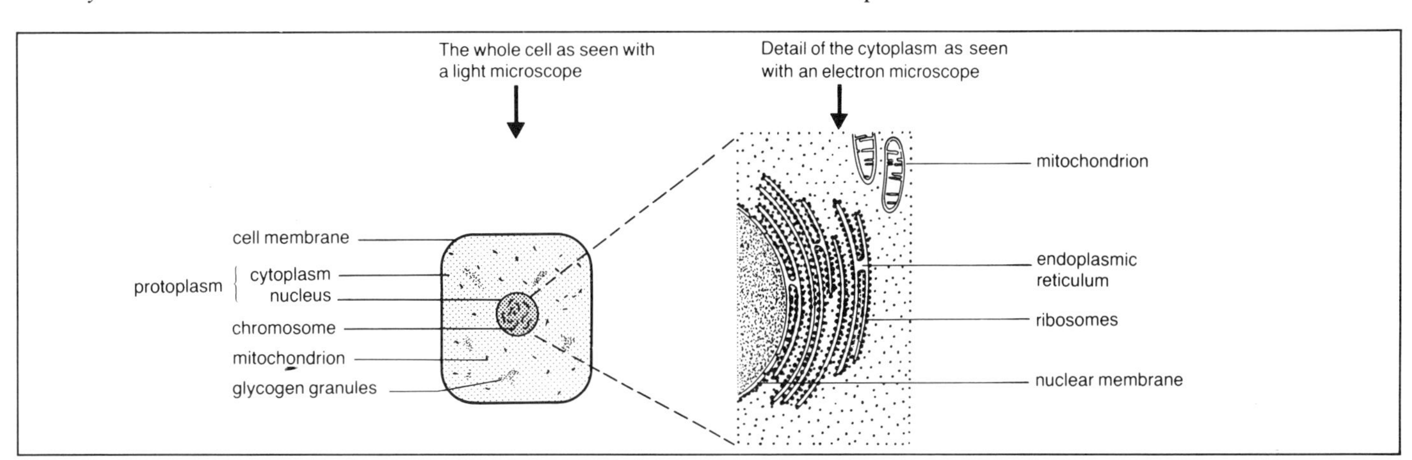

Figure 7.1 *Structure of a typical animal cell. The chromosomes can only be seen distinctly when the cell is dividing, or about to divide. At other times they take the form of a network of particles called chromatin granules.*

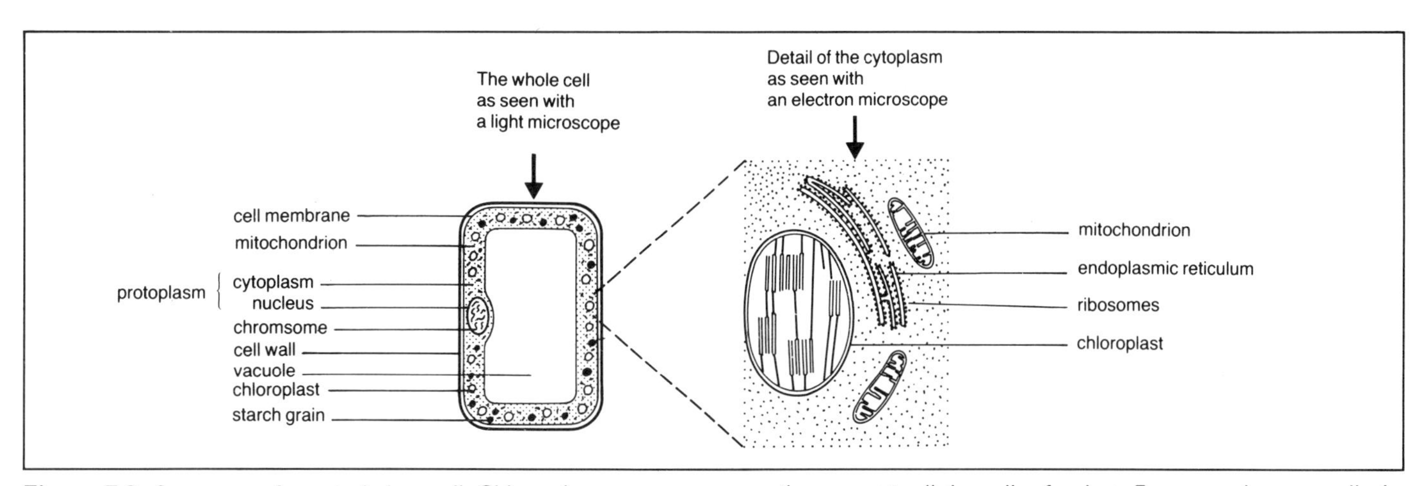

Figure 7.2 *Structure of a typical plant cell. Chloroplasts are not necessarily present in all the cells of a plant. For example, root cells do not have them.*

Cell	Special features	Function
Smooth muscle cell	 Long and thin, can shorten	Brings about movement of gut etc.
Nerve cell	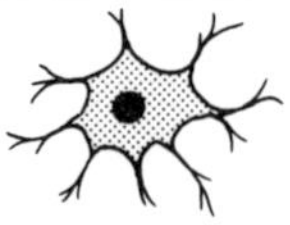 Slender arms project from main part of cell	Conducts electrical messages
White blood cell	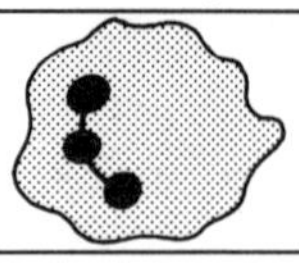 Irregular shape which changes constantly	Ingests harmful bacteria

Figure 7.3 *Three examples of specialised cells found in the human body.*

How do plant cells differ from animal cells?

The plant cell has the same basic features as the animal cell. It differs from the animal cell in these ways:

1. It has a **cell wall** as well as a cell membrane.
2. It has **starch grains** in the cytoplasm instead of glycogen granules.
3. It has **chloroplasts** in the cytoplasm.
4. It has a **vacuole**.

Summary of the structures found in cells

Table 7.1 summarises the various structures found in animal and plant cells.

Different cells for different jobs

Although cells share the same basic features, they are not all alike. They differ, particularly in shape, according to their specific functions. This is called **cell specialisation**.

Figure 7.3 shows three specialised cells which are found in the human body. You will find more examples of specialised cells, both animal and plant, in other parts of this book.

Table 7.1 *Summary of structures found in cells.*

Structure	**Description**	**Functions**
Chromosomes	Thread-like bodies in the nucleus	Carry *genes* which determine the organism's *characteristics*
Cell membrane	Thin lining round the cytoplasm	Holds the cell together and *controls* what enters and leaves it
Cell wall	Layer of *cellulose* surrounding plant cells	Strengthens and protects the cell
Chloroplasts	Round bodies in the cytoplasm of plant cells containing *chlorophyll*	Needed for *photosynthesis*
Glycogen granules	Small particles in the cytoplasm of animal cells	*Carbohydrate* food store
Starch grains	Round bodies in the cytoplasm of plant cells containing starch	Carbohydrate food store
Vacuole	Fluid-filled cavity in the cytoplasm of plant cells	Helps to maintain the shape of the cell
Mitochondria	Small sausage-shaped bodies in the cytoplasm	Release *energy* for use by the cell
Endoplasmic reticulum	Network of membrane-lined channels in the cytoplasm	Transports chemical substances in the cell
Ribosomes	Tiny particles attached to the membranes of the endoplasmic reticulum	Control the *synthesis* of *proteins*

Tissues

Cells are usually massed together into **tissues**. *A tissue is a group of cells which together perform a particular function.*

In simple tissues the cells are all the same kind. In more complex tissues there may be several kinds of cell.

Tables 7.2 and 7.3 summarise the main tissues found in animals and plants.

Table 7.2 *Summary of the main tissues found in animals.*

Name of tissue	What it consists of	Main functions
Epithelial tissue	Sheet of cells	To line tubes and spaces and form the skin
Connective tissue	Tough flexible fibres	To bind other tissues together
Skeletal tissue	Hard material	To support the body and permit movement
Blood tissue	Runny fluid	To carry oxygen and food round the body
Nerve tissue	Network of threads with long cable-like extensions	To conduct and coordinate messages
Muscle tissue	Bundles of elongated cells	To bring about movement

Table 7.3 *Summary of the main tissues found in plants.*

Name of tissue	What it consists of	Main functions
Epidermal tissue	Sheet of cells	To line the surface of plants
Photosynthetic tissue	Cells with chloroplasts	To feed the plant
Packing tissue	Round balloon-like cells	To fill in spaces inside the plant
Vascular tissue	Long tubes	To transport water and food substances
Strengthening tissue	Bundles of tough fibres	To support the plant

Table 7.4 *Some examples of plant organs.*

Name of organ	Functions
Leaf	To feed the plant by photosynthesis
Flower	To produce offspring by sexual reproduction
Bulb	To store food and produce offspring by asexual reproduction

Organs

Different tissues are often combined together to form **organs**. *An organ is a distinct part of the body which carries out one or more specific functions.*

The way cells are grouped into tissues and tissues into organs is summarised in figure 7.4 and the main organs in the human body are summarised in figure 7.5.

Plants are composed mainly of tissues and there are not many organs as such. Examples of plant organs are given in table 7.4.

Organ systems

Certain organs work together to fulfil a particular function. These are called **organ systems**. Table 7.5 summarises the main organ systems in the human body.

Table 7.5 *Summary of the systems in the human body.*

Name of system	Main organs in the system	Main functions
Digestive system	Gut, liver and pancreas	To digest and absorb food
Respiratory system	Windpipe and lungs	To take in oxygen and get rid of carbon dioxide
Blood (circulatory) system	Heart, blood vessels	To carry oxygen and food round the body
Excretory system	Kidneys, bladder, liver	To get rid of poisonous waste substances
Sensory system	Eyes, ears, nose	To detect stimuli
Nervous system	Brain and spinal cord	To conduct messages from one part of the body to another
Musculo-skeletal system	Muscles and skeleton	To support and move the body
Reproductive system	Testes and ovaries	To produce offspring

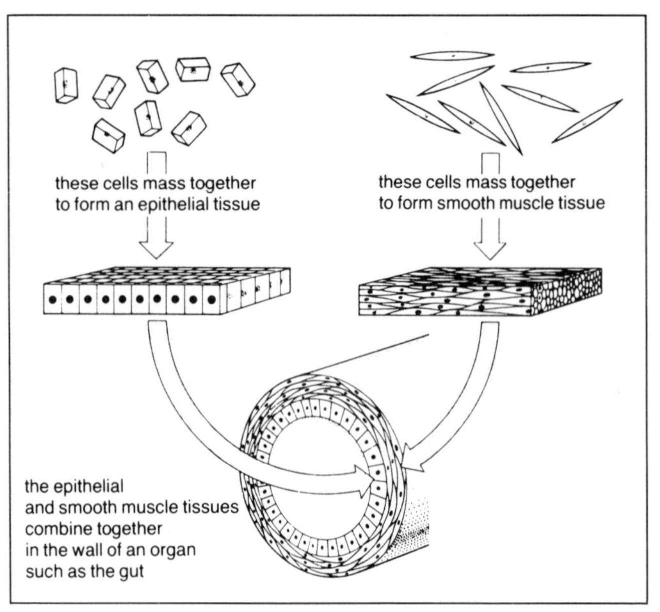

Figure 7.4 *This diagram illustrates how cells are grouped into tissues and tissues into organs.*

Shape and symmetry

Most animals, including the human, are **bilaterally symmetrical**: the various structures are mainly arranged symmetrically on either side of a line drawn down the middle of the body. Examples include the earthworm, locust and goldfish.

Some animals and most plants are **radially symmetrical**: the various structures are arranged in a circle round a central point, like the spokes of a wheel. Examples include *Hydra*, sea anemones and jellyfish.

Most animals have a leading end, back end, top side and bottom side. The scientific names for these four parts are: **anterior**, **posterior**, **dorsal** and **ventral**. The sides of the animal are described as lateral (figure 7.6).

If a **head** is present, it is situated at the anterior end. This is where the sense organs and feeding structures are usually found.

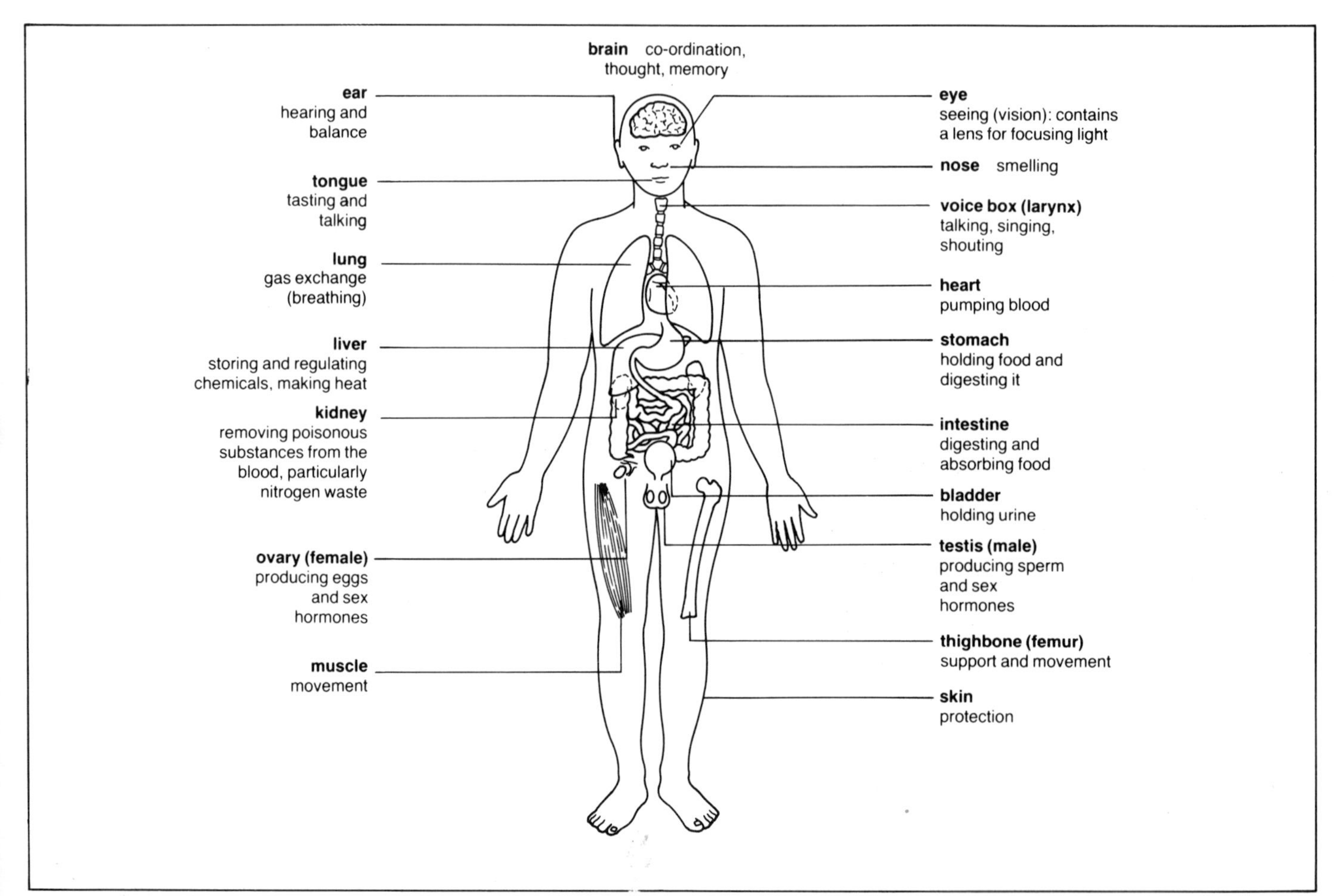

Figure 7.5 *The main organs of the human body (male), and their functions, (From* Biology 11–13, *Mawby and Roberts, Longman).*

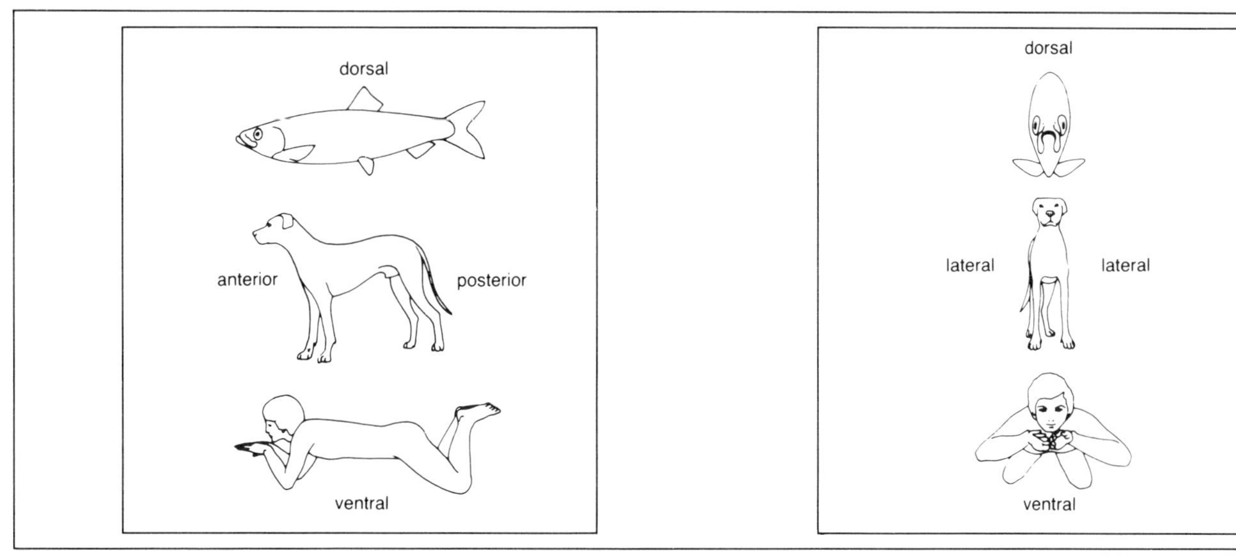

Figure 7.6 *The terms which are used to describe the different parts of the body.*

• Questions

1 The photograph below shows a cell from the inner surface of a person's cheek, viewed under a light microscope.

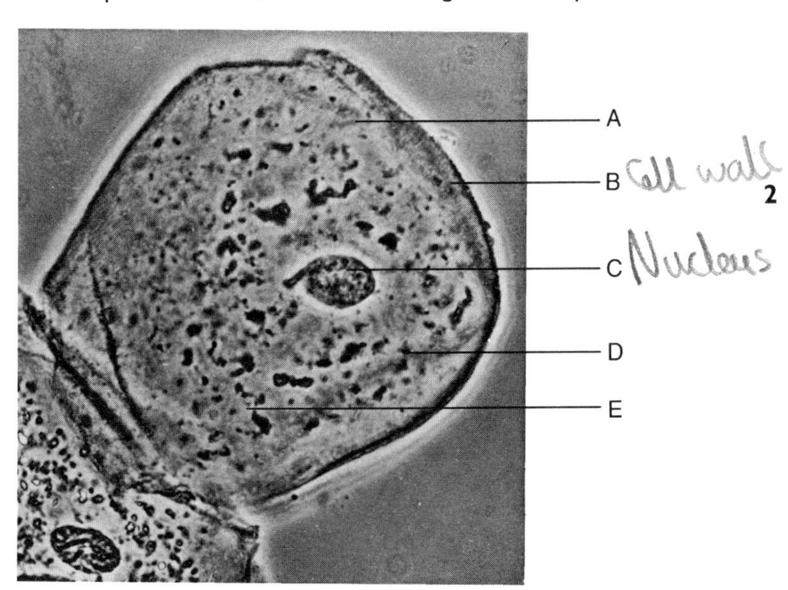

(a) Name the structures labelled A, B and C.

(b) What *might* structures D and E be?

(c) What kind of tissue did this cell come from?

(d) Briefly describe how this cell would have been:
  (i) removed from the cheek;
  (ii) mounted on a microscope slide.

(e) From the scale provided, calculate the magnification of the cell in the photograph.

2 (a) Name *four* structures which are present in a typical plant cell but absent from a typical animal cell.

(b) Name *one* structure which is present in a typical animal cell but absent from a typical plant cell.

(c) Chloroplasts are not present in *all* plant cells. Give two examples of cells in a flowering plant which do *not* contain chloroplasts.

(d) Name two structures in the cytoplasm of a cell which are visible in the electron microscope but *not* under the light microscope.

(e) Name one structure in the cytoplasm of a cell which is *just* visible under the light microscope but is seen in more detail in the electron microscope.

3 Look carefully at table 7.1, then answer the following questions:

(a) Explain, as far as you can, the meaning of the words in italics. Be as precise as possible.

(b) From the information given in the table, state the *overall* function of (i) the nucleus and (ii) the cytoplasm.

4 (a) What is the difference between a tissue and an organ?

(b) Which of the tissues listed in tables 7.3 and 7.4 would you expect to find in each of the following:
(i) the human heart; (ii) the human brain;
(iii) a plant stem; (iv) a plant root?

5 (a) What is meant by (i) bilateral symmetry and (ii) radial symmetry?

(b) Name *one* organ system in the human body which is *not* bilaterally symmetrical. Explain your answer.

(c) Name *one* invertebrate animal which is bilaterally symmetrical and *one* which is radially symmetrical.

(d) Name *two* plant structures which are radially symmetrical and *two* which are bilaterally symmetrical.

8

Movement in and out of cells

How do particles move?

Particles move in three main ways: by **diffusion**, **osmosis** and **active transport**. The particles may be molecules or ions.

What is diffusion?

Diffusion is the net movement of particles from a region where they are more concentrated to a region where they are less concentrated. The difference in concentration between the two regions is called the **concentration gradient** or **diffusion gradient**. Diffusion always takes place *down* a concentration gradient.

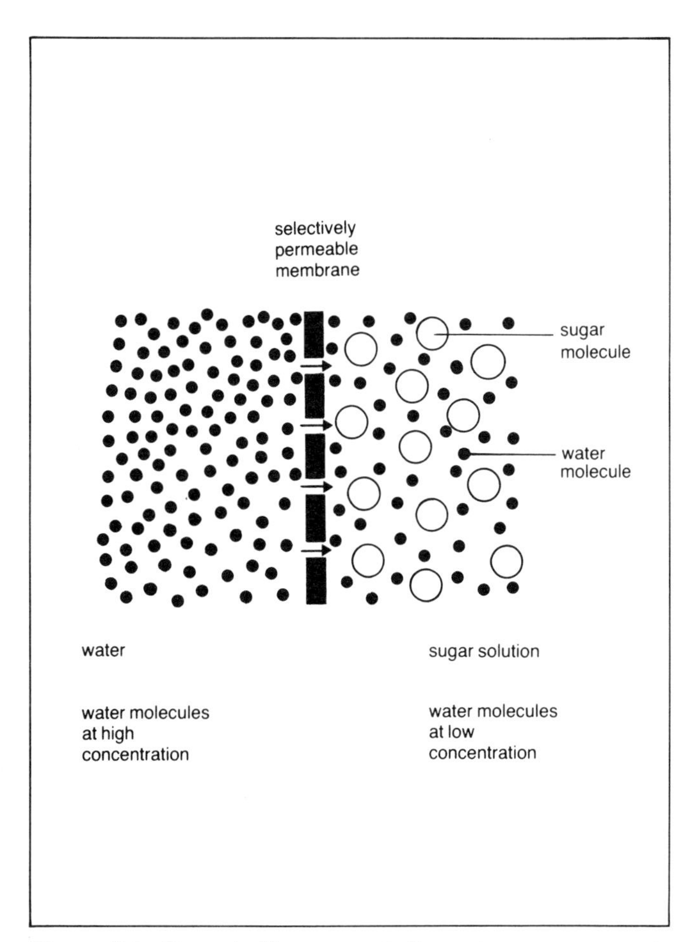

Figure 8.1 *Osmosis. The arrows indicate the net movement of water molecules through the selectively permeable membrane.*

Examples of diffusion in biology

1. The movement of oxygen from the lungs to the bloodstream, and of carbon dioxide from the bloodstream to the lungs in the human (see page 66).
2. The movement of oxygen and carbon dioxide into and out of a leaf (see page 77).
3. The movement of nitrogenous waste matter from the inside of an amoeba to the outside.

Membranes and diffusion

In the above examples, diffusion takes place across a **permeable membrane**. A permeable membrane is one which will let the particles through. For example, in the lungs the membrane between the alveoli and the bloodstream is permeable to oxygen and carbon dioxide.

How does a membrane's surface area affect diffusion?

The larger the surface area of the membrane, the greater will be the amount of diffusion across it. In biology, membranes through which diffusion takes place are often folded so as to increase their surface area.

The surface-volume ratio

The **surface-volume ratio** of an object (e.g. an organism) is its surface area relative to its volume. The larger an object, the smaller is its surface-volume ratio, i.e. the smaller is its surface area relative to its volume.

How does the surface-volume ratio affect organisms?

The surface-volume ratio affects organisms in two main ways:

1. A small organism with a large surface-volume ratio can get all the oxygen it needs by diffusion across the body surface. A larger organism with a small surface-volume ratio requires a special respiratory surface (e.g. lungs, gills) for obtaining oxygen, and a transport system for carrying oxygen quickly to the tissues.
2. A small organism with a large surface-volume ratio loses and/or gains heat more quickly than a large organism with a small surface-volume ratio. This is important in temperature regulation.

What is osmosis?

Osmosis occurs when pure water is separated from a solution of, e.g. sugar by a **selectively permeable membrane** (figure 8.1). The membrane is permeable to the water molecules but impermeable to the sugar molecules. As a result, the water molecules diffuse through the membrane into the sugar solution.

The fluid on the left hand side of the membrane in figure 8.1 does not have to be pure water. It could be another sugar solution. Osmosis will still occur provided that the solution on the left hand side is weaker (i.e. the concentration of sugar molecules is lower) than on the right hand side.

What happens if a red blood cell is put in pure water?

Water enters the cell by osmosis through the selectively permeable cell membrane (**endosmosis**). As a result the cell swells up and bursts.

What happens if a red blood cell is put in a strong solution?

Water leaves the cell by osmosis through the selectively permeable cell membrane (**exosmosis**). As a result the cell shrinks and the cell membrane crinkles.

Note: for this to happen the external solution must be more concentrated than the cell contents.

What happens if a plant cell is put in pure water?

Water enters the vacuole by osmosis through the selectively permeable cell membrane. The cell swells up until the cellulose cell wall resists further expansion. The cell is then fully **turgid**. (*Note* that the cellulose cell wall prevents the cell bursting.)

What happens if a plant cell is put in a strong solution?

Water leaves the vacuole by osmosis through the selectively permeable cell membrane. The cell shrinks and eventually the cytoplasm pulls away from the cellulose cell wall. This is called **plasmolysis** (figure 8.2).

Note: for this to happen the external solution must be more concentrated than the fluid in the vacuole (cell sap).

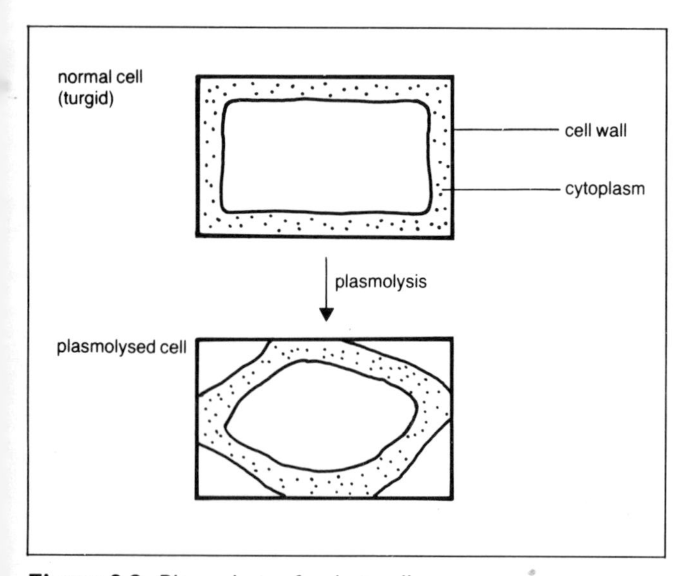

Figure 8.2 *Plasmolysis of a plant cell.*

Why does a gap develop between the cell wall and cytoplasm in plasmolysis?

The reason is that the cellulose cell wall is permeable to both the water and the molecules in the external solution. It is not a selectively permeable membrane. So, as plasmolysis proceeds, the external solution diffuses through the cell wall and fills the gap between the cell wall and the cytoplasm.

Why is it useful for plant cells to be turgid?

When turgid, the cells are tightly packed within the plant and they help to keep the stem erect and the leaves expanded. If the cells lose their turgidity, the stem and leaves droop. This is what happens when a plant **wilts**.

What causes wilting?

Wilting occurs when water evaporates from a plant more quickly than it can be replaced from the soil.

What happens to a plant cell during wilting?

Water passes out of the cell. As a result, the cell shrinks and crumples. The cell wall is pulled in with the cytoplasm. A wilted plant cell does not have a gap between the cell wall and cytoplasm (figure 8.3).

What is active transport?

Active transport is the movement of particles (molecules or ions) through a membrane from a region where they are less concentrated to a region where they are more concentrated, i.e. *against* a concentration gradient.

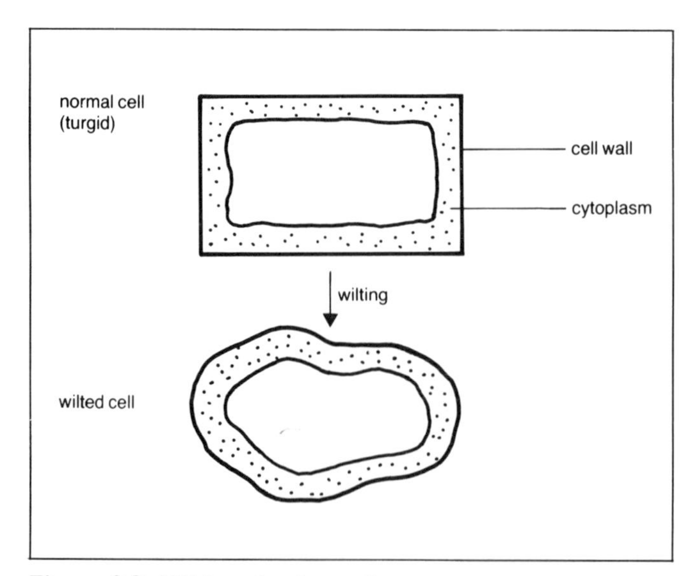

Figure 8.3 *Wilting of a plant cell.*

Unlike diffusion and osmosis, active transport requires **energy from respiration**. Active transport ceases if oxygen is not present, or if the organism is treated with a chemical substance which stops respiration.

In active transport molecules and ions are conveyed across the cell membrane by **protein carriers** within the membrane itself.

Examples of active transport

1. The uptake of mineral salts from the soil water by the roots of plants (see page 80).
2. The absorption of certain products of digestion by the cells lining the small intestine of a mammal (see page 57).

Gas exchange

Gas exchange is the diffusion of oxygen and carbon dioxide between the inside and outside of an organism.

Two processes are associated with gas exchange: **respiration** and **photosynthesis**. Respiration uses oxygen and produces carbon dioxide. Photosynthesis uses carbon dioxide and produces oxygen.

Gas exchange in organisms

Animals take up oxygen and give out carbon dioxide all the time; that is because they respire all the time.
Gas exchange in plants depends on the time of day:

- **At night** plants take up oxygen and give out carbon dioxide. That is because in the dark they respire but do not photosynthesise.
- **During the day** plants take up carbon dioxide and give out oxygen. That is because in bright light they photosynthesise faster than they respire.
- **At dawn and dusk** they do not take up or give out either gas. That is because in dim light they photosynthesise at the same rate as they respire: the carbon dioxide produced by respiration is used for photosynthesis and the oxygen produced by photosynthesis is used for respiration.

• Questions

1 The diagram below shows two cubes. The surface-volume ratio can be expressed as the surface area divided by the volume.

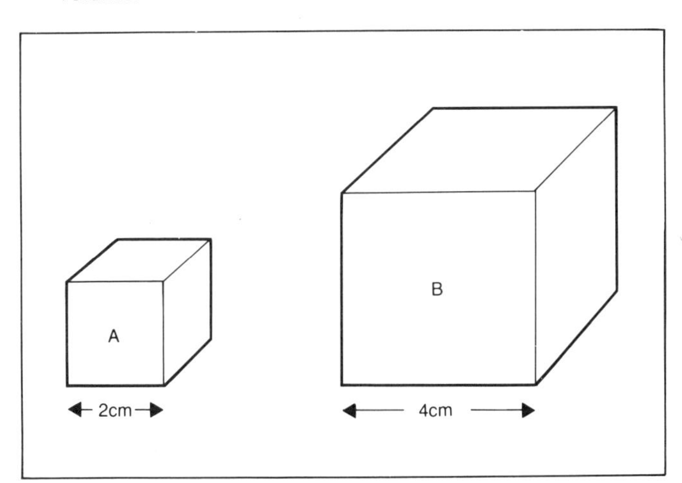

(a) (i) Calculate the surface-volume ratio of cube A.
(ii) Calculate the surface-volume ratio of cube B.

(b) If the two cubes were warmed to the same temperature and then put in a cool place, which one would lose heat more quickly and why?

(c) (i) Suppose the two cubes were organisms. Which one would you expect to require a special surface for gas exchange and why?
(ii) Give two examples of special surfaces which are used for gas exchange in animals.

2 The diagram below shows a simple osmometer. An osmometer is an apparatus for demonstrating osmosis and measuring its rate.

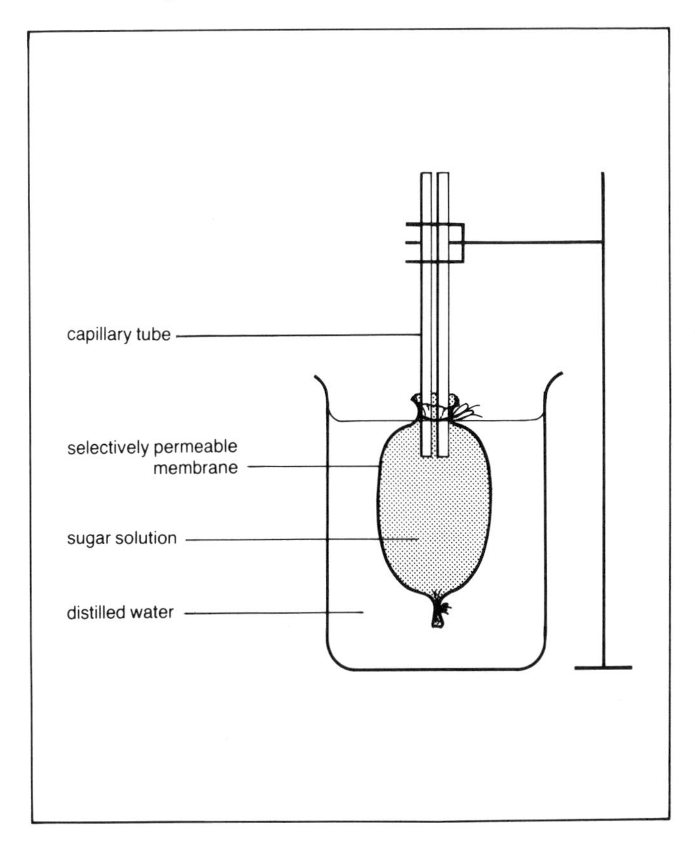

(a) Explain the meaning of the term *selectively permeable membrane.*

(b) After being set up, the fluid rises in the capillary tube.
  (i) What causes the fluid to rise in the capillary tube?
  (ii) What substance could you add to the beaker to prevent the fluid rising?
  (iii) Why should adding this substance prevent the fluid rising?

(c) Describe a simple experiment which you could do, *without using a microscope*, to demonstrate osmosis in a living tissue.

3 The diagrams below show what happens if you put a plant cell in distilled water or a strong sugar solution.

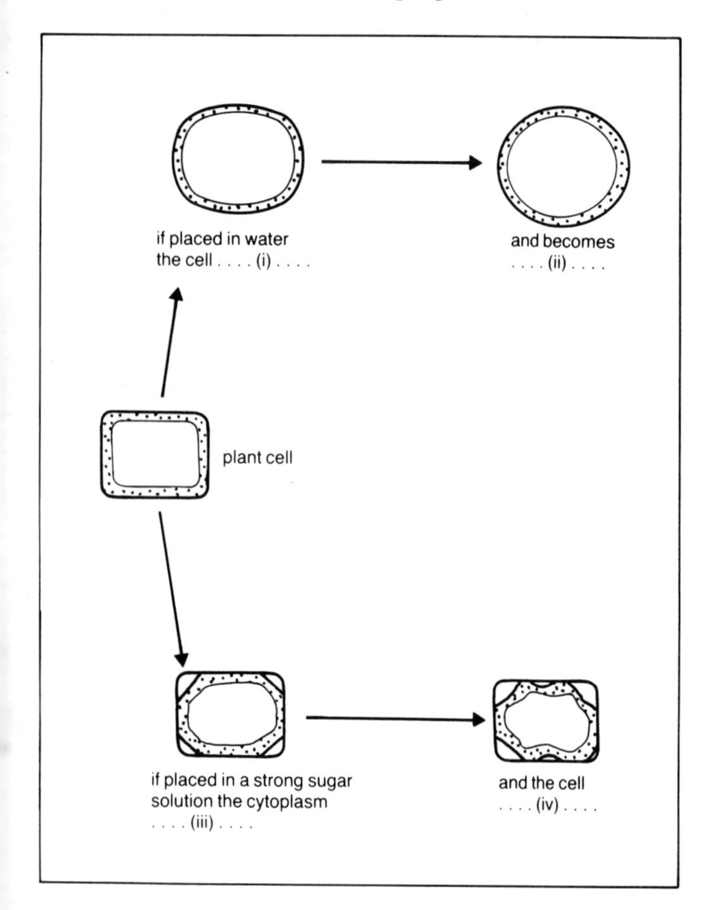

(a) Write down the missing words (i) to (iv) in the labels on the diagrams above.

(b) What prevents the cell which is put in distilled water from bursting?

(c) (i) What would happen if you put a red blood cell in distilled water?
  (ii) Give a reason for your answer.

(d) Plant cells are usually in the condition shown by the top right hand diagram. Why is this important to a plant such as a geranium?

4 Some living organisms are placed in a series of sealed tubes, together with a small quantity of hydrogencarbonate indicator solution. A change in the colour of the indicator solution signifies a change in the concentration of carbon dioxide in the tube: yellow means that the carbon dioxide concentration has increased, purple that it has decreased and red that it has remained unchanged. Details of the experiment and the results are shown below.

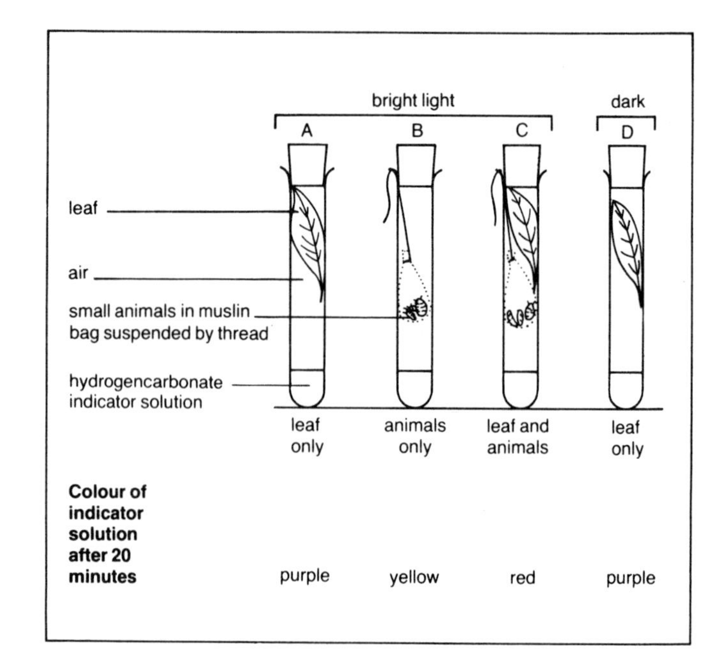

(a) In which tube or tubes does the concentration of carbon dioxide (i) increase, (ii) decrease, (iii) remain unchanged?

(b) Explain the reason for the carbon dioxide concentration in each tube.

(c) What do you think would happen to the carbon dioxide concentration in a sealed tube containing leaf and woodlice in the dark.

(d) A bottle garden is a sealed bottle or jar containing plants growing in soil. Such plants may live for years. How do they survive?

9 The chemistry of life

What does the human body consist of?

The human body consists of the following chemical substances:

1. **Carbohydrate** 5%
2. **Fat** 10%
3. **Protein** 18%
4. **Water** 65%
5. **Other substances** 2%

The proportions are approximate: they vary from person to person and depend on circumstances. 'Other substances' include **vitamins** and their derivatives, and **salts** such as sodium chloride.

Which are organic and which are inorganic?

Carbohydrate, fat, protein and vitamins are all **organic substances**. Water and salts are **inorganic substances**. (Organic substances are complex carbon-containing chemicals, which often have long and complex chemical formulae. Inorganic substances are much simpler, e.g. water, H_2O.)

What is a carbohydrate?

A carbohydrate is an organic substance which contains carbon, hydrogen and oxygen, with twice as much hydrogen as either carbon or oxygen. One of the simplest carbohydrates is **glucose**. Its chemical formula is $C_6H_{12}O_6$.

Types of carbohydrate

There are three types of carbohydrate:

1. **Monosaccharide**: 'single-sugar' molecule o
2. **Disaccharide**: 'double-sugar' molecule o-o
3. **Polysaccharide**: 'multi-sugar' molecule o-o-o-o-o-o

Monosaccharides can be built up into disaccharides and polysaccharides by a type of chemical reaction in which water is taken away (**condensation**).

Polysaccharides can be broken down into disaccharides and monosaccharides by a type of chemical reaction in which water is added (**hydrolysis**).

Table 9.1 shows the three types of carbohydrate. Their properties are summarised and examples of each are given. Notice the way one can be converted into another by either condensation or hydrolysis.

What are the functions of carbohydrates?

1. Glucose (and other sugars) **provide energy** when respired (see page 46). Glucose is found in solution in nearly all cells.

Table 9.1 *The three types of carbohydrate, their properties and examples of each type. The arrows show how one carbohydrate can be converted into another. Downward arrows indicate condensation (water removed), upward arrows indicate hydrolysis (water added).*

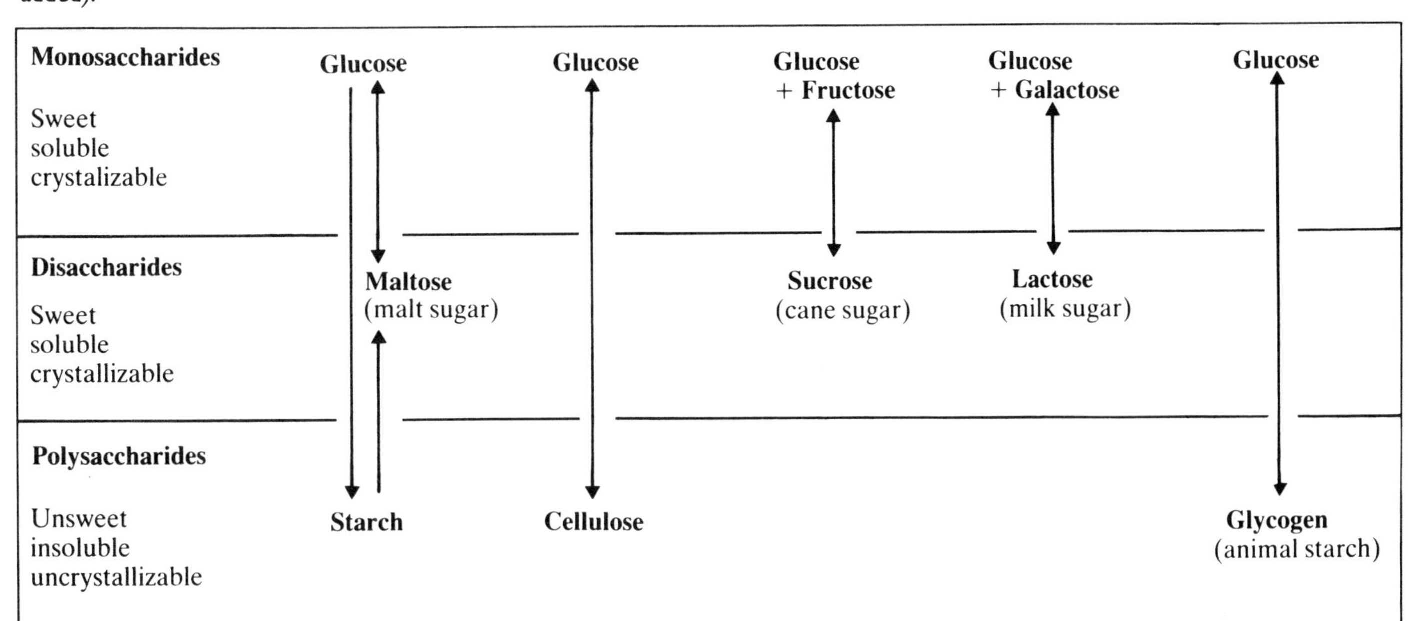

Monosaccharides Sweet soluble crystalizable	**Glucose**	**Glucose**	**Glucose** **+ Fructose**	**Glucose** **+ Galactose**	**Glucose**
Disaccharides Sweet soluble crystallizable	**Maltose** (malt sugar)		**Sucrose** (cane sugar)	**Lactose** (milk sugar)	
Polysaccharides Unsweet insoluble uncrystallizable	**Starch**	**Cellulose**			**Glycogen** (animal starch)

② Starch and glycogen **store energy** for use when required. Starch is found as **starch grains** in plant cells. Glycogen is found as **glycogen granules** in animal cells (see page 31).
③ Cellulose is the material of which **plant cell walls** are made. It is tough but flexible. It is used in manufacturing paper and other important products.

What is fat?

A fat is an organic substance which contains carbon, hydrogen and oxygen, with much more carbon and hydrogen than oxygen. Above a certain temperature fat goes into a fluid state called an **oil**.

What are the chemical components of fat?

A fat is made up of two parts: a molecule of **glycerol** linked with three molecules of **fatty acid**.

Fat can be broken down into glycerol and fatty acids by hydrolysis. And glycerol and fatty acids can be joined together and built up into a fat by condensation.

What are the functions of fat?

① Fat **provides energy** when respired.

② Fat **stores energy** for use when required. Humans and other animals possess fatty tissue which consists of densely packed fat-filled cells.

③ Fat is an important component of **cell membranes**.

④ Oil on the surface of the body helps to **keep water out**. Fats and oils repel water and this makes them good for waterproofing.

⑤ Fat beneath the skin helps to **keep mammals warm**. Fat is a good insulating material.

What is a protein?

A protein is an organic substance which contains carbon, hydrogen, oxygen and nitrogen. Sometimes sulphur is present too.

What are the building blocks of a protein?

A protein consists of one or more chains of **amino acids**. The amino acids are joined together by **peptide links**. Approximately 20 types of amino acid exist in nature.

How do proteins differ from each other?

Proteins differ from each other in three main ways:

① The number and types of amino acids present in their chains.
② The order in which the different amino acids are joined together.
③ The way the chains are folded and arranged.

Types of protein

There are two main types of protein:

① **Globular proteins**, which are soluble: they occur in solution in cells, blood etc., e.g. albumen (egg white), enzymes.
② **Fibrous proteins**, which are insoluble: they are tough and fibre-like, found in structures such as bone, muscle and connective tissue.

What happens if you heat a globular protein?

If a gobular protein is heated much above 40°C, the shape of the molecule changes. The protein is **denatured**. This alters its properties and ruins its function. For example, if you boil an egg, the albumen coagulates (solidifies). This may be nice for humans, but it is a disaster for hens.

What are the functions of proteins?

① Proteins **provide energy** when respired.

② Proteins are the main **structural material** of the body. (Fibrous proteins are particularly important in this respect.)

③ Proteins are an important component of **cell membranes**.

④ Globular proteins **regulate processes** that occur in cells. (Enzymes are particularly important in this respect.)

Breaking down and building up proteins

A protein can be broken down into its amino acids by hydrolysis. First the protein is split into short chains of amino acids called **polypeptides**. The polypeptides are then broken down into single amino acids. Conversely, amino acids can be joined together by condensation to form a polypeptide or protein.

Protein in our food is broken down into amino acids in the gut. The amino acids are then taken to our cells where they may be built up again into a different protein, i.e. one that humans need.

What is metabolism?

Metabolism is the chemical reactions which take place inside cells. (It does not include the chemical reactions that occur in the gut as a result of digestion, because these reactions are taking place outside the cells.)

There are two types of metabolism:

① **Build-up reactions** (synthesis, anabolism): simple molecules are built up into more complex molecules. These reactions require energy (they are **endothermic**).
② **Break-down reactions** (catabolism): complex molecules are broken down into simpler molecules. These reactions release energy (they are **exothermic**).

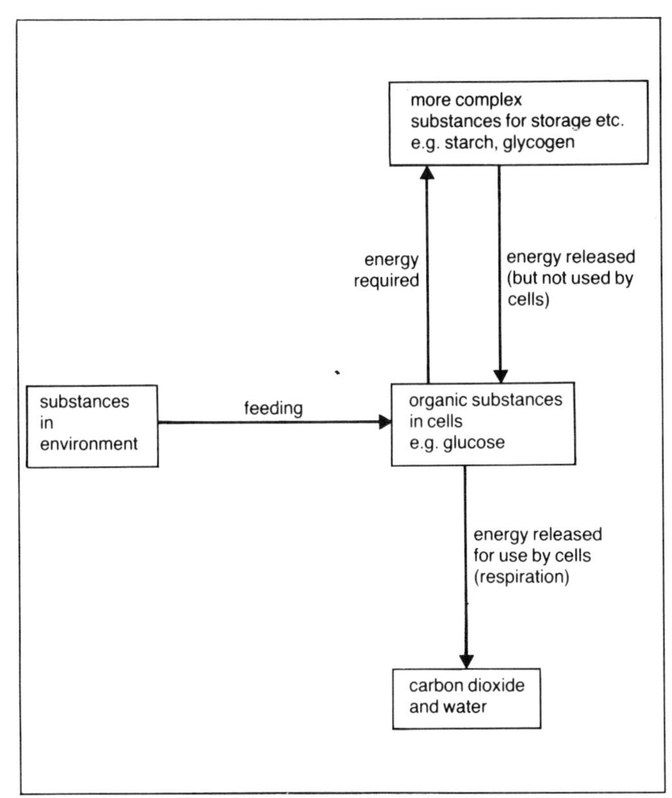

Figure 9.1 *Summary of the main kinds of chemical reactions which occur in organisms.*

Figure 9.1 shows how these two types of metabolism relate to each other.

What are enzymes?

Enzymes are proteins which speed up (**catalyse**) the chemical reactions which occur in organisms. They are 'biological catalysts'.

Why are enzymes important in our bodies?

- Without them the chemical reactions vital to life could not take place quickly enough.
- They control where and when the various chemical reactions take place.

Types of enzymes

Enzymes fall into two main categories:

1. **Intracellular enzymes:** these occur inside cells and catalyse metabolic reactions.
2. **Extracellular enzymes**: these occur outside cells and catalyse reactions involved in digestion (see page 56).

Enzymes are also classified according to the type of substance on which they act. Thus:

- **Carbohydrases** act on carbohydrates.
- **Lipases** act on fats.
- **Proteases** act on proteins.

Within these groups there are many different individual enzymes. For example, in the carbohydrase group, **sucrase** acts on sucrose, **maltase** on maltose. The name of an enzyme usually comes from the name of the substance it acts on with the ending changed to 'ase'.

Reversible reactions

Most metabolic reactions are reversible, that is they will go in either direction. This is shown by a double arrow, like this:

$$\mathbf{A} \rightleftharpoons \mathbf{B}$$

A is converted into **B** if there is more **A** present than **B**.
B is converted into **A** if there is more **B** present than **A**.
The same enzyme catalyses both conversions.

What are the properties of enzymes?

Enzymes have six important properties which affect their functioning:

1. They are always proteins.
2. They are specific in their action.
3. They can be used over and over again.
4. They are destroyed by heat.
5. They are sensitive to pH.
6. They are readily inactivated by poisons.

How do enzymes work?

Suppose this chemical reaction is catalysed by an enzyme:

$$\mathbf{A} \longrightarrow \mathbf{B} + \mathbf{C}$$

A molecule of **A** (the **substrate**) collides with the enzyme molecule and fits into an **active site** on its surface. **A** then splits into the product molecules, **B** and **C**. The active site has a shape which only **A** will fit. This is called the **lock-and-key theory** (figure 9.2).

Many enzymes are helped in their action on substrates by **mineral elements** (e.g. iron, copper) and/or **vitamins**. That is why these substances are needed by organisms (see page 51).

How does the lock-and-key theory explain the properties of enzymes?

Three properties, in particular, can be explained in terms of the lock-and-key theory:

1. **Specificity**: only one type of substrate molecule will fit the active site.
2. **Destruction by heat**: heat alters the shape of the active site so that the substrate molecules will no longer fit.
3. **Inactivation by poisons**: the poison molecules fit into the active site, blocking it temporarily or permanently.

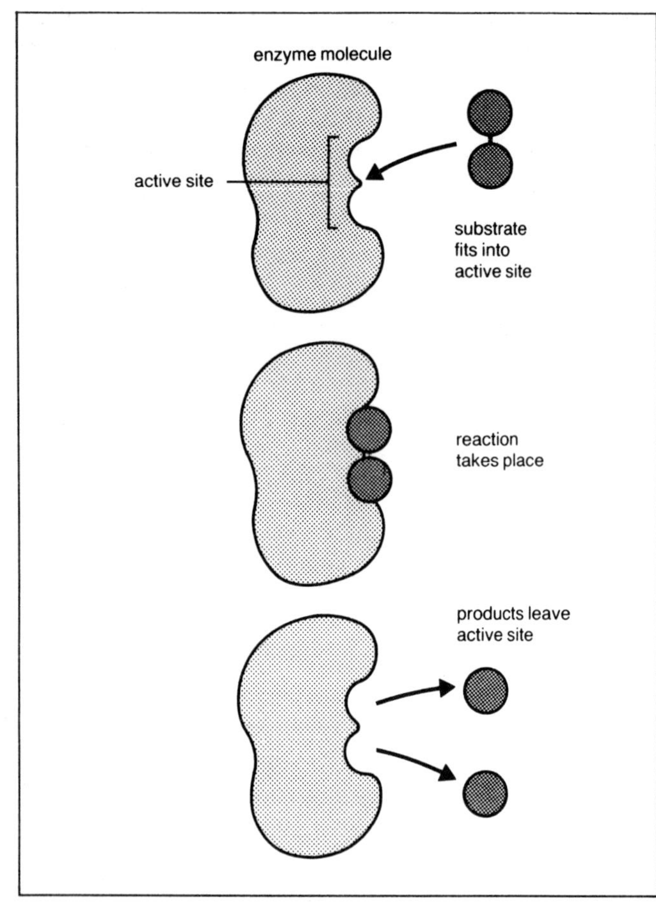

Figure 9.2 *How an enzyme works. The substrate fits into the active site where the reaction takes place.*

What use can we make of enzymes?

Here are four ways enzymes are used in the manufacturing industry:

- Proteases are used as stain-removers in biological washing powders.
- Proteases are used for tenderising meat.
- Cellulase (cellulose-splitting enzyme) is used for softening vegetables.
- Amylase (starch-splitting enzyme) is used for making syrups and chocolates.

Why are enzyme poisons important?

Enzyme poisons stop enzymes working, i.e. they **inhibit** them. Their most important use is as **pesticides**. For example, organophosphate compounds kill insects by inhibiting an enzyme in the nervous system.

How does temperature affect metabolic reactions?

As with chemical reactions in general, raising the temperature increases the rate of metabolic reactions: a rise of 10°C approximately doubles the rate of the reaction. This continues until the temperature reaches approximately 40°C.

Why does raising the temperature increase the rate of metabolic reactions?

Molecules are in random motion. Raising the temperature increases the speed of their movement: as a result, the substrate and enzyme molecules collide more often.

What happens when the temperature reaches 40°C?

Above this temperature the enzyme molecules are denatured. The result is that the rate of the metabolic reaction falls quickly to zero. Because of this, it is dangerous for an organism's body temperature to get much above 40°C. Most organisms have ways of preventing their body temperature from getting too high (see pages 83–4).

Why is water needed by organisms?

There are many reasons for this. One of the main reasons is that water is the substance in which biological molecules are dissolved and in which all the chemical reactions occurring inside the organism take place.

• Questions

1 (a) Why is a polysaccharide such as starch a convenient way of storing energy? (*Hint*: what does a starch molecule consist of?)

(b) Insects have a very thin layer of grease on the surface of their cuticle. In what way is this useful to them?

(c) Protein is the main structural material of animals. Which is the main structural material of plants, and what does it consist of?

(d) Seals have a very thick layer of fatty tissue (blubber) under the skin. In what way is this useful to them?

(e) The proteins we eat may not be the same as the ones in our body. How do we get our proteins from the ones we eat?

2 Read the section headed 'What use can we make of enzymes?' on page 44.

(a) The biological action of a washing powder is reduced if the temperature of the water is too high. Explain the reason for this.

(b) How does a protease tenderise meat?

(c) How does cellulase soften a vegetable such as a cabbage?

(d) In making syrup, starch is mixed with amylase. What is the product of the reaction, and why is it useful for making syrup?

3 The graph below shows the effect of temperature on the rate of a metabolic reaction. The reaction is controlled by an enzyme.

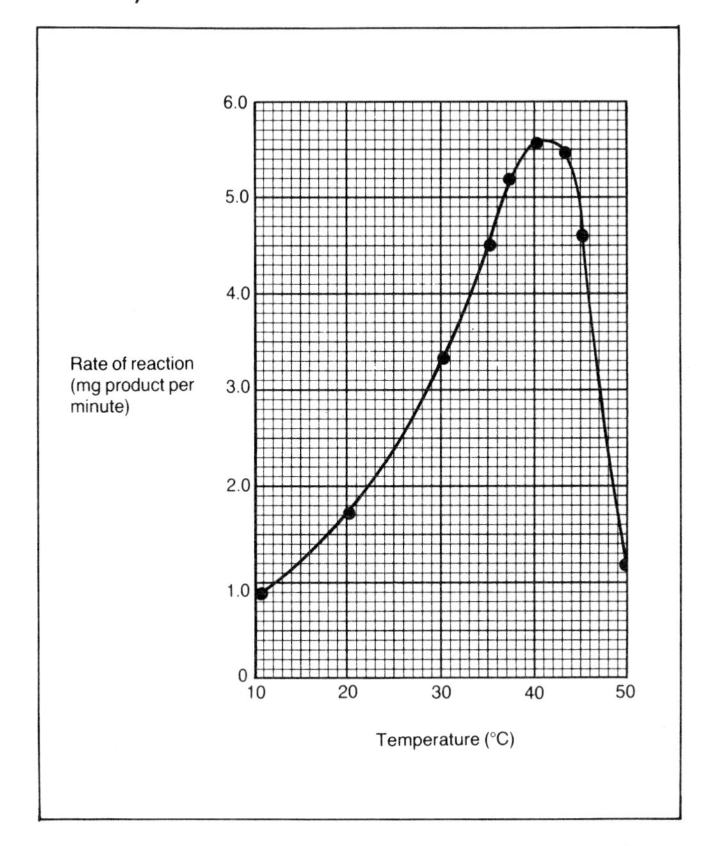

(a) (i) What is the rate of the reaction at 20°C and 30°C?
(ii) By approximately how many times does the rate of the reaction increase between these two temperatures?

(b) Why does raising the temperature increase the rate of the reaction?

(c) Why does the rate of the reaction decrease above 40°C?

(d) Suggest *two* substances you could add to the enzyme-substrate mixture in order to stop the reaction at 30°C. Explain why each substance stops the reaction.

4 Four different reactions involving glucose are shown below:

(1) glucose → starch
(2) glucose → glycogen
(3) glucose → cellulose
(4) glucose → carbon dioxide and water.

(a) Which reaction or reactions (i) require energy, (ii) release energy?

(b) Where, and for what purpose, do reactions 1 to 4 take place?

(c) Give *two* situations in which reaction 1 proceeds in the reverse direction.

(d) Where, and when, does reaction 4 proceed in the reverse direction?

at 37° best temp. for the enzyme to work.

The release of energy

Where do organisms get energy from?

The ultimate source of energy for all organisms is the **sun** (see page 17). However, an organism's immediate source of energy is its **food**.

What is energy needed for?

- Animals need energy for movement (e.g. muscle contraction), sending messages through nerves, transporting things inside the body, and keeping warm.
- Plants need energy for taking up mineral salts from the soil, opening and closing their stomata, and transporting food substances.
- All organisms need energy for growth, cell division, and moving molecules and ions against a concentration gradient (active transport).

How is energy released in organisms?

Energy is released by **respiration**. *Respiration is the chemical process that takes place in every living cell, by which energy is released to keep the cell alive.*

Types of respiration

There are two types of respiration:

1. **Aerobic respiration**: respiration with the use of oxygen.
2. **Anaerobic respiration**: respiration without the use of oxygen.

What happens in aerobic respiration?

In aerobic respiration glucose is oxidised to give carbon dioxide and water:

$C_6H_{12}O_6$	+	$6O_2$	$\rightarrow$	$6CO_2$	+	$6H_2O$	+	energy
glucose		oxygen		carbon dioxide		water		

Glucose comes from the organism's food. Oxygen comes from the air which the organism breathes in; carbon dioxide is got rid of in the air breathed out. Most of the energy is lost as heat, but some of it is used for keeping the organism alive.

How can you find out if carbon dioxide is present in the air we breathe out?

Breathe out (exhale) through a narrow glass tube or drinking straw into **lime water**. If the lime water turns milky, carbon dioxide is present in the exhaled air.

How can you find out if a small mammal gives out carbon dioxide?

Use a suction pump to draw air through the following, one after the other:

1. soda lime pellets to absorb carbon dioxide from the air;
2. lime water to make sure the carbon dioxide has been absorbed by the soda lime;
3. a bell jar containing the animal;
4. lime water to test if the animal has given out carbon dioxide.

How can you find out if small animals and plants give out carbon dioxide?

Suspend the animals or plants (e.g. woodlice, leaf) above a small quantity of **hydrogencarbonate indicator solution** in a test tube. Seal the tube (it must be airtight). If carbon dioxide is given out by the organisms, the indicator solution will turn from reddish-orange to yellow.

Hydrogencarbonate indicator solution is more sensitive than lime water. It is therefore more suitable for small organisms.

How can you find out if the carbon dioxide given out by an organism comes from glucose?

First, 'label' the carbon atoms in some glucose: this is done by replacing the normal carbon atoms with radio-active carbon atoms. Feed the radio-active glucose to a mouse. Then collect the mouse's exhaled air as described in the experiment before last and pass it through lime water. Test the lime water for radio-activity with a Geiger counter. If it is radio-active, the labelled carbon atoms in the glucose must have got into the carbon dioxide given out by the mouse.

How can you find out if small organisms release energy?

You can do this by finding out if the organisms give out heat. Fill a thermos flask with germinating peas. Insert a thermometer, then seal the thermos flask with cotton wool. Take the temperature at intervals, and see if it rises. Set up a second flask containing killed peas to serve as a control. (A control is particularly important in this experiment as the temperature increase may be very small.)

How can you measure the rate of respiration of small organisms?

By means of a **respirometer** (figure 10.1). This apparatus measures the rate at which oxygen is absorbed by the organisms, i.e. their **metabolic rate**. You can use a respirometer to:

① compare the rate of respiration of different organisms;
② find the effect of varying the temperature on the rate of respiration.

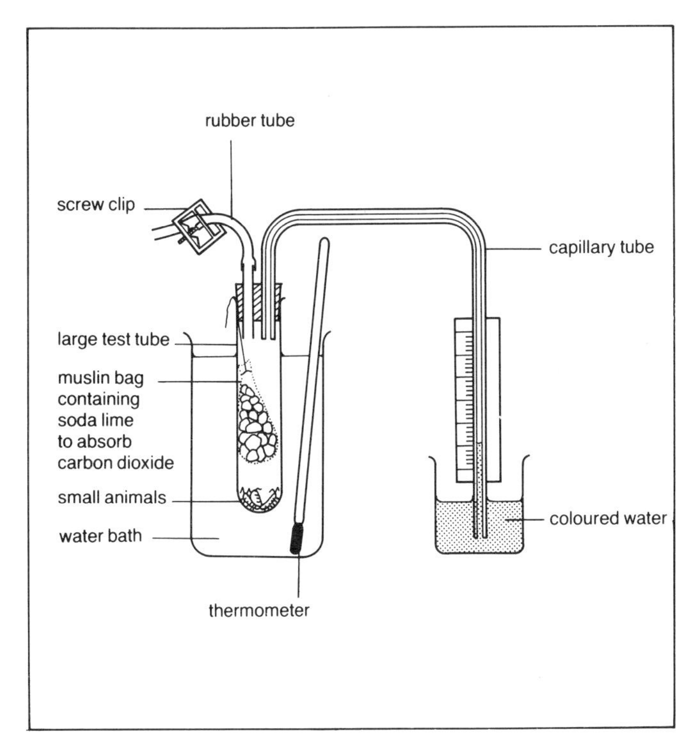

Figure 10.1 *A simple respirometer.*

How does temperature affect the rate of respiration?

A rise of 10°C approximately doubles the rate of respiration. This is true of chemical reactions in general, so it shows that basically respiration is just an ordinary chemical process.

If the temperature rises much above 40°C, the rate of respiration quickly decreases and soon stops. This is because respiration is catalysed by enzymes which are denatured by heat.

Can energy be released from other substances besides carbohydrates?

Energy can be released in cells from fats and proteins, as well as from carbohydrates. In the case of fats, most of the energy comes from oxidation of the fatty acid part of the molecule. In the case of proteins, the energy comes from oxidation of the amino acids after the nitrogen part of the molecule has been removed (see page 84).

What happens in anaerobic respiration?

In anaerobic respiration sugar is broken down, without the use of oxygen, into either **ethanol** (alcohol) or **lactic acid**. Energy is released.

A lot less energy is produced by the anaerobic breakdown of a glucose molecule than is produced by the aerobic breakdown of a glucose molecule.

How is ethanol produced by anaerobic respiration?

Here is a word equation summarising what happens:

sugar → ethanol + carbon dioxide + energy

This process is called **alcoholic fermentation**. It takes place in **yeast**. Yeast is a single-celled fungus which grows on the surface of fruit, feeding on sugar.

How can you demonstrate alcoholic fermentation?

By means of the apparatus shown in figure 10.2. The liquid paraffin is permeable to carbon dioxide but not to oxygen. If alcoholic fermentation occurs, the lime water turns milky and you can smell alcohol.

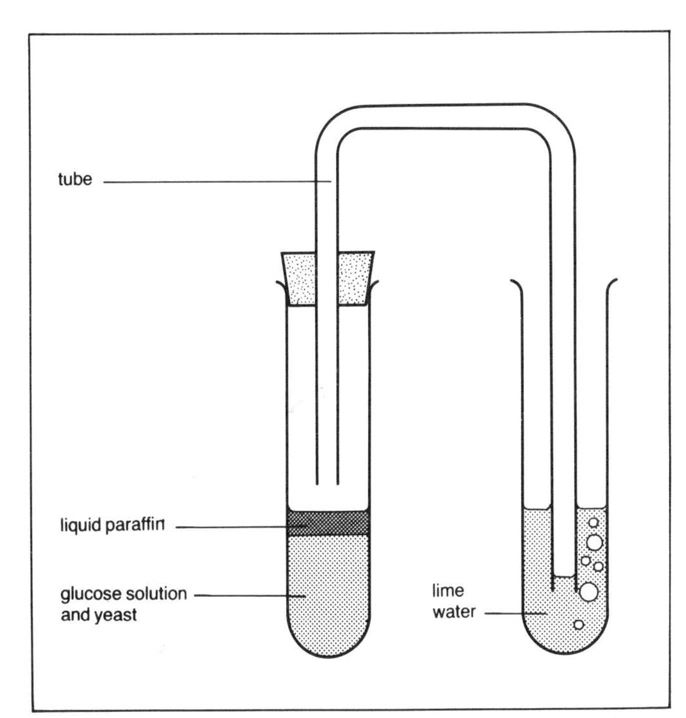

Figure 10.2 *Apparatus for demonstrating alcoholic fermentation.*

How do humans make use of alcoholic fermentation?

Alcoholic fermentation has two uses:

① Making alcoholic drinks, e.g. wine, beer.

② Making bread.

How is wine made?

Juice is extracted from grapes which contain fruit sugar (fructose). Yeast is added. The yeast ferments the sugar, turning it into alcohol. (For home-made wine other plant materials may be used, e.g. elderberries, turnips etc.).

How is beer made?

Juice is extracted from barley grain which contains malt sugar (maltose). Hops are added to give the right flavour. Then yeast is added. The yeast ferments the sugar, turning it into alcohol. Beer-making is called **brewing**.

How is bread made?

Mix together flour, water, sugar and yeast. This makes dough. Leave the dough in a warm place. The yeast ferments the sugar, giving off carbon dioxide gas. The gas makes the dough rise. Then bake the dough: the heat kills the yeast and the alcohol evaporates.

How is lactic acid produced by anaerobic respiration?

Here is a word equation summarising what happens:

$$\text{glucose} \rightarrow \text{lactic acid} + \text{energy}$$

Note that in this case no carbon dioxide is produced.

When and where is lactic acid produced?

Lactic acid is produced in situations where there is little or no oxygen. Here are some examples:

- Animals living in stagnant water or mud, e.g. certain types of worm.
- Parasites living in the gut, e.g. tapeworm.
- Mammals which dive for long periods, e.g. whale, seal.
- Muscles of humans and other mammals during strenuous exercise.

Why is lactic acid produced by muscles during strenuous exercise?

The breathing and circulatory systems cannot deliver oxygen to the muscles quickly enough to keep pace with their needs.

What happens to lactic acid produced by muscles during strenuous exercise?

After the exercise is over, the lactic acid is broken down into carbon dioxide and water. This requires oxygen, and it is why we pant after exercise. The oxygen required to get rid of lactic acid after exercise, is called the **oxygen debt**. It is important for lactic acid to be removed because it is poisonous. It causes the muscles to ache and get tired.

In prolonged exercise which is not too severe, lactic acid builds up to begin with, but later it is removed while the exercise is still taking place. When this happens the person is said to have got his or her 'second wind'.

How do bacteria respire anaerobically?

Some types of bacteria produce ethanol, others produce lactic acid. For the lactic acid types, see page 19.

How long can organisms respire anaerobically?

Organisms which can respire anaerobically fall into two groups:

① Those which respire anaerobically all their lives and never use oxygen, e.g. certain bacteria. Some anaerobic organisms of this type are poisoned by oxygen.

② Those which respire anaerobically only when oxygen is in short supply, e.g. yeast, human (muscle). Organisms of this type are poisoned by excessive amounts of ethanol or lactic acid.

The chemistry of respiration

The equation on page 46 is very simplified. Respiration really takes place in a series of small steps, each catalysed by a different enzyme. Some of the steps release energy which enables **adenosine triphosphate (ATP)** to be synthesised. The ATP then provides energy for muscle contraction and all other biological activities.

ATP is the link between the oxidation of glucose and all energy-requiring activities in organisms.

Where is ATP made?

ATP is made in all living cells. Most of it is made inside the **mitochondria** (see page 32). Cells which need a lot of energy (e.g. muscle cells and sperm cells) contain a particularly large number of mitochondria.

Questions

1 Read the account on page 46 of how you can find out if small animals and plants give out carbon dioxide.

(a) Why must the test tube be airtight?

(b) (i) What should the control be in this experiment?
(ii) Why is a control necessary?

(c) For this experiment it is better to detect carbon dioxide with hydrogencarbonate indicator solution than with lime water. Why?

(d) If you do this experiment with a leaf, the test tube should be placed in the dark. Why?

2 Look at the diagram of a respirometer in figure 10.1, page 47.

(a) What should the control be in this experiment?

(b) Why is it necessary to absorb carbon dioxide in the test tube?

(c) Describe in detail how you would use this apparatus to measure the rate of respiration of small animals at the following temperatures: 10°C, room temperature and 30°C.

3 Two muscle fibres were teased out of a muscle. After suitable preliminary treatment, a drop of glucose was added to one of the muscle fibres, and a drop of ATP was added to the other. Here is what happened:

Substance added	Response of the muscle fibre
Glucose	No change
ATP	Shortens

(a) What do the letters ATP stand for?

(b) Interpret the results of this experiment in terms of how energy is made available in cells.

(c) What do you think the 'suitable preliminary treatment' consisted of, and why was it necessary?

(d) Write down two other processes in humans, besides muscle contraction, for which energy is required.

4 An investigation was carried out into the concentrations of lactic acid in the blood of a human before, during, and after a two-minute period of vigorous exercise. The results are summarised below.

	Time (minutes)	Blood lactic acid (mg per 100 cm^3)
	−10	9
Exercise (2 minutes) →	0	14
	10	80
	15	95
	20	70
	40	35
	60	22
	80	18

(a) Plot the data on graph paper, putting blood lactic acid concentration on the vertical axis, and time on the horizontal axis.

(b) (i) Where in the human body is lactic acid produced during exercise?
(ii) Why is lactic acid produced during exercise?

(c) (i) Why does the lactic acid concentration rise after the exercise is over?
(ii) Why does the lactic acid concentration then fall?

(d) Name one other organism, besides the human, which produces lactic acid, and explain why the organism produces it.

11

Nutrition in animals

What is nutrition?

Nutrition is the study of food. The food that we eat makes up our **diet**. The chemical substances which we require in our diet are called **nutrients**. A diet which contains all the necessary nutrients in the right proportions is called a **complete diet** or **balanced diet**.

What are the constituents of a complete diet?

A complete diet should contain the following substances:

1. **Carbohydrates**
2. **Fats**
3. **Proteins**
4. **Water**
5. **Minerals**
6. **Vitamins**

For the chemistry of these substances see page 41.

Substances needed in bulk

Carbohydrates, fats, proteins and water are needed in bulk. Table 11.1 gives examples of each substance, where they are found and why we need them.

Why is starch a good source of energy?

Starch is the storage carbohydrate of plants. It consists of folded chains of glucose molecules packed tightly together inside **starch grains** (see page 31). A starch grain is therefore a concentrated source of glucose molecules from which energy may be obtained when required.

Saturated and unsaturated fats

A **saturated fat** contains the full number of hydrogen atoms in its fatty acid chains. An **unsaturated fat** has room for more hydrogen atoms in its fatty acid chains.

Animal fats are mainly saturated, e.g. butter, lard. Plant fats (oils) are mainly unsaturated, e.g. margarine, sunflower oil.

Which sort of fat is better for us?

Some scientists claim that for good health we should eat mainly unsaturated fat. A diet too rich in saturated fat may lead to heart disease.

Essential and non-essential amino acids

Essential amino acids are needed in the diet, because the body cannot make them.
Non-essential amino acids are not needed in the diet, because the body can make them from other amino acids.

Table 11.1 *Summary of substances required by humans in bulk.*

Name of substance	Main sources	Why needed
Carbohydrates		
Sugars		
Glucose (dextrose)	Sweet drinks, etc.	Provide energy
Fructose	Fruit	Provide energy
Sucrose (table sugar)	Sugar cane and beet	Provide energy
Lactose	Milk	Provide energy
Maltose	Barley grain	Provide energy
Starch	Bread, potatoes, cereals	Provides concentrated source of glucose molecules for energy.
Cellulose	All unrefined plant foods	Provides fibre (roughage) for maintaining healthy gut and preventing constipation.
Fats and oils	In animals as animal fat. In plants as vegetable oil.	Provide energy, insulation against heat loss, and fatty part of cell membrane.
Proteins	Milk, eggs, meat.	Provide body structure, enzymes and some energy.
Water	In all foods and drinks.	Provides biological solvent and prevents blood getting too concentrated.

Why are some proteins more useful than others?

Dietary proteins can be divided into two groups:

1. **High biological value proteins** provide all the essential amino acids. They are mainly animal proteins, e.g. in milk, eggs, meat.
2. **Low biological value proteins** lack certain essential amino acids. They are mainly plant proteins, e.g. in wheat, peas, beans.

What is textured vegetable protein?

Textured vegetable protein (TVP) is 'artificial meat' which has been manufactured from plant protein. Soya beans, with their relatively high biological value, are much used for making TVP.

What happens if a child does not get enough protein?

The child develops a condition called **kwashiorkor**: growth is retarded and the child is weak.

Substances needed in small amounts

Minerals and vitamins are needed in the diet in relatively small amounts.

Minerals are inorganic substances, mainly salts, whose chemical elements perform a variety of functions in the body (table 11.2). Check your syllabus to see which ones you need to know.

Vitamins are an assortment of organic substances which perform specific functions in the body, mainly as enzyme-helpers (table 11.3). Check your syllabus to see which ones you need to know.

Food tests

Foods can be tested for their nutrients. Here are some of the more important tests:

1. **Test for sugar**. If the food is not already in liquid form, mash it up and make a suspension or solution of it in water. Add an equal quantity of Benedict's or Fehling's solution. Heat to boiling (care!). Precipitate (usually green or brown) indicates sugar.[1]
2. **Test for starch**. Add a little dilute iodine solution to the food. A blue-black colour indicates starch.
3. **Test for fat/oil**. *Either*: Rub the food onto thin paper. A translucent mark on the paper indicates fat.
   Or: Mix the food with absolute ethanol, then add an equal quantity of water. A white precipitate indicates fat.

[1]Benedict's and Fehling's solutions contain copper sulphate. The sugar reduces the copper sulphate on heating, forming a precipitate. Sugars which do this include glucose, fructose and lactose: we call them **reducing sugars**.
Sucrose is not a reducing sugar. It will give a precipitate with Benedict's or Fehling's solution only if you first break it down into glucose and fructose by boiling it with a little dilute hydrochloric acid.

Table 11.2 *Some important mineral elements required by humans.*

Name of element	Main sources	Why needed	Effect of lack or shortage
Major elements			
Sodium	Common salt (sodium chloride) in many foods.	Nerve transmission, muscle contraction.	Cramp
Calcium	Milk, cheese, fish	Hardening of bones and teeth. (Also muscle contraction and blood-clotting.)	Soft bones in children (rickets).
Iron	Liver, kidneys	Constituent of haemoglobin in blood.	Anaemia, tiredness
Phosphorus	All foods	Consituent of nucleic acids (e.g. DNA) and ATP.	Unknown
Minor (trace) elements			
Iodine	Drinking water	Constituent of thyroid hormone (thyroxine).	Goitre
Fluorine	Drinking water	Helps to harden teeth and prevent tooth decay.	Poor teeth (?)

Table 11.3 *Some important vitamins required by humans.*

Vitamin	Main sources	Why needed	Effect of lack or shortage
Fat soluble			
Vitamin A	Liver, carrots	Functioning of eye.	Poor dark adaptation, dry cornea.
Vitamin D	Fish liver oil	Hardening of bones and teeth.	Soft bones (rickets in children).
Vitamin K	Cabbage, spinach	Blood clotting.	Long clotting time.
Water soluble			
Vitamin B1 (thiamin)	Cereals, yeast	Enzyme-helper in respiration.	Wasting of muscles, paralysis (beri-beri).
Vitamin B2 (riboflavin)	Leafy vegetables, fish, eggs	Enzyme-helper in respiration.	Sore mouth, ulcers.
Niacin	Meat, fish, wheat	Enzyme-helper in respiration	Diarrhoea, skin sores, mental disorder (pellagra).
Vitamin C	Citrus fruits, green vegetables	Helps epithelial cells to stick together.	Bleeding of gums and lips (scurvy).

④ **Test for protein**. If the food is not already in liquid form, mash it up and make a suspension or solution in water. Add a little sodium of potassium hydroxide till the suspension/solution clears, then add a few drops of dilute copper sulphate and shake. A purple colour indicates protein.

⑤ **Vitamin C**. Extract juice from the food. Add the juice, drop by drop, to a drop of blue DCPIP solution. Disappearance of the blue colour indicates vitamin C.

The smaller the number of drops required to decolorise the DCPIP, the greater is the concentration of vitamin C in the food. This enables us to *compare* the concentration of vitamin C in different foods.

Why do we cook food?

Cooking involves heating the food. The heat

- softens cellulose, thereby tenderising vegetables,
- bursts open starch grains, releasing the starch molecules,
- softens fibrous proteins, thereby tenderising meat,
- coagulates globular (soluble) proteins, e.g. egg albumen,
- sterilises the food by killing germs.

A chef could add many more points to this list!

Note: Over-cooking reduces the nutritional value of certain foods. For example, excessive heat destroys vitamin C.

Food additives

A **food additive** is a chemical substance, natural or synthetic, which is added by the manufacturer to food. Additives sweeten, flavour, colour, stabilise or preserve the food.

- **Stabilisers** prevent chemical reactions taking place in the food which might spoil its consistency.
- **Preservatives** include anti-bacterial and anti-fungal agents which help to prevent the food going bad.
- **Sweeteners** include saccharine, a non-fattening synthetic substance about 300 times sweeter than sugar.

Advantages and disadvantages of food additives

- **Advantages**: they make the food look and taste more attractive, and prolong its life.
- **Disadvantages**: some people are allergic to certain additives, some additives make children hyperactive, and some may have long-term damaging effects on health.

Vegetarians and vegans

A **vegetarian** eats plant foods and animal products, but not meat. A vegetarian on a well balanced diet can get all the nutrients he or she needs for a healthy life.

A **vegan** eats plant foods, but not meat *or* animal products. Vegans must ensure that they have a complete diet by making the diet very varied and by supplementing it with certain vitamins.

How can we find out how much energy food contains?

A known mass of food is ignited and made to heat up a known quantity of water. The rise in temperature of the water is measured. The energy content (energy value) of the food is then calculated in kilojoules (kJ) or in calories (cal). (4.2 kJ (1 cal) of energy are required to raise the temperature of 1 kg of water through 1°C.)

The experiment can be done crudely by burning a small sample of food under a test tube containing water, or more accurately by means of a **food calorimeter** (figure 11.1).

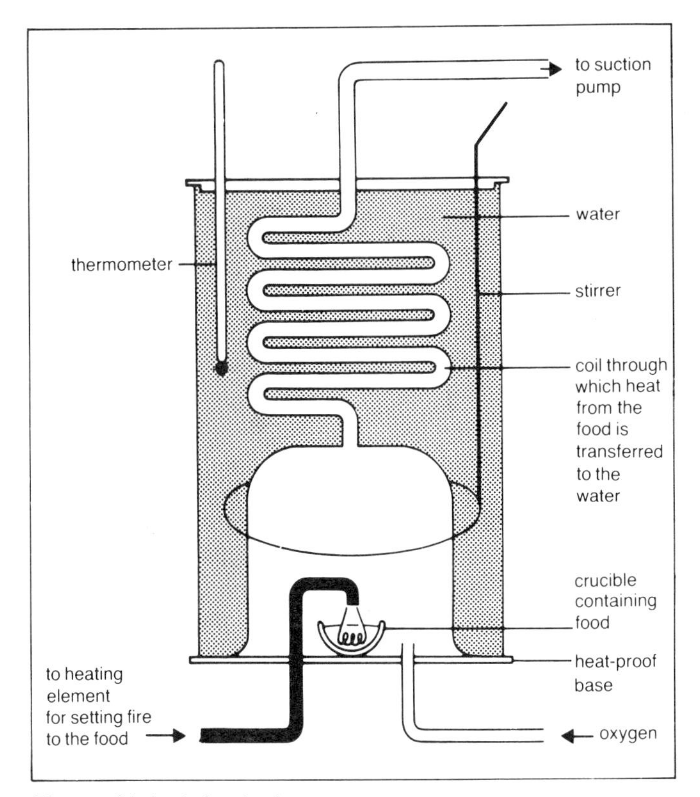

Figure 11.1 *A food calorimeter.*

What are the energy values of carbohydrate, fat and protein?

Carbohydrate: 1 gram contains 17 kJ.
Fat: 1 gram contains 39 kJ.
Protein: 1 gram contains 18 kJ.

The energy value of a particular food such as meat or potatoes depends on the relative amounts of carbohydrate, fat and protein which it contains, and how concentrated each is.

Which substances do we actually get energy from?

We get energy from (i.e. we *respire*) carbohydrate, fat *and* protein. However, the amount of energy which a person gets from each depends on circumstances.

In general, a healthy person on a well balanced diet respires mainly carbohydrate (about 50%), to a lesser extent fat (about 30%), and relatively little protein (about 20%).

How much energy does a person need?

Everyone needs sufficient energy to:

- maintain the **basal metabolic rate (BMR)**: this is the energy needed to keep the body alive when completely at rest;
- sustain such additional activities as are required by the individual according to his or her occupation.

The basal metabolic rate depends on the person's age, mass and sex. As a rough guide, approximately 7000 kJ/day are needed to maintain the BMR of an average adult.

A total of 15000 kJ/day are needed to fulfil the requirements of a heavy manual worker.

What happens when a person does not get enough energy food?

Energy output exceeds energy input. Carbohydrate and fat reserves are respired, and the person's body mass decreases. When carbohydrate and fat reserves have been used up, tissue protein is respired and the body 'wastes away'.

What happens when a person gets too much energy food?

Energy input exceeds energy output. Excess carbohydrate is turned into fat, and the person's body mass increases. This may lead to 'overweight' (**obesity**).

For everyone there is a correct body mass which depends on age, height and build.

Why is obesity a disadvantage?

Here are four reasons:

- With the extra mass to carry, you get tired more quickly.
- It can shorten your life: an obese person has a greater chance of having a stroke or heart attack.
- It can lower your self-esteem.
- It is more difficult to get life insurance!

How can an obese person lose 'weight'?

- By eating less energy food: this decreases energy input.
- By taking more exercise: this increases energy output.

What is malnutrition?

A person whose diet lacks one or more essential constituents suffers from **malnutrition**. This may give rise to various deficiency diseases, e.g. kwashiorkor, rickets, scurvy.

• Questions

1 Explain the reason for each of the following:

(a) A person engaged in heavy manual work requires a diet containing a lot of carbohydrate.

(b) Citrus fruits are good for you.

(c) A young child may be given cod liver oil.

(d) Margarine is considered better for health than butter.

(e) Milk protein is more useful to us than wheat protein.

2 To fulfil his or her dietary needs, a vegan requires a bulky and varied diet.

(a) Suggest *one* reason why a vegan's diet needs to be bulky.

(b) What is meant by a 'varied diet', as applied to a vegan?

(c) Name *two* nutrients which are likely to be in short supply in a vegan's diet.

(d) If a vegan was willing to eat animal products, how would this affect his or her diet?

3 Analysis of a potato produced the following results:

Substance	**Percentage composition by mass**
Starch and sugars	11.0
Water	82.0
Vitamins	0.3
Other substances	6.7

(a) Where, and in what form, would you find starch in a potato?

(b) (i) What is the function of starch in the human diet?
(ii) Why do potatoes fulfil this function particularly well?

(c) How would you test a potato for the presence of (i) starch; (ii) reducing sugar; (iii) vitamin C?

(d) Give *one* function of water in (i) a potato and (ii) the human.

4 The table below gives approximate figures for the daily requirements of energy and protein for males of different ages:

Age (years)	**Body mass** (kg)	**Energy** (kJ)	**Protein** (g)
1	6	3 300	15
11	34	10 500	70
18	62	16 000	100
25	75	13 500	65

(a) (i) Calculate the energy requirement of each male as kJ per kg body mass.
(ii) Calculate the protein requirement of each male as g per kg of body mass.
(iii) Why is it useful to express the energy and protein requirements per kg body mass?

(b) (i) Account for the difference in energy requirements per kg body mass of a one-year and 25-year-old male.
(ii) Account for the difference in protein requirements per kg body mass of a one-year and 25-year-old male.

(c) (i) State *one* function of the protein consumed by the 25-year-old male.
(ii) Name *two* foods which would provide substantial amounts of protein.

(d) The 25-year-old went on to develop a 'weight' problem. He decided to remedy this by cutting out all starchy and fatty foods. Why do you think his plan was likely to lead to poor health?

12

Feeding and digestion

What is feeding?

Feeding is the process in which food substances are taken in by an organism and incorporated into its body.

In the human feeding is carried out by the **gut** (**alimentary canal**). The gut and associated structures are shown in figure 12.1.

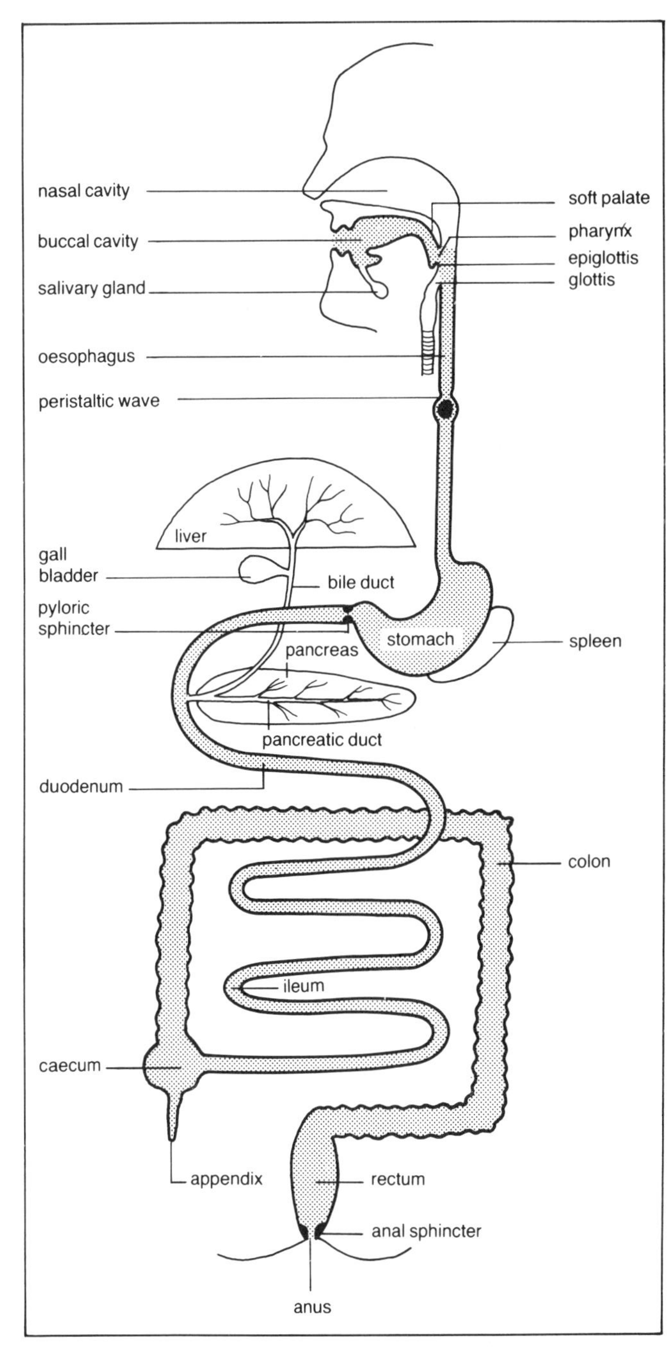

Figure 12.1 *Structure of the human gut (alimentary canal) and associated structures.*

What does feeding involve?

In the human and most other animals feeding involves the following processes:

- **Ingestion**. Food is taken into the gut through the mouth.
- **Digestion**. The food is broken down into soluble substances.
- **Egestion**. Components of the food which cannot be digested are got rid of through the anus.
- **Absorption**. The soluble products of digestion are absorbed into the bloodstream and transported round the body.
- **Assimilation**. The soluble products of digestion are taken into the cells.

What is meant by 'soluble products of digestion'?

A soluble substance is one which can dissolve. The substances formed as a result of digestion can dissolve in the gut fluid. This has to happen before they can be absorbed.

How is digestion achieved?

Digestion is achieved in two ways:

① The food is broken up *physically* by the **teeth** and by contractions of the **stomach**.

② The food is broken down *chemically* by means of **digestive enzymes**: large, complex molecules are broken down (hydrolysed) into smaller, simpler molecules.

Why is physical breaking up of the food important?

It helps the digestive enzymes to work. This is because it increases the surface area of the food on which the enzymes can work.

What happens to food as it passes down the gut?

① **In the buccal cavity and pharynx**

- chewed (masticated);
- mixed with water and mu
- partially digested by an e
- formed into a ball (bolus
- swallowed.

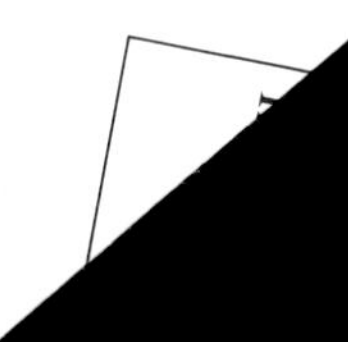

② **In the oesophagus food is:**
- transported to the stomach by peristalsis.

③ **In the stomach food is:**
- acidified by hydrochloric acid in gastric juice;
- churned up (kneaded) by contractions of the muscular stomach wall;
- further digested by enzymes in gastric juice;
- turned into semi-fluid chyme;
- let through into the duodenum bit by bit (controlled by pyloric sphincter).

④ **In the small intestine:**
- the gut contents are made alkaline by sodium hydrogencarbonate in bile, pancreatic and intestinal juice;
- fat is emulsified by bile salts;
- digestion is completed by enzymes in pancreatic and intestinal juice;
- the products of digestion are absorbed.

Note: Glucose, ethanol (alcohol) and water are absorbed in the stomach as well as the small intestine.

⑤ **In the caecum and appendix** of certain herbivores but not the human
- cellulose (fibre, roughage) is digested by the enzyme cellulase to form soluble glucose.

Note: In the human cellulose is not digested. It helps to keep the colon and rectum healthy and prevents constipation.

⑥ **In the colon:**
- water is absorbed;
- indigestible matter becomes compacted into faeces;
- faeces are moved towards the rectum by peristalsis.

⑦ **In the rectum:**
- faeces are stored before being egested through the anus (**defaecation**, controlled by the anal sphincter).

How does swallowing take place?

Swallowing is a complex reflex. The sequence of events is as follows:

① The back of the tongue rises.

② The soft palate is pushed against the back of the pharynx, so food cannot get into the nasal cavity.

③ The glottis becomes closed by the epiglottis, so food cannot get into the larynx and trachea.

④ Food passes into the oesophagus.

How does peristalsis take place?

The gut constricts by a localised contraction of the circular muscle in the gut wall. The constriction then moves along the gut, sweeping the contents before it.

Which enzymes digest food?

The main enzymes involved in digestion are summarised in table 12.1. They fall into three types: **carbohydrases, lipases** and **proteases** (see page 43).

What is the function of hydrochloric acid in the stomach?

The hydrochloric acid has two functions:

① It helps to sterilise the food by killing germs.

② It makes the stomach contents acidic (low pH): this is necessary for pepsin to work.

What is the function of sodium hydrogencarbonate in the small intestine?

The sodium hydrogencarbonate neutralises the acid from the stomach and makes the contents of the small intestine alkaline (high pH): this is necessary for trypsin (and other intestinal enzymes) to work.

Table 12.1 *Summary of the main digestive enzymes found in the human gut. Bile salts are included though they are not really enzymes. The stomach of calves produces an additional enzyme called rennin. Rennin turns soluble milk protein into a solid which is then attacked by pepsin. Rennin is not produced by the human stomach.*

Juice	Where it comes from	Where it works	Name of enzyme	Food acted on	Substances produced
Saliva	salivary glands	mouth cavity	amylase	starch	maltose
Gastric	stomach wall	stomach	pepsin	protein	polypeptides
Bile	liver	small intestine	bile salts (not enzymes)	fat	fat droplets
Pancreatic	pancreas	small intestine	amylase trypsin lipase	starch protein fat	maltose polypeptides fatty acids and glycerol
Intestinal	wall of small intestine	small intestine	maltase sucrase peptidases	maltose sucrose polypeptides	glucose glucose and fructose amino acids

(Pancreatic and intestinal products: can be absorbed)

What is bile?

Bile is a fluid produced by the liver. It consists of a mixture of chemical substances. It is stored in the gall bladder from which it passes into the duodenum when required.

What is the function of bile in the gut?

Bile **emulsifies fat**: the fat breaks up into lots of tiny droplets. This increases the surface area of the fat, thereby helping the action of the enzyme lipase.

Structure of the wall of the small intestine

Figure 12.2 shows part of the wall of the small intestine as seen in transverse section. These two structures are particularly important in digestion:

① The **intestinal glands** (also called crypts of Leiberkühn): the cells lining these glands secrete digestive enzymes.

② The **circular muscle**: contraction of this muscle constricts the intestine and may make a peristaltic wave.

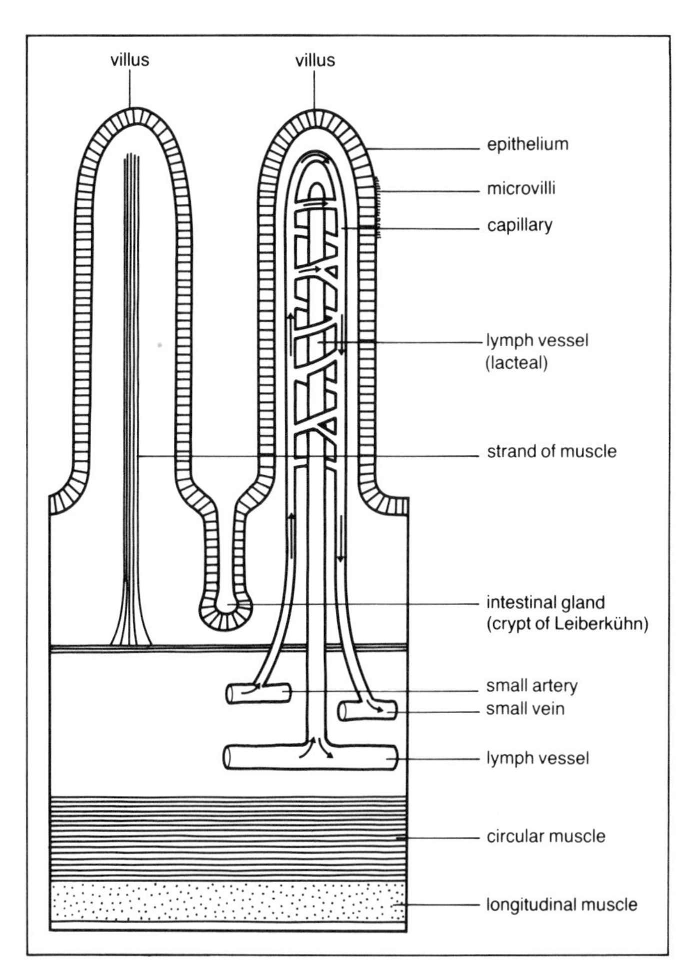

Figure 12.2 *Transverse section of part of the wall of the small intestine. The arrows indicate the direction in which blood and lymph flow.*

Absorption

Once they have dissolved, the products of digestion pass through the epithelium lining the villi and enter the capillaries.

Some of the fat enters the lymph vessels in the villi. These particular lymph vessels are called **lacteals**. Lacteal means milky. They were given this name because after a meal they fill up with fat droplets and this makes them look milky.

How is the wall of the small intestine adapted for absorption?

① The **finger-like villi** increase the surface area.

② The **hair-like microvilli** projecting from the epithelial cells increase the surface area still further.

③ The **strand of muscle** in each villus enables the villi to wave around, bringing them in contact with the digested food.

④ The **capillary network** in each villus is close to the epithelium through which the food substances are absorbed.

What happens to the absorbed substances?

The absorbed substances are carried to the liver by the **hepatic portal vein** (see page 67). The liver may store them, break them down for energy release, or shed them into the hepatic vein.

Substances shed into the hepatic vein will be carried by the circulation to all parts of the body and assimilated by the cells.

What sort of teeth do we have?

An adult human has the following teeth on each side of the upper and lower jaws:

- **Incisors**: flat, chisel-shaped for cutting food.
- **Canines**: cone-shaped also for cutting food.
- **Premolars** } **cheek teeth:** broad top with bumps (cusps)
- **Molars** } for crushing food.

How many teeth does an adult human have?

An adult human has, on each side of the upper and lower jaws: 2 incisors, 1 canine, 2 premolars and 3 molars. This is summed up in the **dental formula**:

$$i\tfrac{2}{2} \quad c\tfrac{1}{1} \quad pm\tfrac{2}{2} \quad m\tfrac{3}{3}$$

The top figures are the number of teeth on one side of the upper jaw; the bottom figures are the number of teeth on one side of the lower jaw. So the total number of teeth in an adult human is **32**.

How do our teeth work?

The teeth belonging to the upper and lower jaws are brought together by the **jaw muscles**. The main jaw muscles run from the lower jaw (mandible) to the sides of the skull.

The external structure of our teeth

A tooth consists of a **crown** and one or more **roots** (figure 12.3). The roots fit into sockets in the jaw bone.

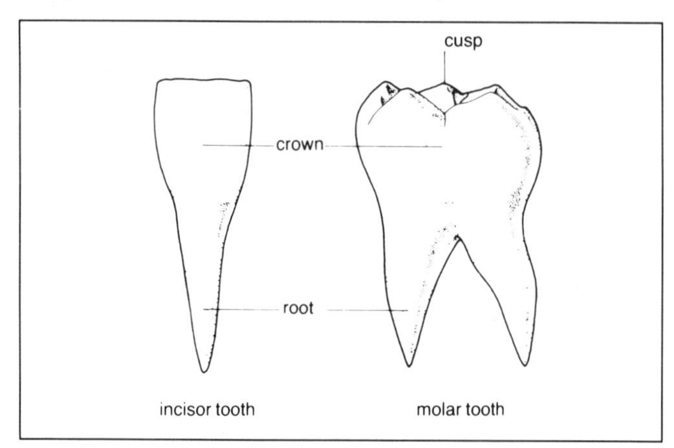

Figure 12.3 *External structure of human teeth.*

The internal structure of our teeth

Figure 12.4 shows the internal structure of an incisor tooth. Notice the different layers of the tooth, and the way the roots are attached to the jaw bone.

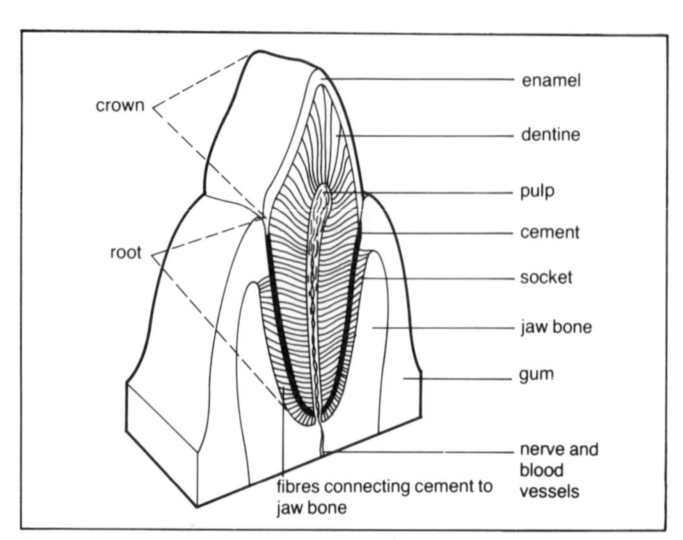

Figure 12.4 *Internal structure of an incisor tooth. The tooth has been sectioned at right angles to the line of the jaw.*

What causes tooth decay?

Tooth decay (**dental caries**) is caused by bacteria in the mouth. The bacteria form a layer of **plaque** over the teeth. This is what happens when a tooth decays:

- After a meal the bacteria metabolise sugar and produce acid. (Saliva is slightly alkaline but it cannot neutralise the acid quickly enough to prevent decay getting underway.)
- The acid gradually eats into the enamel and dentine of the tooth, causing toothache.
- Eventually the acid reaches the pulp, enabling the bacteria to get inside the tooth (severe toothache).
- The bacteria spread to the base of the tooth (below the root), causing an **abscess** (excruciating pain).

Two other conditions associated with tooth decay are:

- **Gum disease**: infection of the gums, particularly between the teeth.
- **Pyorrhoea**: infection of the fibres which hold the tooth in its socket.

How can we prevent tooth decay?

① Clean teeth regularly, so as to remove plaque.

② Finish meals with a rough vegetable, e.g. carrot or celery.

③ Do not eat sweet things or drink sugary drinks between meals.

④ Use fluoride toothpaste and/or drink fluoridised water.

⑤ Visit the dentist every six months for a check-up.

How can you tell if there is any plaque on your teeth?

Dissolve a **disclosing tablet** in your mouth. The resulting solution will stain the plaque red.

How does fluoride help to prevent tooth decay?

Fluoride helps to harden the teeth of growing children, making them more resistant to decay. It also helps to prevent plaque-formation.

Feeding in other mammals

Other mammals feed in basically the same way as we do. However, herbivores and carnivores have special features to help them with their diets.

What special features do herbivores have?

The problem facing a herbivore is to digest cellulose and break open plant cells. Herbivorous mammals are helped by these features:

① They have **sharp, chisel-like incisor teeth** for cutting and gnawing, e.g. rabbit, squirrel.

② Their cheek teeth have **enamel ridges** for grinding, e.g. horse.

③ Ruminants, e.g. cattle, sheep, have a special stomach (**rumen**) containing bacteria which produce the cellulose-splitting enzyme cellulase.

④ They have a **long intestine**: this provides a greater surface for digestion and absorption.
⑤ Some species (e.g. horse, rabbit) have a **large caecum and appendix** containing cellulase-producing bacteria.

What is the relationship between a herbivore and its gut bacteria?

This relationship is an example of **mutualism** (see page 15). The herbivore (host) gains cellulase from the bacteria. The bacteria gain shelter and food from the herbivore. (*Note*: animals cannot produce cellulase themselves.)

What special features do carnivores have?

The problem facing a carnivore is to kill and dismember its prey. Carnivorous mammals are helped by these features:

① They have **sharp claws** for bringing down prey, e.g. lion, tiger.
② They have **long, dagger-like canine teeth** for tearing flesh, e.g. lion, tiger.
③ Their cheek teeth have **large cusps** for crushing bone.
④ Dogs and their relatives have an enlarged cheek tooth (**carnassial**) on both sides of each jaw for scraping flesh off bones.

How do bacteria and fungi feed?

Bacteria and fungi secrete digestive enzymes across their body surface onto the food. The food is digested and the soluble products are absorbed.

The food may be of two kinds:
- a living organism, in which case the bacteria or fungi are **parasites**;
- a dead organism, in which case the bacteria or fungi are **saprotrophs**.

Why is the feeding of bacteria and fungi important to humans?

There are two reasons:

① Parasitic bacteria and fungi cause **disease** (see page 26).
② Saprotrophic bacteria and fungi are decomposers and bring about **decay** (see page 18).

How do other organisms feed?

Check your syllabus and find out what you need to know!
- *Amoeba* engulfs micro-organisms (e.g. green protists) and takes them into a **food vacuole** where they are digested.
- *Euglena* feeds by **photosynthesis** in the light, using its chloroplasts. In the dark it feeds by absorbing soluble organic substances across its surface.
- *Hydra* catches its prey (e.g. a water flea) with its **tentacles** and poisons it with its **sting cells**. The prey is then drawn to the mouth by the tentacles and digested in the **digestive cavity**.
- *Insects* have **mouth parts** operated by muscles. The mouth parts of different species are adapted for feeding on different types of food. Thus the locust's are adapted for cutting and chewing leaves, the mosquito's for piercing skin and sucking blood, the butterfly's for probing flowers and sucking nectar, and the housefly's for dissolving solids and sucking up the resulting solution.

• Questions

1 Look at table 12.1 (page 56) which summarises the main digestive enzymes found in the human gut.

(a) Which enzyme or enzymes, if any, would be involved in the digestion of each of the following?
(i) A piece of lean meat.
(ii) A slice of bread.
(iii) A lump of sugar.
(iv) A dextrosol (glucose) tablet.
(v) A pat of butter.

(b) Why are bile salts *not* regarded as enzymes?

(c) Describe an experiment which you could do to find out if bile salts enable lipase to digest fat more quickly.

2 Gastric juice is produced by glands in the wall of the stomach. The graph here shows the amount of gastric juice produced by the stomach of an individual who had just chewed some food. The food was spat out after being chewed, none was swallowed.

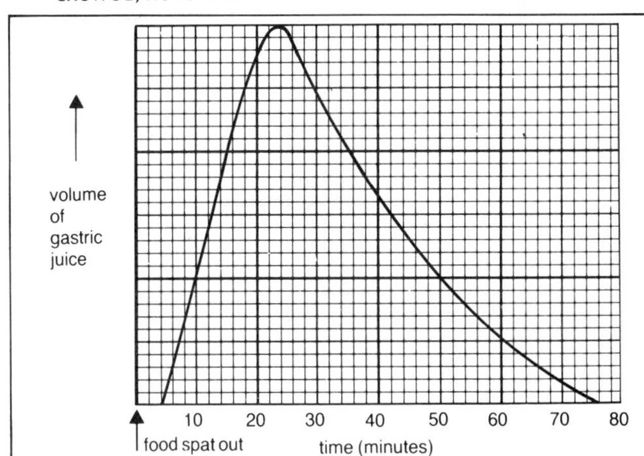

(a) Name *two* constituents of gastric juice.

(b) Assuming that no traces of food got down into the stomach, suggest how the glands in the stomach wall 'knew' when to produce gastric juice.

(c) (i) How much time elapsed between the moment the food was spat out and the moment gastric juice started to be produced?
(ii) Account for the delay.

(d) Mr Smith has the whole of his stomach removed in an operation. What sort of diet would you recommend for Mr Smith after he has recovered from the operation. Give reasons for your recommendations.

3 A student carried out an experiment to test the hypothesis that saliva contains an enzyme which breaks down starch to sugar. She set up ten test tubes as follows:

Test tube 1	sugar only.
Test tube 2	sugar only.
Test tube 3	starch only.
Test tube 4	starch only.
Test tube 5	saliva only.
Test tube 6	saliva only.
Test tube 7	starch plus saliva.
Test tube 8	starch plus saliva.
Test tube 9	starch plus boiled saliva.
Test tube 10	starch plus boiled saliva.

The student left all the test tubes at room temperature for fifteen minutes. She then tested the contents of tubes 1, 3, 5, 7 and 9 for starch with iodine solution; and she tested the contents of tubes 2, 4, 6, 8 and 10 for sugar by heating with Benedict's reagent.

(a) Why did the student set up test tubes 1 to 6?

(b) Why did the student set up test tubes 9 and 10?

(c) What safety precautions should be taken when carrying out the Benedict's test?

4 Tooth decay is caused by bacteria which cause a thin film (plaque) to develop on the surface of the teeth. The graphs show the pH of plaque at different times of the day for two people, Pat and Ann. The shaded areas in each graph show the pH values at which tooth decay occurs.

(a) For approximately how many hours, between 6 a.m. and 11 p.m., is tooth decay likely to occur in (i) Pat and (ii) Ann?

(b) How would you explain the pattern of the curves in each graph?

(c) Suggest *two* ways by which Ann might reduce the rate at which her teeth decay.

(d) The bacteria that cause tooth decay respire anaerobically.
(i) What does the term anaerobic mean?
(ii) Explain as precisely as you can how these bacteria bring about tooth decay. (*Hint*: what happens when bacteria respire anaerobically?)

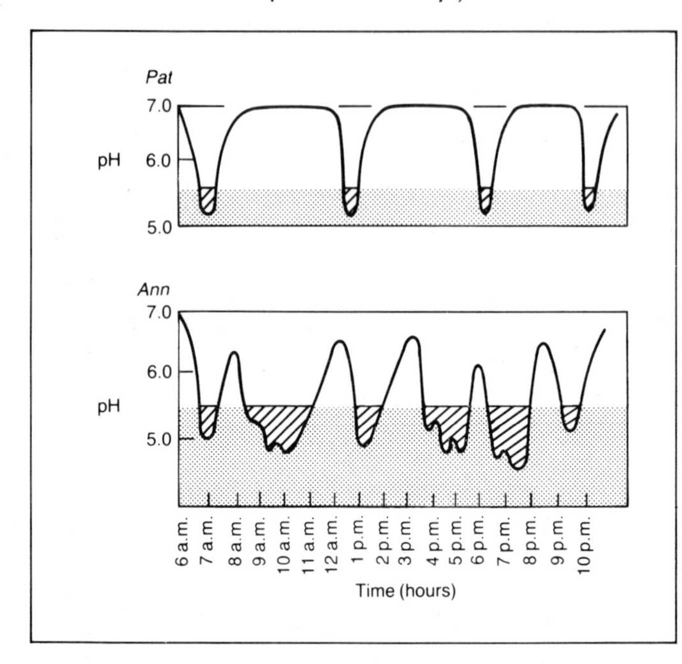

5 The diagram below shows the jaws of a rabbit, a mammal which feeds on plants.

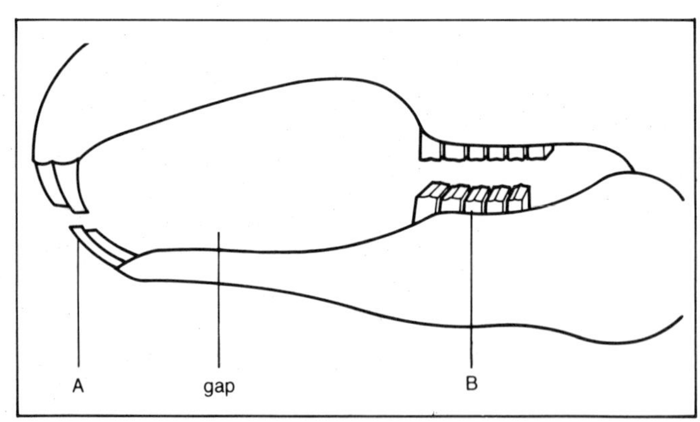

(a) (i) Name the type of teeth labelled A.
(ii) What function do these teeth perform, and how is their structure suited for this function?

(b) (i) Name the type of teeth labelled B.
(ii) What function do these teeth perform, and how is their structure suited for this function?

(c) What type of teeth, present in humans, appears to be absent in the rabbit?

(d) Suggest a function for the large gap between the front and back teeth.

Breathing

What is breathing?

Breathing is the process by which animals bring the external medium (air or water) into contact with their gas exchange surface.

In mammals and other land vertebrates the gas exchange surface is the inner lining of the **lungs**. In such animals breathing can be divided into two parts: **Inspiration** (breathing in) and **expiration** (breathing out).

How does inspired air differ from expired air?

Expired (exhaled) air contains less oxygen and more carbon dioxide than inspired (inhaled, atmospheric) air. The amount of nitrogen remains about the same.

How can we find out the amount of oxygen and carbon dioxide in air?

By **gas analysis**. First you mix a known quantity of the air with a solution which absorbs carbon dioxide, and you note the decrease in the size of the air sample. You then mix the remaining air with a solution which absorbs oxygen, and you note the further decrease in the size of the air sample. From the decrease in the size of the air sample, you can work out the percentage of carbon dioxide and oxygen in the air.

The reagent for absorbing carbon dioxide is **potassium hydroxide**, and the reagent for absorbing oxygen is **potassium pyrogallate**.

How is the composition of expired air affected by exercise?

If a sample of expired air is collected during, or immediately after, exercise, the amount of carbon dioxide is higher, and oxygen lower, than in a sample of air collected from a person at rest. This is because active muscles use up more oxygen, and produce more carbon dioxide, than resting muscles.

Structures involved in breathing

The structures involved in breathing in humans are shown in figure 13.1. They make up the **breathing system** (also known as the **gas exchange system** or **respiratory system**). The functions of the various structures are summarised in table 13.1.

How does inspiration take place?

The diaphragm flattens, the ribs swing upwards and outwards and the sternum moves forwards. *Result*: the volume of the chest increases, creating a negative pressure, and air is sucked into the lungs.

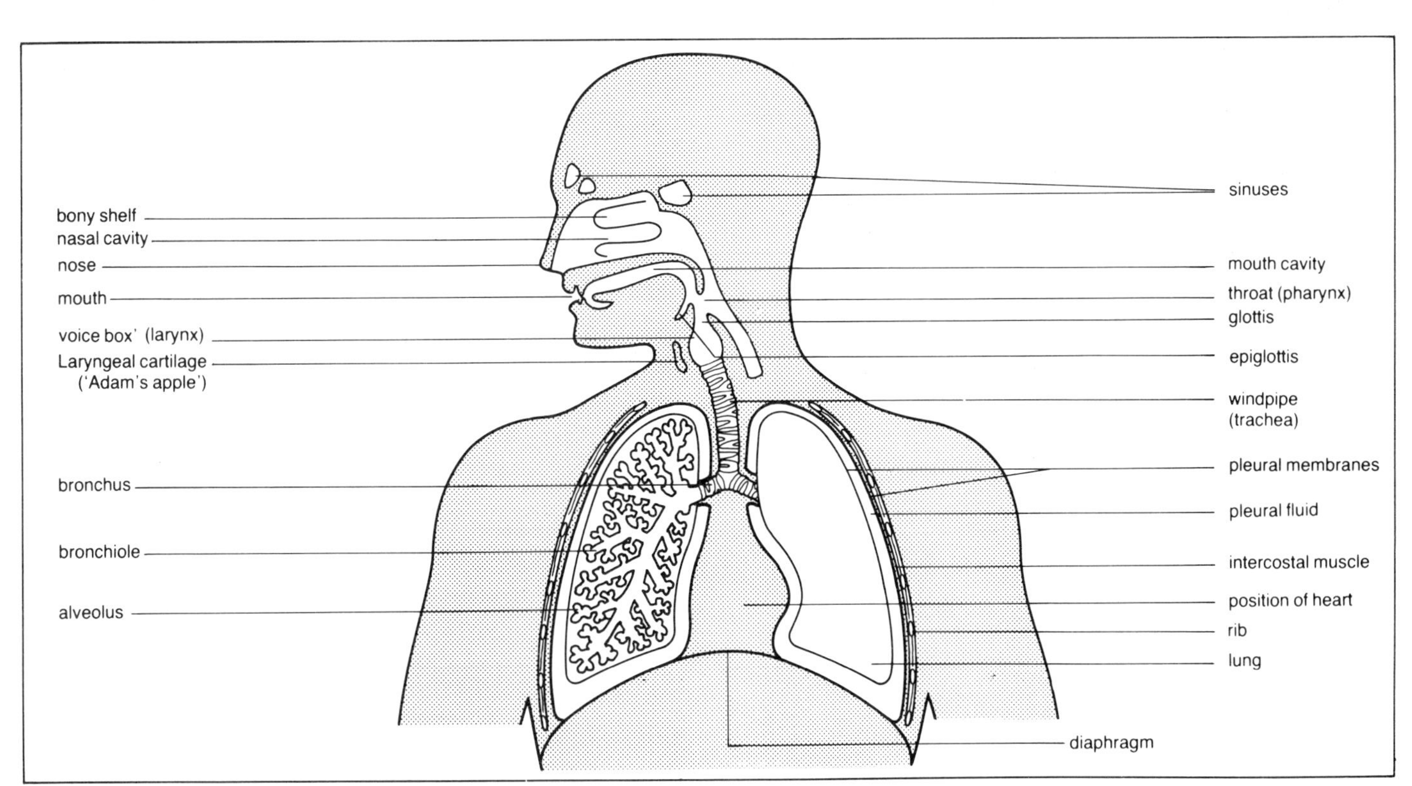

Figure 13.1 *The breathing system of the human.*

Table 13.1 *The main structures that make up the breathing system.*

Structure	Description and function
Nasal cavity	Cavity divided up by shelves (turbinates) which warms, moistens, cleans and tests the air breathed in. Foreign particles get caught in mucus and are wafted by cilia towards throat.
Pharynx (throat)	Point where the breathing and alimentary pathways cross.
Glottis	Small hole through which air enters larynx.
Epiglottis	Flap which closes over the glottis during swallowing, thereby preventing food 'going down the wrong way'.
Larynx (voice box)	Contains vocal cords which, when vibrated, make sounds.
Trachea (windpipe)	Tube by which air passes to and from the lungs. Incomplete rings of cartilage keep it permanently open. Foreign particles get caught in mucus and are wafted by cilia towards throat.
Bronchi	Similar to trachea, carry air to and from left and right lungs.
Bronchioles	Narrow, tree-like branches of bronchi through which gases diffuse to and from alveoli.
Alveoli	Tiny air sacs across whose thin, moist walls gas exchange takes place. Very close association with capillaries. The vast number of alveoli provide a large surface area for gas exchange.

How does expiration take place?

The diaphragm bows upwards, the ribs swing downwards and inwards and the sternum moves backwards. *Result*: the volume of the chest decreases, creating a positive pressure, and air is forced out of the lungs.

How are the chest movements brought about?

- During inspiration, the ribs are moved by contraction of the **intercostal muscles**, and the diaphragm is flattened by contraction of the **diaphragm muscles**.
- During expiration, the intercostal and diaphragm muscles relax.
- During inspiration and expiration, the **pleural membranes** slide easily against each other without friction. This is due to the presence between them of **pleural fluid** which acts as a lubricant.

How can you increase the rate at which oxygen is taken into the body?

You can do this by increasing the **depth** and/or **frequency** of your breathing. The depth is the volume of air drawn into the lungs each time you inspire. The frequency is the number of times you inspire in a given time.

When does a person's breathing rate increase?

The depth and/or frequency of breathing increases in the following circumstances:

- During, and immediately after, exercise when a lot of oxygen is being used by the muscles.
- At high altitudes where the concentration of oxygen in the atmosphere is relatively low.
- When the lungs are not allowing efficient gas exchange, e.g. in emphysema (see below).
- When the circulation is not delivering oxygen to the tissues efficiently, e.g. in coronary heart disease.

What is a person's vital capacity?

The **vital capacity** is the maximum volume of air which you can take into your lungs. It is measured like this: first you breathe in as much as you can; then you breathe out as much air as you can into a calibrated chamber and measure the volume of the air.

The vital capacity is generally greater for:

- males than females;
- adults than children;
- fit people than unfit people.

Diseases of the breathing system

Here are some important examples, some more serious than others:

- **Bronchitis**: inflammation of the bronchial tubes.
- **Pleurisy**: inflammation of the pleural membranes.

- **Pneumoconiosis**: inflammation of the lungs due to, e.g. asbestos or silica dust.
- **Pneumonia**: infection of the lungs by bacteria or viruses.
- **Tuberculosis**: infection of the lungs by bacteria.
- **Emphysema**: destruction of walls of the alveoli caused by, e.g. smoking.
- **Lung cancer**: growths in walls of the bronchial tubes caused by, e.g. smoking.

What is the evidence that smoking causes lung cancer?

The evidence comes from two types of research:

1. Correlating the incidence of lung cancer in a large group of people with the number of cigarettes smoked per day.
2. Laboratory experiments investigating the effect of cigarette smoke on animals.

The conclusion is that tobacco smoke contains 'tar' which causes cancer of the lung.

Gas exchange in other organisms

Check your syllabus to find out how much you need to know!

- In **plants and simple organisms** (e.g. *Amoeba, Hydra*, earthworm) gas exchange takes place by **diffusion** across the body surface.
- **Fish** breathe by maintaining a flow of water of much folded, vascularised **gills** on either side of the pharynx.
- In **frogs and toads** air is breathed in and out of lungs, but gas exchange also occurs by diffusion across the moist, vascularised skin and the lining of the mouth cavity.
- **Insects** have a **tracheal system** through which oxygen diffuses (or in some cases is pumped) to the tissues. **Spiracles** (pores in the cuticle) lead to **tracheae** which branch into **tracheoles**.

Questions

1 The nose cleans, moistens, tests and warms the air which passes through it. These functions are made possible by the presence in the lining of the nose of these structures: blood capillaries, cilia, mucus glands and sensory cells. Which structures are connected with which functions?

2 (a) Make a list of the structures through which a molecule of oxygen passes as it travels from outside your body to an alveolus in your lungs.

(b) What kind of force propels the oxygen molecule, and how is the force brought about?

3 The diagrams below show a small part of the lung from a healthy person and from a person suffering from the effects of air pollution. Both are drawn to the same scale.

(a) Write down *two* differences which can be seen in the diagrams between the healthy lung and the diseased lung.

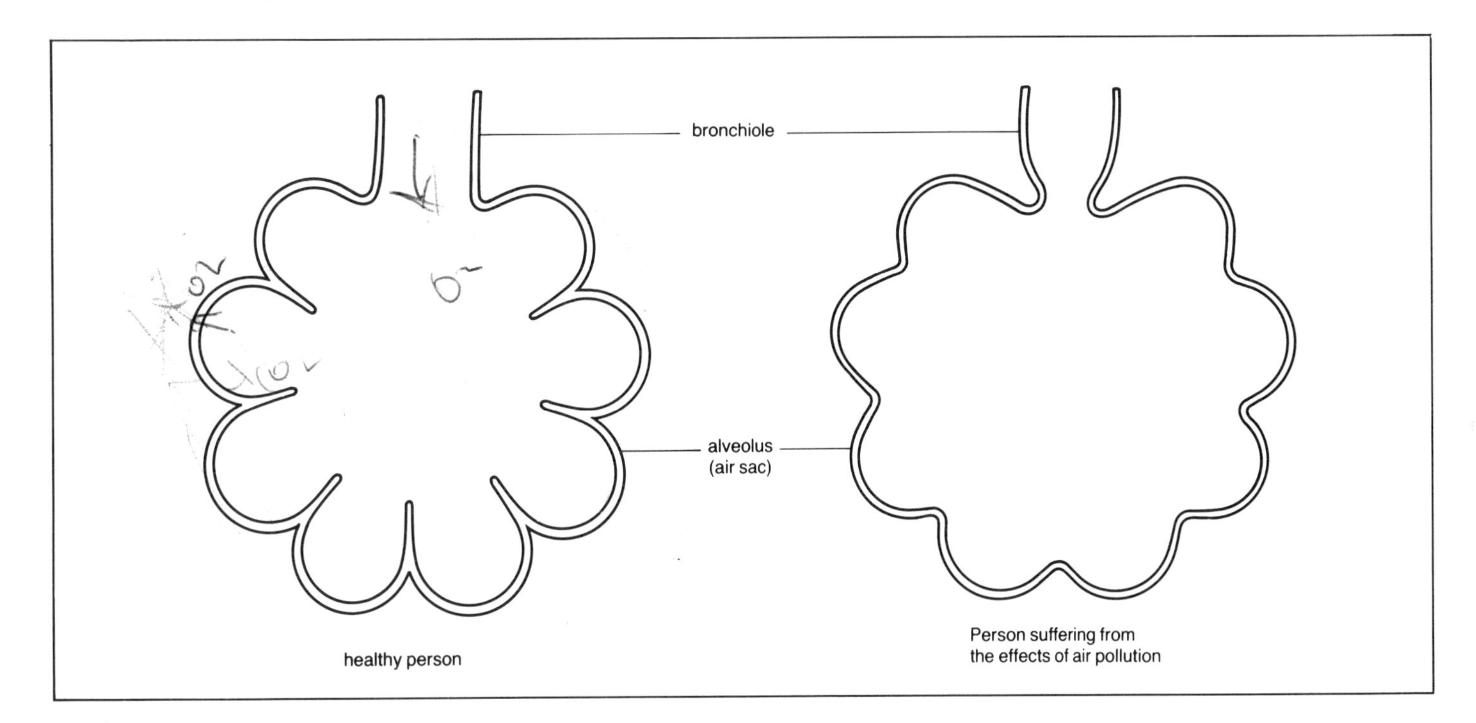

(b) Suggest why each of the differences which you have described would make the diseased lung work less well than the healthy lung.

(c) What effect would you expect the diseased lung to have on the person's breathing, and why?

(d) Suggest *two* types of air pollution which might cause this condition.

(NEA, extended)

4 A spirometer is an apparatus for recording a person's breathing movements. One is shown in diagram 1 below. It consists of an air chamber which moves up and down as the person breathes in and out. The movements are recorded by a pen which writes on calibrated paper attached to a slowly revolving cylinder.

Some recordings which were made with a spirometer of this type are shown in diagram 2 below.

(a) What volume of air is inspired in a single breath at rest?

(b) What volume of air is the person capable of expiring after a maximum inspiration?

(c) Express the volume of air inspired in a single breath at rest as a percentage of the maximum volume of air which could be inspired in a single breath.

(d) Why is it an advantage for only a small proportion of the lung's capacity to be used when breathing at rest?

(e) (i) In what way does the pattern of breathing in the recordings change during exercise?

(ii) Suggest *one* other way in which the pattern of breathing might change during exercise.

(iii) Why does the pattern of breathing change during exercise?

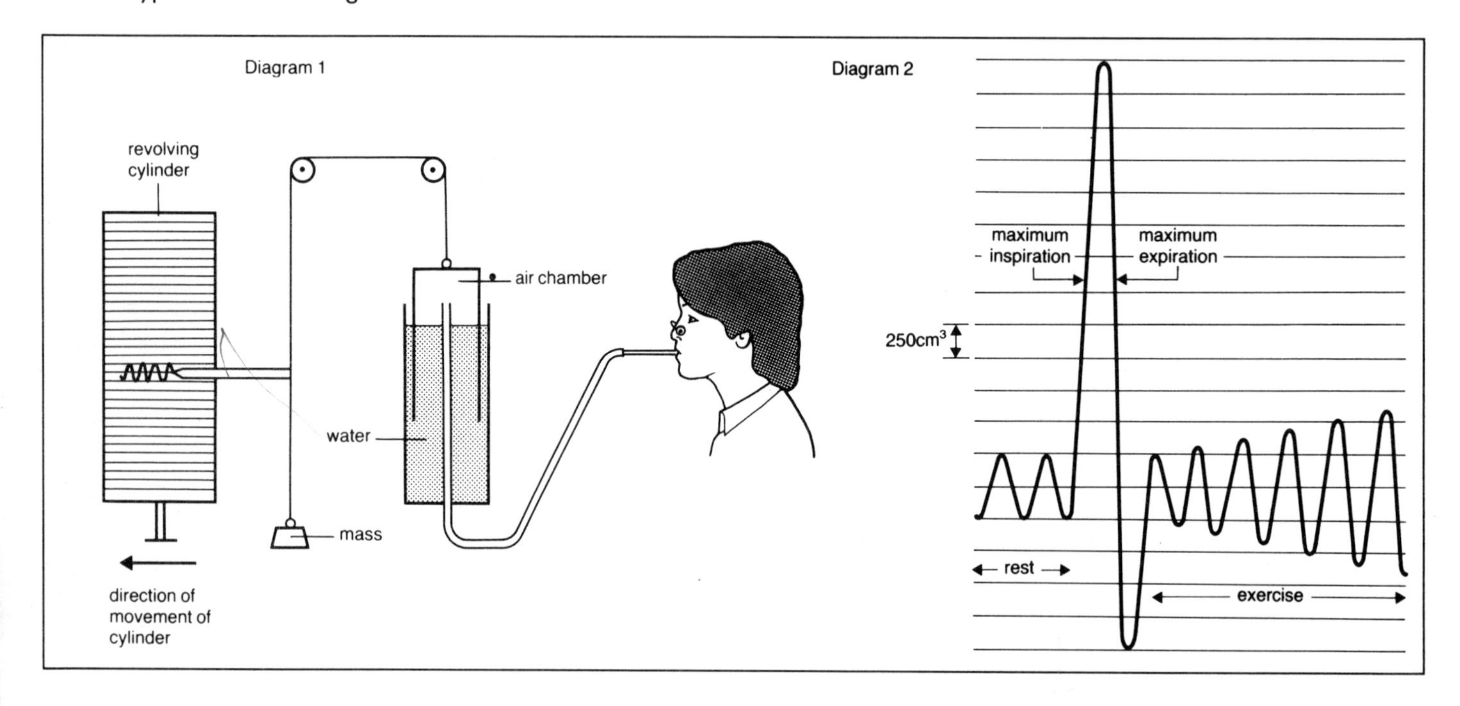

14

Circulation

What is the circulation?

The circulation is the movement of **blood** around the body. The structures which propel and carry the blood make up the **circulatory system**.

What is blood?

Blood is a fluid-like tissue consisting of **blood cells** and **platelets** suspended in a watery solution called **plasma** (figure 14.1).

Plasma proteins

Plasma proteins are soluble proteins in blood plasma. There are three kinds:

- **Albumen** makes the blood slightly viscous.
- **Globulin** is the protein of which antibodies are made.
- **Fibrinogen** is responsible for clotting of blood.

What is serum?

Serum is blood plasma after the fibrinogen has been removed. The fibrinogen can be removed by making blood clot and then removing the clot.

Where are blood cells formed?

Blood cells (and platelets) are manufactured in **bone marrow** in the centre of certain bones. Red blood cells have a life-span of about 120 days and are constantly being replaced.

The functions of blood

Transport

1. It carries oxygen from lungs to tissues, and carbon dioxide from tissues to lungs.
2. It carries dissolved food substances from the gut to other parts of the body.
3. It carries unwanted substances, including urea, to the kidneys which get rid of them.
4. It carries hormones from endocrine glands to other parts of the body.
5. It carries antibodies from one part of the body to another.

Protection

1. By clotting, it prevents fluid being lost from cuts and wounds.
2. It protects us against disease by killing germs.

Regulation

1. It helps to control the amount of water in the tissues.
2. It helps to control the amounts of various chemical substances in the tissues.

Figure 14.1 *Chart summarising the components of human blood.*

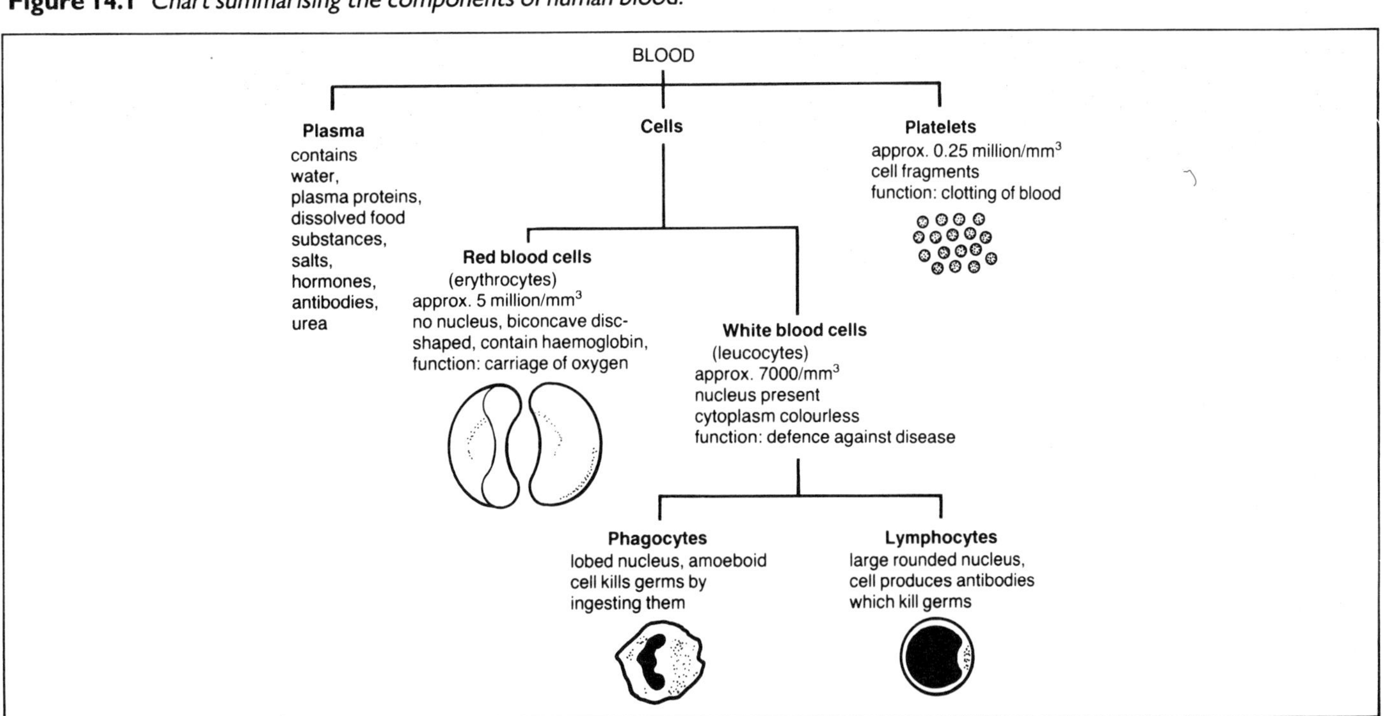

③ It helps to keep our body temperature constant by absorbing and/or losing heat at the surface of the body, and distributing it evenly around the body.

How does blood carry oxygen?

In the lungs, oxygen combines rapidly with **haemoglobin** in the red blood cells to form **oxyhaemoglobin**. In the tissues, oxyhaemoglobin splits into haemoglobin and oxygen, and the oxygen diffuses into the tissue cells.

How does blood carry carbon dioxide?

Carbon dioxide is taken up rapidly by the red blood cells and then carried in solution as **sodium hydrogencarbonate**, partly in the red blood cells but mainly in the plasma.

How does blood clot?

Damage of tissues and/or contact of platelets with an unfamiliar surface triggers a chain reaction which results in the conversion of fibrinogen to **fibrin** (**clot**). For the chain reaction to occur, numerous factors are required including **calcium ions** and **vitamin K** (figure 14.2).

One of the factors required for blood-clotting is called **factor VIII**: absence of it results in **haemophilia**.

Blood is prevented from clotting inside our blood vessels by **anticoagulants**.

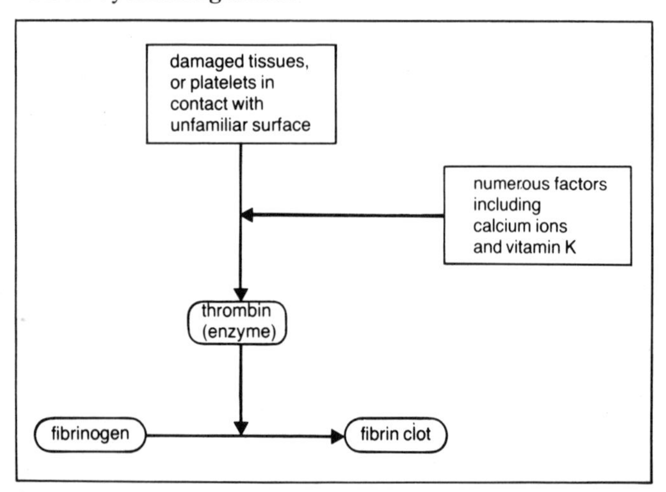

Figure 14.2 *Simplified scheme showing how blood-clotting is brought about.*

Why is blood-clotting important?

- It prevents excessive loss of blood from cuts and wounds.
- It is the first step in the **healing** of cuts and wounds.

What are antibodies?

Antibodies are proteins (globulins), produced by lymphocytes. They combine with **antigens** on the surface of germs, thereby destroying them.

Types of antibodies and how they work

- **Lysins** make germs burst.
- **Agglutinins** make germs clump together, after which they may be ingested by phagocytes.
- **Opsonins** stick onto germs, making it easier for phagocytes to ingest them.
- **Precipatins** produce a precipitate with the antigens.
- **Antitoxins** combine with, and render harmless, poisons produced by germs.

Helper cells

Not all lymphocytes produce antibodies. Some, called **helper cells**, assist the antibody-producing lymphocytes in their action. The virus that causes AIDS attacks helper cells.

How are antibodies produced?

Figure 14.3 shows what happens.
Note: Antibodies are specific, i.e. they will only work against the type of germ whose antigens triggered their production.

Why do we get certain diseases only once?

The first time a particular type of germ attacks you, your lymphocytes 'learn' how to produce antibodies against it. If the same type of germ attacks you again, the lymphocytes produce antibodies more quickly than before.

Once your lymphocytes have 'learned' how to produce antibodies against a particular type of germ, you are **immune** to that disease.

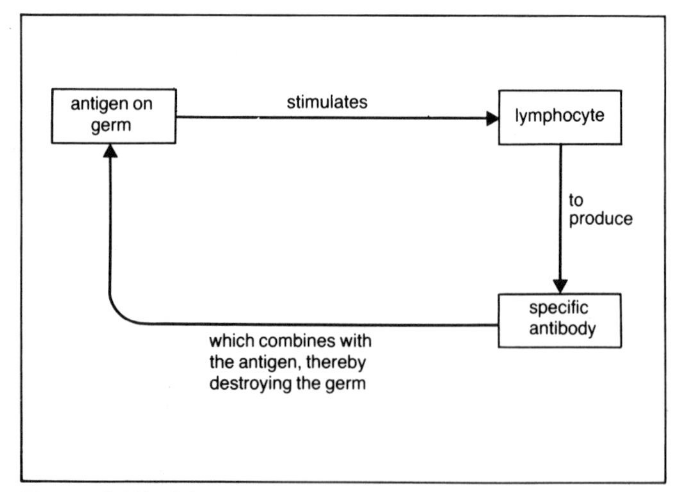

Figure 14.3 *Scheme showing how antibodies are produced.*

Blood groups

The human population can be divided into four groups according to the presence or absence of certain substances on the surface of the red blood cells, and of corresponding 'anti' substances in the plasma (table 14.1).

A person's plasma never contains an 'anti' substance corresponding to one of the substances on the red blood cells. If it did, the 'anti' substance would combine with the red blood cell substance, causing the red blood cells to clump together (**agglutination**).

Table 14.1 *The ABO blood group system. The percentages denote the proportion of people belonging to the different groups in Great Britain.*

Name of blood group	Substances on the red blood cells	Substances in the plasma
A (40%)	A	anti-B
B (10%)	B	anti-A
C (3%)	A and B	none
O (47%)	none	anti-A and anti-B

Blood transfusions

In a blood transfusion, care must be taken to ensure that the recipient's plasma does not contain 'anti' substances corresponding to the donor's red blood cell substances. If it does, the donor's red blood cells will agglutinate, blocking the recipient's blood vessels: such bloods are incompatible.

Normally blood transfusions are carried out using donor blood which belongs to the same group as that of the recipient.

Components of the circulatory system

The components of the circulatory system are:

- The **heart**, which pumps blood round the body.
- **Arteries**, which carry blood from the heart to the tissues.
- **Capillaries**, which carry blood through the tissues where exchanges take place.
- **Veins**, which carry blood from the tissues back to the heart.

Arteries, veins and capillaries are all **blood vessels**. Their main features are summarised in figure 14.4.

The circulatory system in more detail

Figure 14.5 summarises the human circulatory system. Note that the heart is divided into left and right halves: blood flows from the heart to the lungs (where it is oxygenated), then back to the heart, and then to the rest of the body. This is called a **double circulation** and is typical of mammals.

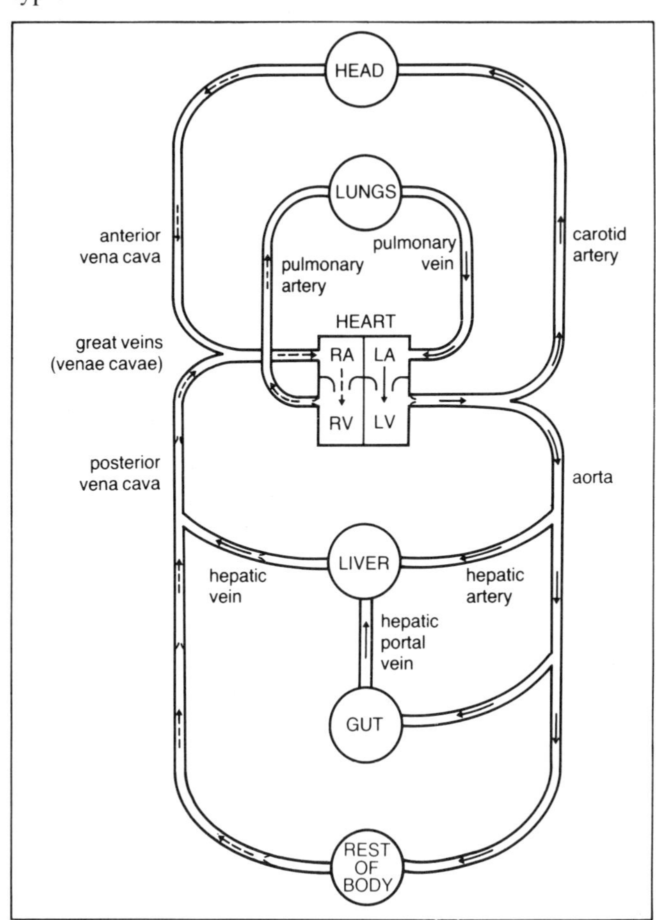

Figure 14.5 *General plan of the human circulatory system. Solid arrows, oxygenated blood; broken arrows, deoxygenated blood.*

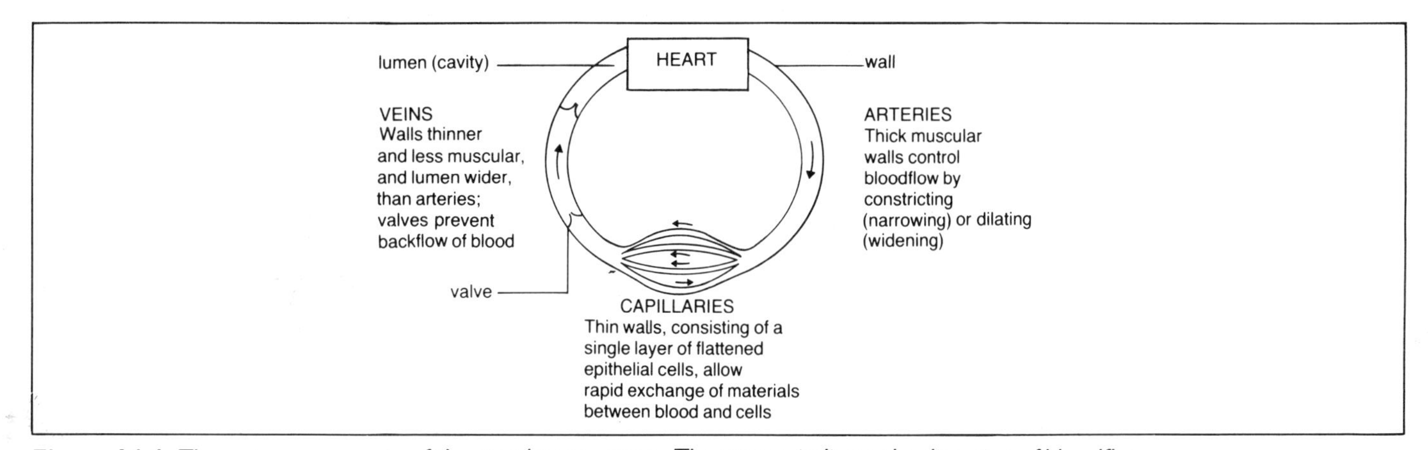

Figure 14.4 *The main components of the circulatory system. The arrows indicate the direction of bloodflow.*

Structure of the heart

The heart is shown in detail in figure 14.6. The main pumping action is achieved by the **ventricles** whose walls are thick and muscular. **Valves** between the atria and ventricles, and at the openings into the arteries, prevent backflow of blood.

Contraction (beating) of the heart is called **systole**; relaxation of the heart is called **diastole**. The heart muscle is supplied with oxygen by **coronary arteries**.

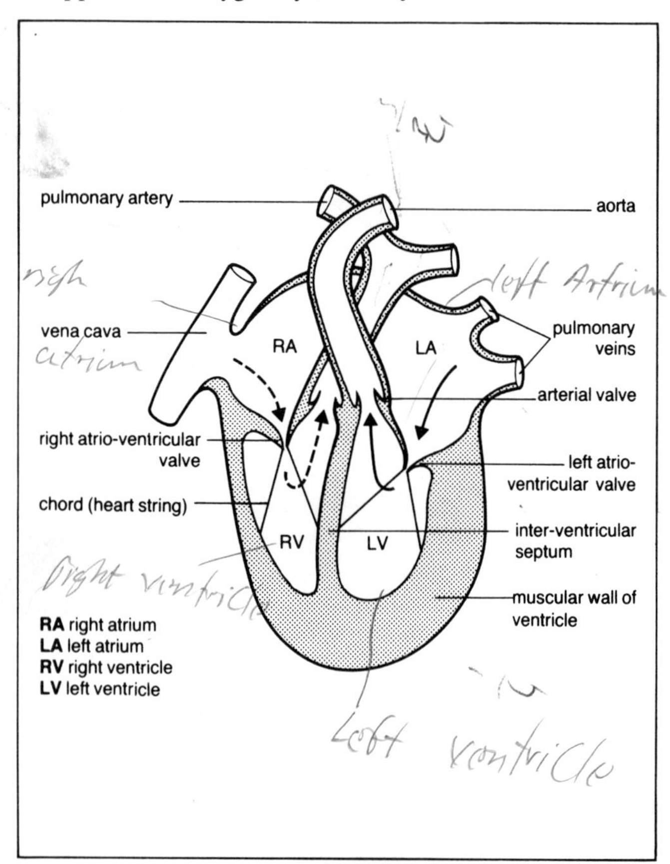

Figure 14.6 *The heart in ventral view. The atrio-ventricular valves are also known as the cuspid valves. The arterial valves at the openings of the pulmonary artery and aorta are also known as pocket valves or semi-lunar valves, on account of their shape. Solid arrows, oxygenated blood; broken arrows, deoxygenated blood.*

What makes the heart beat?

The human heart beats approximately 70 times per minute. Each beat is initiated by a **pacemaker** in the wall of the right atrium. The pacemaker's frequency can be modified by nerve impulses: in this way the heart is 'told' to beat faster when necessary, for example during exercise.

What is meant by blood pressure?

Blood pressure is the pressure exerted by the blood against the walls of the blood vessels as a result of the pumping action of the heart.

Where and when is the blood pressure highest?

Blood pressure is highest in the arteries and lowest in the veins. The arterial pressure rises when the heart contracts (systole) and falls when the heart relaxes (diastole).

Defects of the circulation

Here are three particularly important defects:

- **Hypertension**: high blood pressure.
- **Stroke**: brain damage resulting from blockage, or bursting, of a blood vessel in the brain.
- **Heart attack**: failure of the heart muscle (or part of it) to contract, caused by blockage of the coronary artery or one of its branches (figure 14.7).

What is tissue fluid and how is it formed?

Tissue fluid is a watery fluid found between the cells in the tissues. It is formed by filtration from the capillaries and consists of blood plasma minus the plasma proteins.

Excess tissue fluid drains into the lymphatic system where it becomes **lymph**.

The lymphatic system

The **lymphatic system** consists of numerous thin-walled vessels containing **lymph**. The lymph vessels lead to the veins, so eventually lymph returns to the bloodstream.

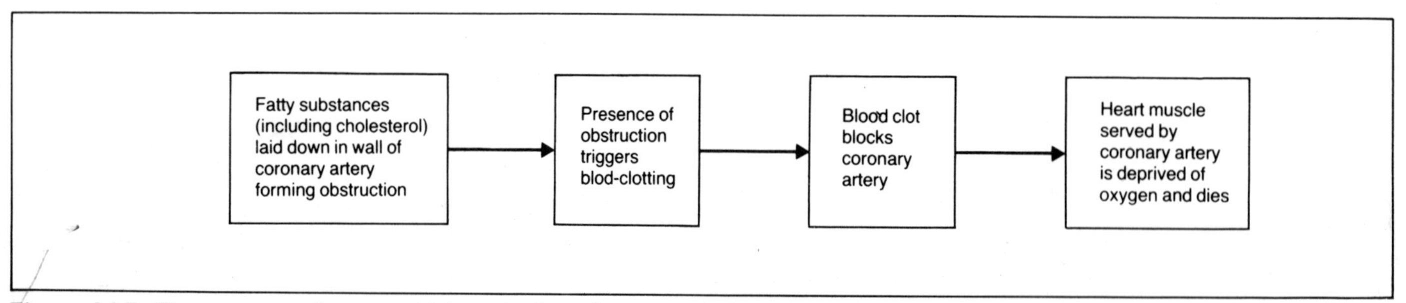

Figure 14.7 *The sequence of events which can lead to a heart attack. Smoking, and possibly excessive eating of animal (saturated) fats, can cause fatty substances to be laid down in the walls of the coronary arteries.*

Valves in the lymph vessels keep the lymph moving towards the veins.

Lymph nodes (lymph glands) occur at intervals along the length of the lymph vessels. Lymph nodes contain a meshwork of fibres with numerous phagocytes and lymphocytes. Germs get trapped in the meshwork and destroyed by the phagocytes and lymphocytes.

Where are the main places where lymph nodes occur?

The main places are the sides of the neck, armpits and groins. Lymph tissue is also found in the tonsils and adenoids.

The circulation in other animals

Check your syllabus to find out how much you need to know!

- In **simple organisms** (e.g. *Amoeba, Hydra*) there is no blood or circulation. Substances are transported by diffusion.
- The **earthworm** has a simple circulation with blood containing haemoglobin. Blood is pumped by contractions of the dorsal blood vessel.
- **Fish** have a single circulation: blood (containing haemoglobin) is pumped from the heart to the gills whence it flows to the rest of the body and then returns to the heart. (Compare this with the double circulation of mammals.)
- **Insects** have an open circulation: colourless blood (no haemoglobin) is pumped by a tubular heart into the body cavity whence it seeps round the body before returning to the heart. The blood does not carry oxygen; this function is performed by the tracheal system.

• Questions

1 Carbon monoxide combines with haemoglobin even more readily than oxygen does. Use this information to explain why carbon monoxide is a fatal poison.

2 (a) Blood has difficulty returning to the heart from the legs. Suggest *two* reasons for this.

(b) Suggest as many ways as you can think of by which blood is helped to get back to the heart. (There are five possible ways.)

3 Explain the reason for each of the following observations.

(a) When a red blood cell is viewed under a light microscope, the central part of the cell appears paler than the outer part.

(b) When you have an infectious disease, the number of white cells in your blood is greater than at other times.

(c) The plasma protein fibrinogen can be removed from a sample of blood by stirring the blood vigorously with a glass rod and then slowly withdrawing the rod from the blood.

(d) The blood pressure in the aorta is approximately six times higher than in the pulmonary artery.

4 A person's blood group can be found by mixing small samples of his or her blood with dried sera in test panels on a card. Copy the following table:

	Test panels		
	anti-A	anti-B	neither anti-A nor anti-B (control panel)
Blood group A			
Blood group B			
Blood group AB			
Blood group O			

(a) Fill in the table, writing a tick if agglutination occurs and a cross if agglutination does not occur.

(b) Sarah's blood agglutinates in the anti-A and anti-B panels, but Tom's blood does not agglutinate in any of the panels.
(i) Which blood groups do Sarah and Tom belong to?
(ii) Whose blood group is the rarer?

(c) Why is the control panel needed?

(d) When sampling a person's blood, certain safety precautions have to be taken.
(i) State two such safety precautions.
(ii) Why are they necessary?

5 The diagram below shows the mammalian heart and major blood vessels in ventral view.

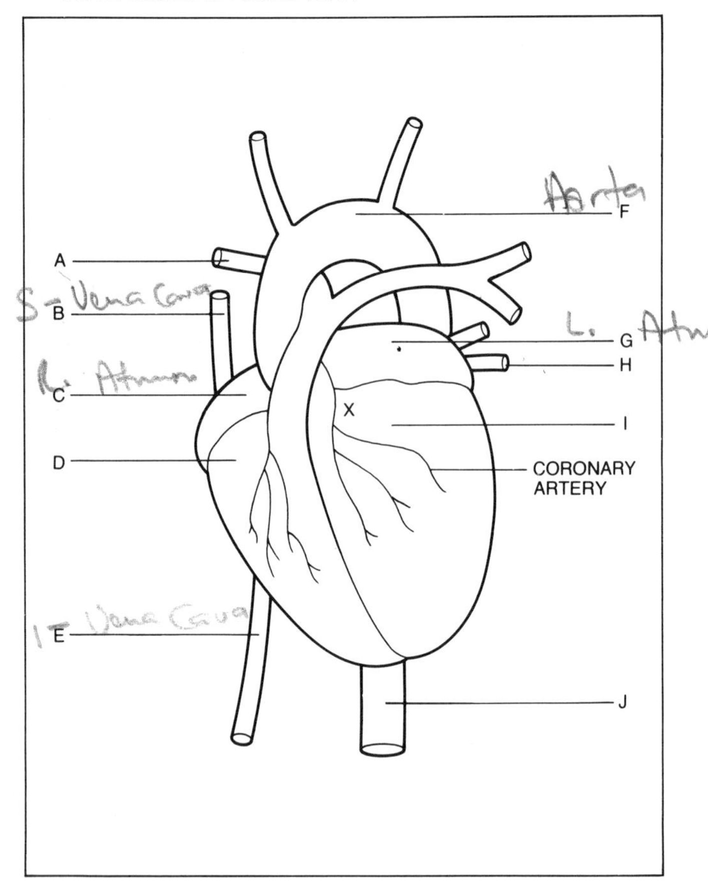

(a) Give the names of each of the parts labelled A to J.

(b) Write down the letters in the order in which blood passes through the parts.

(c) Give the letter of the part which, when it contracts, sends blood to the brain.

(d) Suppose a blood clot forms in the coronary artery at the point marked X. How is this likely to affect the heart as a whole?

(e) Describe two human activities which might cause a clot to form in the coronary artery.

6 The graph below shows the blood pressure in the left ventricle and the aorta throughout two cycles of the heart. (The heart cycle is the time interval between one heartbeat and the start of the next heartbeat.)

Use the graph to answer the following questions.

(a) How long is a heart cycle?

(b) For how long during each heart cycle is the valve between the left ventricle and the aorta closed. How did you work out your answer?

(c) What is the difference between the maximum and minimum pressure in the aorta, and why should this difference occur?

(d) Explain how blood pressure is increased in the left ventricle.

(SEG, with slight addition)

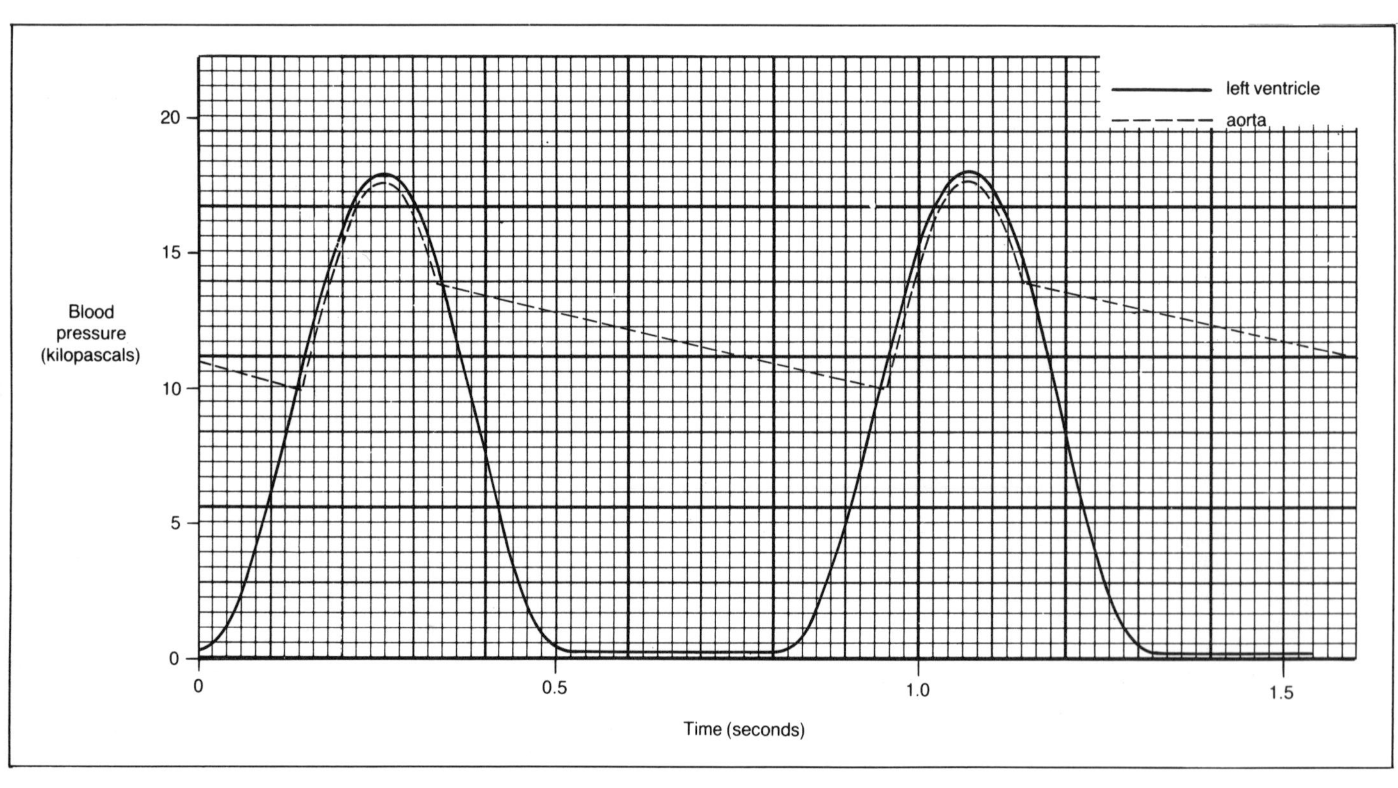

Nutrition in plants

What do plants need?

For successful growth, plants need:
- carbon dioxide (for photosynthesis);
- water (for photosynthesis and other functions);
- mineral elements (for various purposes – see below);
- oxygen (for respiration).

What mineral elements do plants need?

Table 15.1 summarises the main mineral elements needed by plants. Many of them are obtained as salts, e.g. potassium nitrate and potassium phosphate. They are called **nutrients**.

How can we find out the effect of depriving a plant of a mineral element?

By carrying out a laboratory experiment involving **water culture** or **sand culture**. A plant (or group of plants) is grown in distilled water or purified sand containing all the required mineral elements in sufficient amounts. This is the control. A second plant (or group of plants) is grown in an identical medium except that it lacks the mineral element whose absence you wish to investigate. The two plants (or groups of plants) must be kept in identical conditions.

Where do plants get mineral salts from?

Water plants get their mineral salts from the surrounding water. They are mainly in the form of **ions**.

Land plants get their mineral salts from the **soil** (see page 11). Fertile soil contains all the necessary mineral elements for plant growth.

Where do the mineral elements in the soil come from?

The mineral elements come from two sources:

1. Humus, the decayed remains of animals and plants.
2. Soil particles, which are derived from rocks.

How is soil kept fertile?

In nature, soil is kept fertile by the cycling of elements such as nitrogen (see page 20).

In crop-growing fields, soil is kept fertile by:
- Leaving the field empty of crops (fallow) once every few years. This allows the chemicals which have been taken out of the soil to build up again.
- Rotating crops: Different crops remove different chemicals from the soil at different rates, so rotation helps to prevent complete depletion.

Table 15.1 *The main mineral elements needed by plants, why they are needed and what happens if they are absent (deficiency effects). Yellowing of leaves is known as chlorosis.*

Element	Why needed	Deficiency effects
Major elements (needed in relatively large amounts)		
Nitrogen (N)	Contained in amino acids and proteins	Poor growth, yellow leaves
Phosphorus (P)	Contained in important chemicals	Poor growth, leaves dull green with curly brown edges
Potassium (K)	Increases hardiness	Yellow edges to leaves, die early
Sulphur (S)	Contained in proteins	Yellow leaves
Calcium (Ca)	Needed for cell formation	Poor buds, stunted growth
Magnesium (Mg)	Contained in chlorophyll	Yellow leaves
Iron (Fe)	Needed for chlorophyll formation	Yellow leaves
Manganese (Mn)	Helps certain enzymes	Yellow leaves with grey spots
Trace elements (needed in very small amounts)		
Copper (Cu)	Helps certain enzymes	Shoots die back
Molybdenum (Mo)	Helps certain enzymes	Poor growth
Zinc (Zn)	Helps certain enzymes	Malformed leaves

- Growing a leguminous crop in the field from time to time. Leguminous plants enrich the soil in nitrates (see page 20).
- Putting the chemicals which have been removed from the soil back into it. This is achieved by applying a fertiliser.

What are fertilisers?

A **fertiliser** is any substance containing nutrients needed for plant growth. The main types of fertiliser are:

- **Organic fertilisers:**
  (a) Brown manure
    (i) Farmyard manure: dung and urine of animals mixed with straw.
    (ii) Compost: partly decayed remains of garden waste.
  (b) Green manure
  Plants which are ploughed into the soil; leguminous plants are particularly beneficial.
- **Inorganic fertilisers:**
  Commercially produced preparations of inorganic elements in liquid or powdered form.

Growing crops the modern way

Thanks to fertilisers and pesticides (see page 26), it is possible to grow single, weed-free crops at high density. This is called **monoculture**.
Monoculture ensures high productivity but it has these disadvantages:

- If a pest gets into the crop, there are no natural predators to control it. The pest spreads quickly because the plants are so close together. This can be avoided by rotating crops: different plants are infested by different pests, and rotation can break the pest's life cycle.
- Intensive use of the soil can make it fine and dusty, leading to erosion. This can be avoided by leaving fields empty of crops (fallow) every third year or so, and/or ploughing in manure.
- It makes the countryside look dull, particularly if the fields are very large, and it can be destructive of wildlife.

• Questions

1 A biology student carried out an experiment on two groups of broad bean seedlings, A and B.
There were ten seedlings in each group. The seedlings in group A had their roots immersed in distilled water to which had been added all the mineral elements necessary for plant growth minus nitrogen. The seedlings in group B had their roots immersed in distilled water to which had been added all the mineral elements necessary for plant growth including nitrogen.

(a) What was the student trying to investigate?

(b) Why did she add the mineral elements to distilled water rather than tap water?

(c) In what form would the nitrogen have been when it was added to the water for the group B seedlings?

(d) What was the purpose of setting up the group B seedlings?

(e) The student was tempted to have only two seedlings in each group rather than ten. Why was she wise to have ten?

2 (a) Put the following types of fertiliser in order according to their speed of action: compost, green manure, inorganic fertiliser. Explain your answer.

(b) Give two advantages of farmyard manure over inorganic fertilisers.

(c) Why are leguminous plants particularly useful as green manure?

(d) When making compost it is important to keep the compost heap well aerated. Why?

3 When visiting a farm you might see some bags labelled 'N, P and K'.

(a) What do N, P and K stand for?

(b) Why are they needed on the farm?

16 Photosynthesis

What is photosynthesis?

Photosynthesis is the method of nutrition used by plants and some single-celled organisms.

Simple carbon dioxide and water are built up into complex sugars, using energy from sunlight. The green pigment **chlorophyll** plays an important part.

Overall equation:

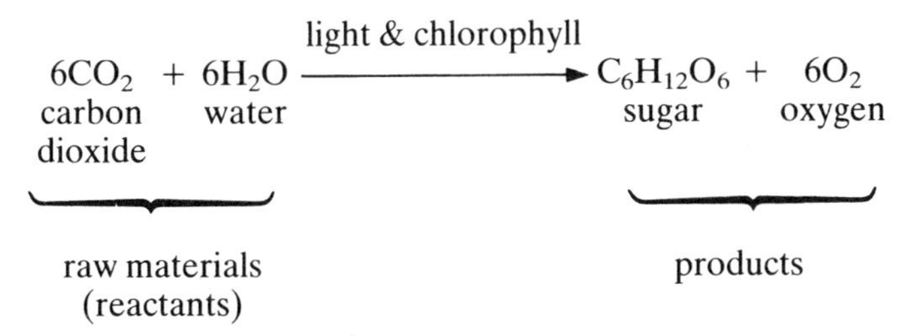

The sugar may be built up into starch for storage. This reaction does *not* require light energy.

How can you tell if a plant has been photosynthesising?

The best way is to test a leaf for sugar or starch. Photosynthesis is the only way a plant can produce these substances, so their presence tells us that the plant must have been photosynthesising.

How can you test a leaf for reducing sugar?

Grind up a piece of leaf with water. Transfer it to a test tube and shake. Add a little **Benedict's** or **Fehling's solution**. Heat to boiling. A red, brown or green precipitate means that reducing sugar is present.

How can you test a leaf for starch?

Boil the leaf in water till flabby (soft), then in ethanol till it is decolorised. Wash it in water, then add **dilute iodine solution**. If the leaf turns blue-black, starch is present. (The purpose of boiling the leaf in water is to enable the iodine solution to penetrate it; the purpose of decolorising it is to allow the blue-black colour to show up, and washing it softens the leaf which is hardened by the ethanol.)

How can you remove the starch from a plant?

Put the plant in the dark for several days. Without light the plant cannot photosynthesise and make starch; any starch already present is turned into sugar and used up. Such a plant is described as **de-starched**.

What do plants need in order to photosynthesise?

They need

- light;
- carbon dioxide;
- chlorophyll;
- water.

How can you show that light is needed for photosynthesis?

Either:

De-starch two potted geranium plants. Leave one of them in the dark. Place the other one (the control) in the light. All other conditions should be the same for both plants. Leave the two plants for 48 hours, then take a leaf from each one and test it for starch. Only the control plant should have made starch.

Or:

De-starch a potted geranium plant. Attach a strip of opaque paper or foil to the upper and lower sides of a leaf. Leave the plant in a well lit place for 48 hours, then test the leaf for starch. Only the exposed (i.e. uncovered) parts of the leaf should give a blue-black colour with iodine: this is called a **starch print**.

How can you show that carbon dioxide is needed for photosynthesis?

De-starch two potted geranium plants. Place one of the plants in a sealed polythene bag containing a dish of **soda lime**: this will absorb carbon dioxide from the air. Place the other plant (the control) in a polythene bag without soda lime. All other conditions, apart from the soda lime, should be the same for both plants.

Leave the two plants for 48 hours, then take a leaf from each one and test it for starch. Only the control plant should have made starch.

How can you tell that chlorophyll is needed for photosynthesis?

De-starch a potted geranium plant which has **variegated leaves**. (A variegated leaf is green in places but white or yellow elsewhere; chlorophyll is present only in the green areas.) Place the plant in the light for 48 hours, then remove a leaf and test it for starch. Only the green parts of the leaf give a blue-black colour with iodine.

How can you show that oxygen is given off in photosynthesis?

Obtain a handful of the water plant *Elodea* (Canadian pondweed) or its tropical relative *Hydrilla* and set it up as shown in figure 16.1. If the plant is well lit, it will give off bubbles of gas which collect at the top of the test-tube as shown. Bubbling ceases if the plant is put in the dark. The gas contains oxygen: it will re-kindle a glowing splint.

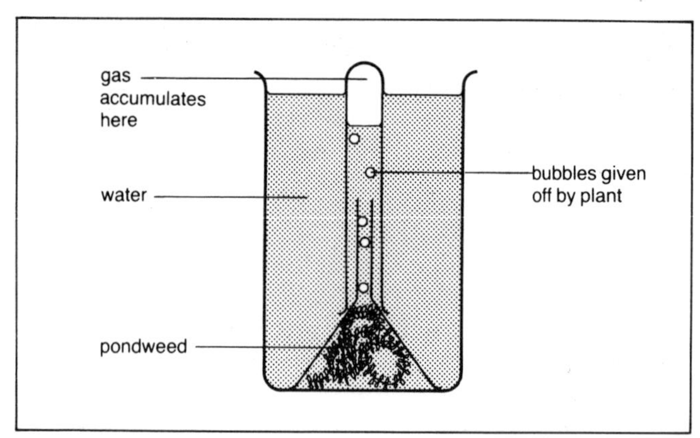

Figure 16.1 *How to collect gas given off by* Elodea *or* Hydrilla.

How can you measure the rate of photosynthesis?

Cut off a short length of *Elodea* or *Hydrilla* and place it in a test tube of water. Count the number of bubbles given off from the cut end of the stem in a given time. If you do this with the plant in different conditions, you can find out which conditions affect the rate of photosynthesis.

What conditions affect the rate of photosynthesis?

The rate of photosynthesis is affected by three main conditions:

- light intensity;
- temperature;
- carbon dioxide concentration.

These conditions vary in different places at different times of the day and year. Light and temperature are particularly variable, carbon dioxide less so.

At any given moment the rate of photosynthesis is controlled by whichever condition is closest to its minimum value. So if, for example, the carbon dioxide concentration and temperature are high but the light intensity is low, then light will be controlling the rate.

The pigments in a leaf

Chlorophyll is one of several pigments (coloured substances) present in leaves. The main pigments are:

- Chlorophyll (green);
- Xanthophyll (yellow);
- Carotene (yellow).

Chlorophyll is the main pigment involved in photosynthesis. Xanthophyll and carotene are supplementary pigments which help to capture light energy.

How can we separate the pigments?

By the technique of **paper chromatography**. First, grind up several leaves with a **solvent** (ethanol or acetone). The pigments come out of the ruptured cells and dissolve in the solvent. Next, make a concentrated spot of the pigment solution near one end of a strip of absorptive paper. Then set up the paper strip as shown in figure 16.2. The solvent rises up the paper strip, carrying the pigments with it. The different pigments travel at different speeds, so they become separated and can be identified individually.

What is chlorophyll?

Chlorophyll is a complex organic molecule containing **magnesium**. The magnesium is at the centre of the molecule and helps the chlorophyll to do its job.

What does chlorophyll do?

Chlorophyll absorbs light energy and transforms it into chemical energy in sugar molecules.

Which wavelengths of light does chlorophyll absorb?

White light is made up of a series of wavelengths (i.e. colours). They can be separated by a prism to give a **spectrum**. The colours of the visible spectrum are:

RED ORANGE YELLOW GREEN BLUE INDIGO VIOLET

The colours that are absorbed by chlorophyll are BLUE and, to a lesser extent, RED. (Green is reflected: that's why chlorophyll looks green.)

How can we show that chlorophyll absorbs blue and red light?

First project a beam of light through a prism so as to produce a spectrum on a white screen. Then insert a thin glass vessel containing a concentrated solution of chlorophyll in the path of the beam. The blue, and part of the red, colours disappear from the spectrum.

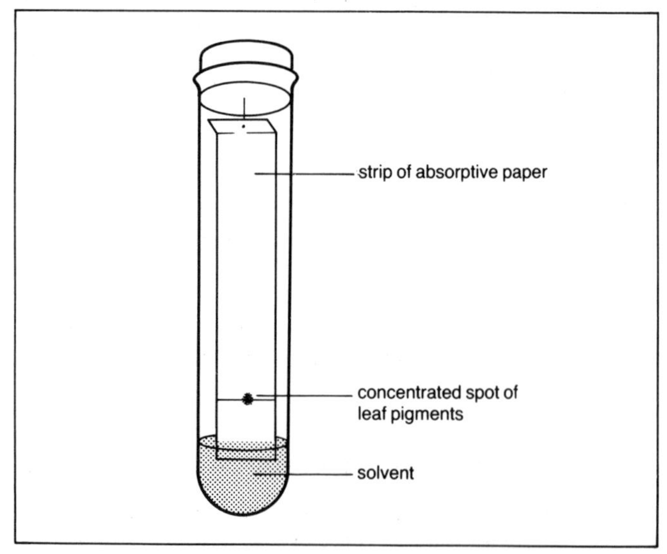

Figure 16.2 *How to separate the different leaf pigments by paper chromatography.*

Which wavelengths are used in photosynthesis?

The wavelengths which are absorbed by chlorophyll (i.e. blue and red) are also *used* in photosynthesis.

This can be shown by illuminating a plant with different coloured lights and finding out which ones give the highest rates of photosynthesis. As expected, the highest rates are produced by blue and red lights, particularly blue.

Where does chlorophyll occur?

Inside **chloroplasts**. A chloroplast is a spherical or ovoid body whose hollow interior is divided up by numerous **membranes** (figure 16.3). The chlorophyll molecules are located in the membranes. The membranes together present a large surface area for the efficient harvesting of light energy.

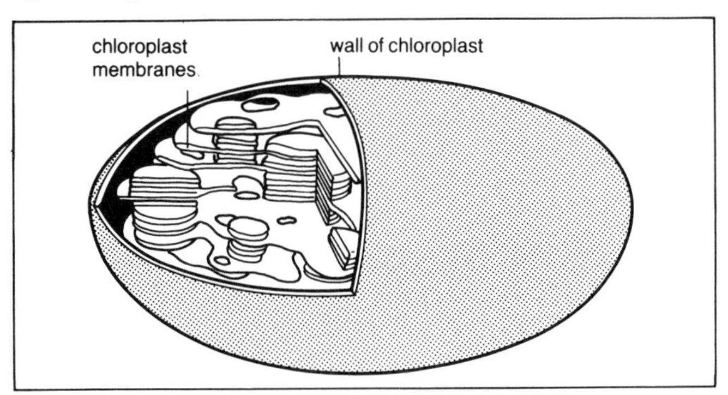

Figure 16.3 *Structure of a chloroplast. The chlorophyll is situated in the membranes.*

The chemistry of photosynthesis

Scientists have confirmed the photosynthesis equation by 'labelling' the carbon and oxygen atoms of the raw materials (reactants) and tracing what happens to them during photosynthesis. They used the unicellular organism *Chlorella*.

The carbon and oxygen in the carbon dioxide supplied to the organism get into the carbohydrate (sugar and starch) which the organism makes; and the oxygen which is given off comes from the water:

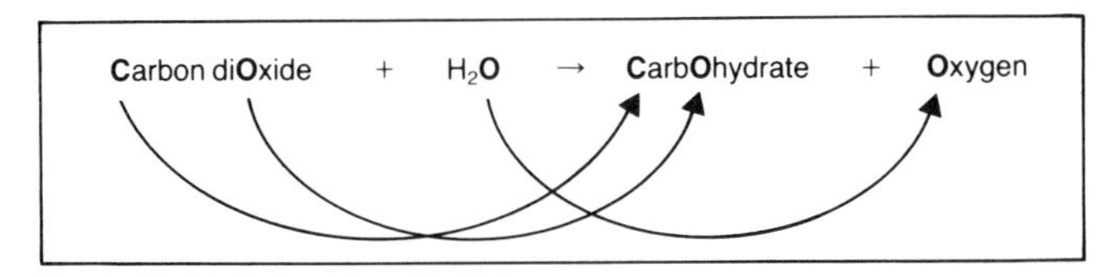

How were the carbon and oxygen atoms labelled?

The carbon in the carbon dioxide supplied to the organism was replaced by its **radio-active isotope**. The radio-active carbon was later detected in the carbohydrate products by means of a Geiger counter.

The oxygen in the carbon dioxide supplied to the organism was replaced by its **heavy isotope**. The heavy oxygen was later detected in the carbohydrate products by means of a mass spectrometer.

The oxygen in the water supplied to the organism was replaced by its heavy isotope. In this case the heavy oxygen was later detected in the oxygen given off by the organism.

Photosynthesis occurs in two stages

- *Light stage* (can only occur in the light): Water is split into oxygen and hydrogen (top part of figure 16.4).
- *Dark stage* (can occur in the light or dark): Hydrogen from the light stage combines with carbon dioxide to form sugar (bottom part of figure 16.4).

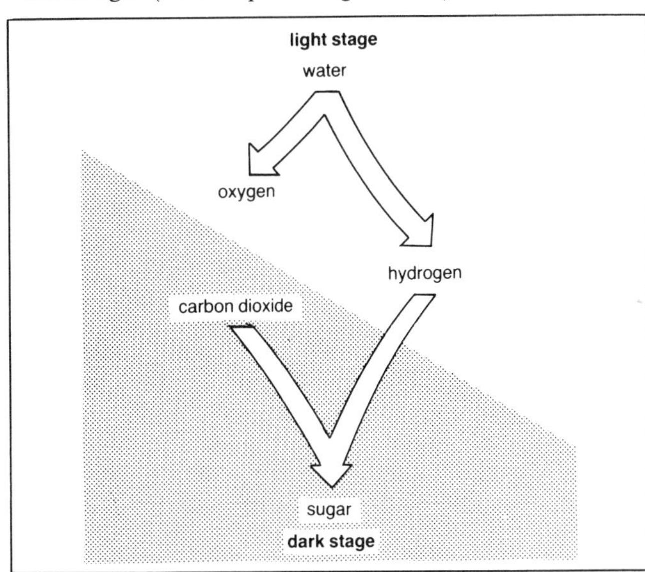

Figure 16.4 *Photosynthesis occurs in two stages. First water is split into oxygen and hydrogen. Then the hydrogen combines with carbon dioxide to form sugar.*

What happens to the sugar formed in photosynthesis?

It may be:

1. Broken down in respiration to give energy (see page 46).
2. Built up into starch for storage.
3. Built up into cellulose to make cell walls.
4. Turned into fats and/or oils for storage and other purposes.
5. Combined with nitrogen to form proteins (see page 42).
6. Moved to other parts of the plant (see page 81).

Why is photosynthesis important?

1. It provides animals, including humans, with a natural source of food.
2. It puts oxygen into the atmosphere and this is used by organisms for respiration.

• Questions

1 In 1692 Van Helmont weighed a young willow tree and planted it in a pot containing a known mass of soil. He then left the tree to grow, giving it nothing but water. After five years he re-weighed the tree, and the soil. He found that the soil's mass had decreased by 56 g whereas the tree's mass had increased by 74 g. Explain Van Helmont's result in the light of modern knowledge about how plants feed.

2 The graph below shows how a plant's rate of photosynthesis is affected by the light intensity at two different concentrations of carbon dioxide. Curve A was obtained with the plant in a low concentration of carbon dioxide. Curve B was obtained at a higher concentration of carbon dioxide.

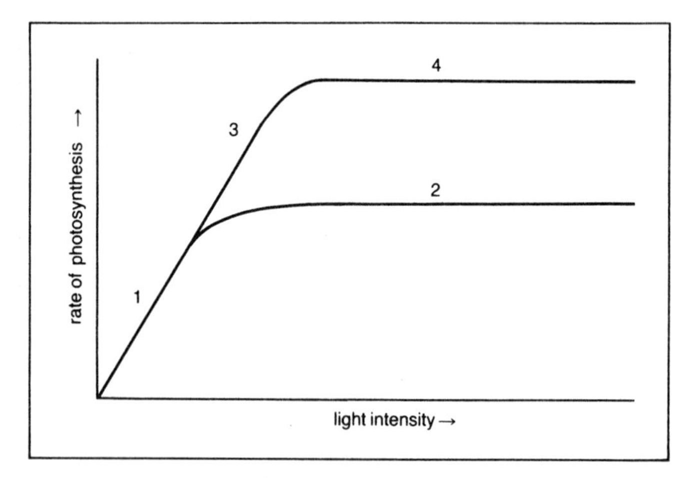

(a) What factor was controlling the rate of photosynthesis at each of the points 1, 2 and 3 in the graph? Give reasons for your answers.

(b) What factors might have been controlling the rate of photosynthesis at point 4 in the graph?

(c) During what part(s) of a summer's day in a temperate region would you expect the rate of photosynthesis of a plant in an open field to be:
  (i) at its maximum;
  (ii) zero;
  (iii) at its minimum but above zero;
  (iv) controlled by the light intensity;
  (v) controlled by the carbon dioxide concentration?

(d) Under what *natural* circumstances would you expect temperature to control the rate of photosynthesis?

3 Describe how you would show which particular wavelengths of light are used for photosynthesis by the pondweed *Elodea*. What assumptions are made in your method?

4 The graph below shows the changes in carbon dioxide concentration measured in a field of grass during a warm day in the summer.

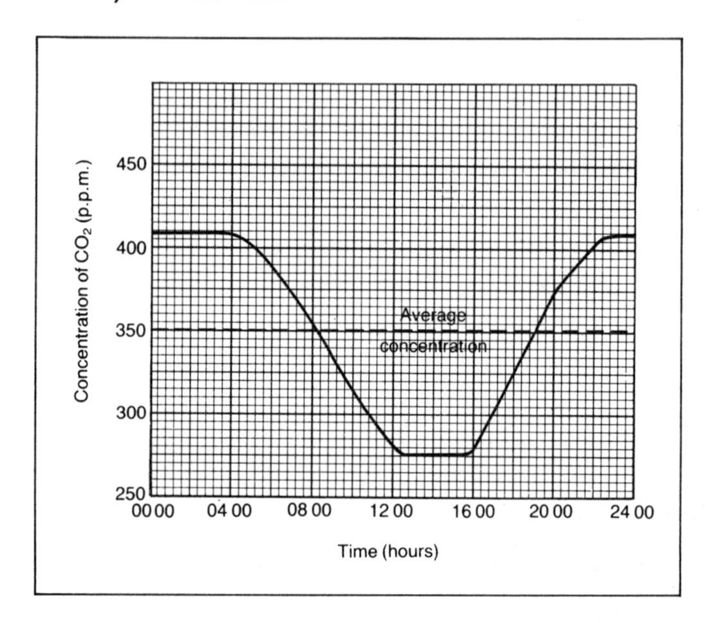

(a) (i) What was the highest concentration of carbon dioxide recorded?
  (ii) Between which hours is the concentration of carbon dioxide below average?

(b) Describe what happens to the carbon dioxide concentration between:
  (i) 0000 to 0400
  (ii) 0400 to 1200

(c) What makes the carbon dioxide concentration change between 0400 and 1200?

(NEA)

Structure and function in the flowering plant

External structure of the leaf

The external structure of a typical leaf (dicotyledon type) is shown in figure 17.1. Most leaves are green because of the presence of chlorophyll in the cells. The veins strengthen the leaf and transport substances to and from it.

Internal structure of the leaf

The internal structure of the leaf is shown in figure 17.2.

- The **epidermis** protects the leaf against damage and excessive water loss.
- The **mesophyll** undergoes photosynthesis and gives the leaf strength (by means of the turgidity of its cells).
- The **xylem** transports water and mineral salts to the leaf and gives it strength.
- The **phloem** transports food substances, which have been made by photosynthesis, away from the leaf.
- The **stomata** are for gas exchange and escape of water vapour.

In addition **woody fibres** beneath the larger veins give the leaf extra strength.

What are the functions of leaves?

The main function of leaves is to feed the plant by photosynthesis (see Chapter 16). In some cases leaves have other functions such as support in climbing plants (where the leaves are modified into tendrils), and feeding on small animals in carnivorous plants (where the leaves may be modified into insect traps).

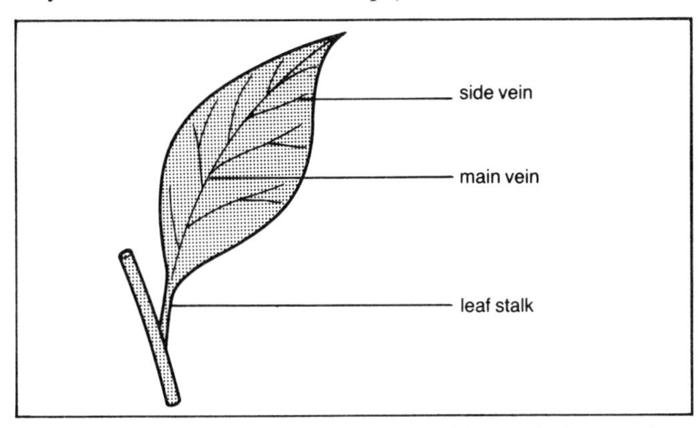

Figure 17.1 *External structure of a leaf (dicotyledon type).*

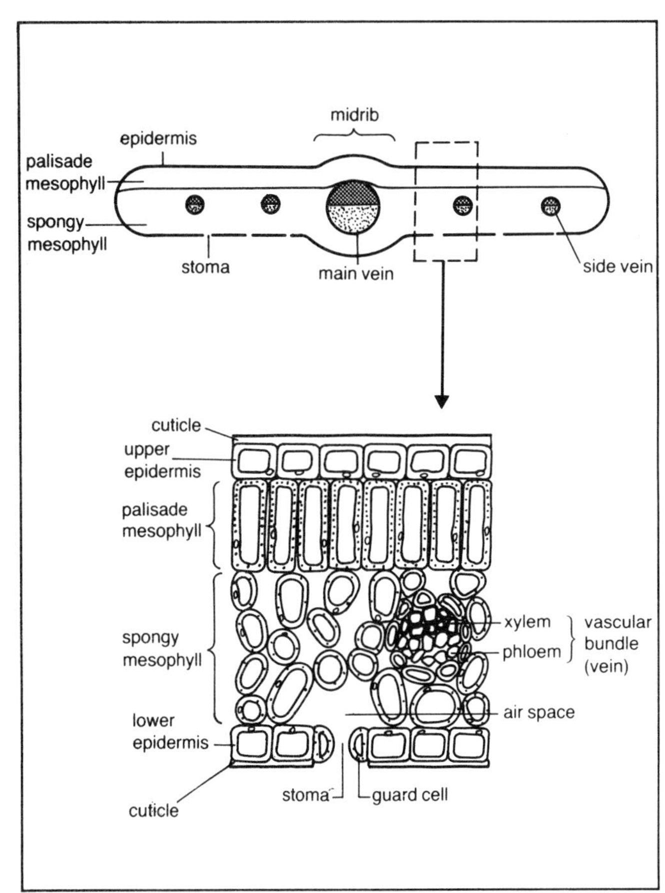

Figure 17.2 *Internal structure of a leaf (dicotyledon type) as seen in a transverse section.*

How are leaves adapted for photosynthesis?

1. They have a **large surface area** for absorbing light and carbon dioxide.
2. They are usually arranged in such a way that they all receive light.
3. They have **numerous stomata** for taking in carbon dioxide.
4. They are **thin**, thus reducing the distance through which carbon dioxide has to diffuse after it has entered the leaf.
5. The **palisade mesophyll**, the main site of photosynthesis, is situated towards the upper side of the leaf, which is usually the most brightly lit side.
6. There are **large air spaces** between the spongy mesophyll cells, allowing gases to circulate.
7. There are **transport (vascular) tissues** for bringing water and mineral salts to the leaf and removing the products of photosynthesis from it.

How can you observe the stomata in a leaf?

Either observe the surface of the leaf under the microscope, illuminating it from above; *or* make a **replica** of the leaf surface with nail varnish, then peel off the dried varnish and observe it under the microscope.

How do stomata open and close?

Figure 17.3 shows how a stoma opens. The mechanism depends on the **guard cells** taking up water by osmosis from the surrounding epidermal cells. The inner wall of the guard cells is thicker and less stretchable than the outer wall, so the guard cells bend as they expand, thereby opening the stoma.

The stoma closes by water being drawn out of the guard cells by osmosis, the reverse of the mechanism described above.

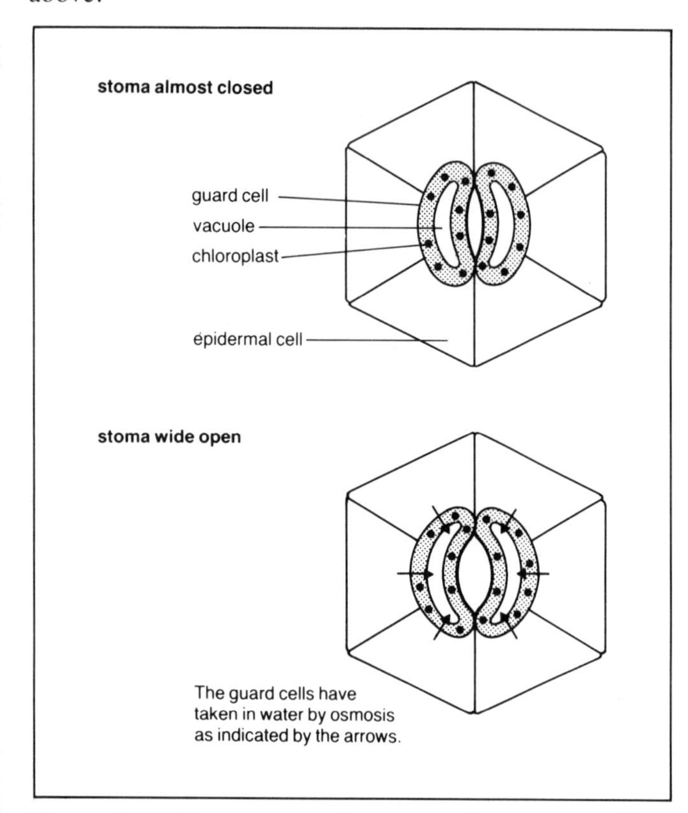

Figure 17.3 *The structure of a stoma and how it opens. The stoma is shown here in surface view.*

What is transpiration?

Transpiration is the evaporation of water from the above-ground parts of a plant, particularly the leaves. Water is lost mainly through the stomata.

How is the transpired water replaced?

Water is taken up by the roots and flows up the stem to the leaves, replacing the water which has been lost by transpiration. The flow of water through the plant is called the **transpiration stream**.

What use is transpiration to a plant?

① The evaporation of water cools the leaves in hot weather.

② The transpiration stream carries mineral salts to the leaves (see below).

How can you show that a leaf loses water?

Attach a piece of dried **cobalt chloride** (or **cobalt thiocyanate**) paper to the leaf surface. Cobalt paper is blue when dry, pink when moist. If the paper changes from blue to pink, you know that water is being lost from the leaf surface.

Generally cobalt paper turns pink more quickly when attached to the lower side, than when attached to the upper side, of the leaf. This is because there are more stomata on the lower side.

How can you measure the rate of transpiration of a plant?

- **Weighing method**. Weigh the plant (or a leaf or leaves thereof) now, and again after a given time. Assuming that a decrease in mass is due to loss of water, calculate the mass of water lost per unit time.
- **Potometer method**. Fill a capillary tube with water and attach it, by rubber tubing, to the cut stem of a leafy plant. Introduce an air bubble into the open end of the capillary tube and find out how far it travels along the tube in a given time. Express the rate of uptake of water as the distance moved by the air bubble per unit time.
  Note: in using a potometer to measure the rate of transpiration, we are assuming that the rate of water uptake is equal to the rate of water loss.

What useful investigations can be carried out with a potometer?

The main use of a potometer is for *comparing* the rate of transpiration in different circumstances. For example you can compare:

- The rate of transpiration of different plant species. (In general, plants with a thin cuticle and/or numerous stomata transpire more rapidly than plants with a thick cuticle and/or fewer stomata).
- The effect of various external conditions on the rate of transpiration of a particular species.

Which external conditions affect the rate of transpiration?

Here are three external (i.e. environmental) conditions which particularly affect the rate of transpiration:

- **Temperature**. (Heat increases the rate of evaporation.)
- **Humidity**. (A dry atmosphere increases the rate of evaporation.)
- **Wind and air currents**. (They increase the rate of evaporation by blowing away the water vapour from the surface of the leaf.)

What happens to a plant if the rate of transpiration is very high?

If water evaporates from a plant faster than it can be replaced from the soil, the cells lose their turgidity and the plant droops (**wilts**). Some plants close their stomata in these conditions and this reduces the rate of water loss.

In what kind of weather do plants wilt?

Plants wilt in two kinds of weather:

- In hot, dry weather when the evaporative power of the atmosphere is great and the rate of transpiration is high.
- In very cold weather when the soil freezes and cannot be taken up by the roots.

How are plants adapted for living in dry places?

In general, plants living in dry places have some or all of the following features:

- They have relatively **few stomata**.
- Their stomata are sunk down into **pits in the epidermis**.
- They have **folded or rolled up leaves**.
- They have a **hairy epidermis**.
- They have **small leaves**.
- They **drop their leaves** in the winter or dry season.
- They have **shallow roots** for rapid absorption of water after a shower of rain.
- They have **deep roots** for absorbing water from deep down in the soil.
- They **store water** in swollen stems and/or leaves. (Plants which do this are called **succulents**.)

The features listed above are particularly well shown by plants which live in dry places such as the desert. Such plants are called **xerophytes** (literally means 'dryness-loving').

Structure of the stem

The internal structure of a young stem (dicotyledon type) is shown in figure 17.4. Many of the tissues are the same as in the leaf and have the same functions. The **packing cells** give the stem strength by means of their turgidity. The cells immediately beneath the epidermis are elongated longitudinally and have thick cellulose walls, creating **cellulose strands** which make the stem strong but flexible. The **cambium** forms secondary tissues in older stems (see below).

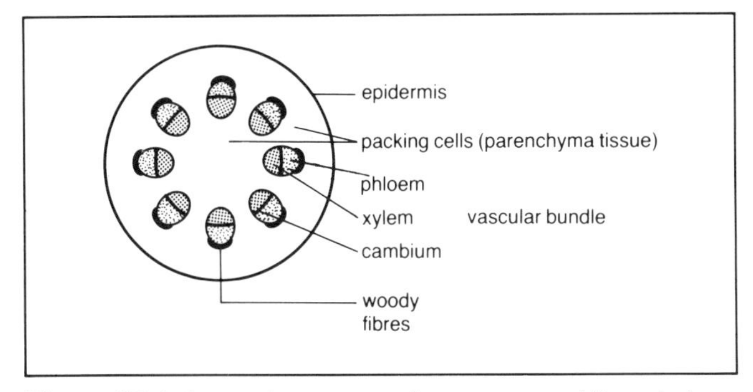

Figure 17.4 *Internal structure of a young stem (dicotyledon type) as seen in a transverse section.*

Xylem and phloem

- **Xylem** is dead tissue consisting mainly of long, tube-like **vessels** with lignified walls (figure 17.5). Lignified means that the cellulose cell walls have had lignin added to them: this makes them hard and impermeable to water. The function of the vessels is to transport water and mineral salts.
- **Phloem** is a living tissue consisting mainly of long **sieve tubes** with cellulose walls (figure 17.6). The function of the sieve tubes is to transport food substances.

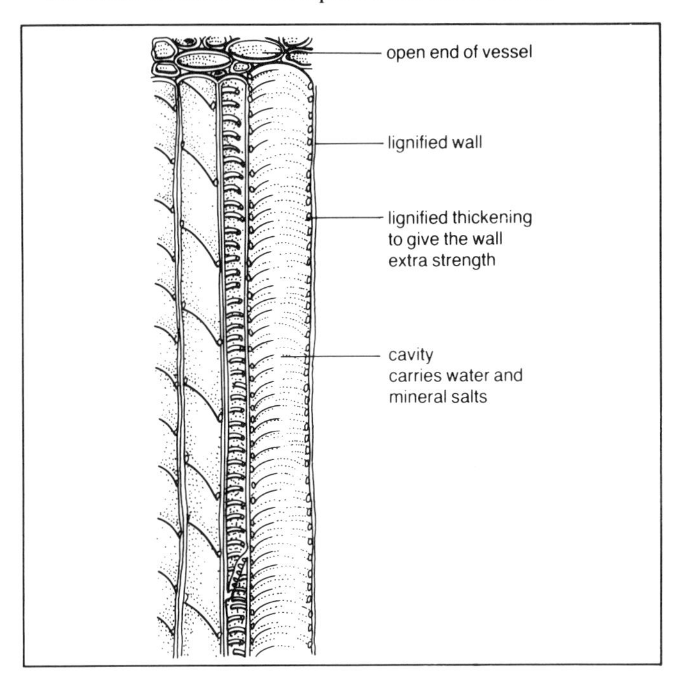

Figure 17.5 *Vessels.*

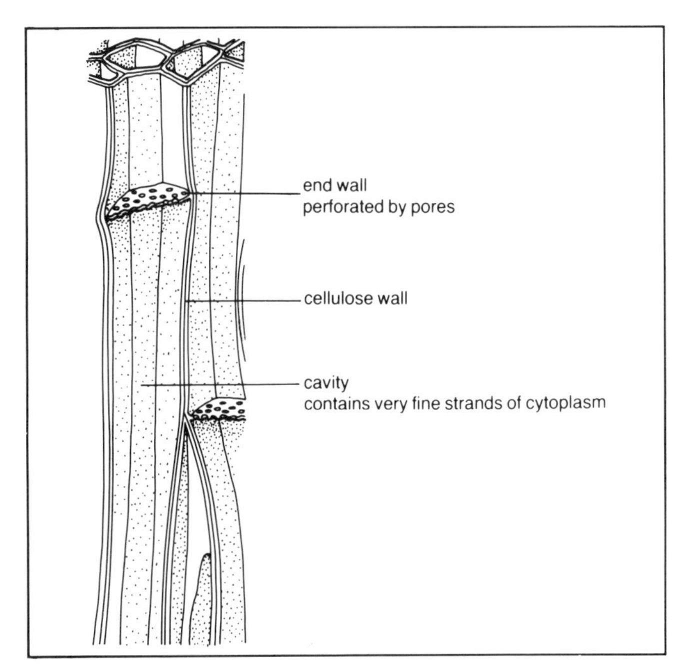

Figure 17.6 *Sieve tubes.*

Older stems

The stems (and branches) of shrubs and trees (woody perennials) contain extra tissues formed by secondary growth (see page 120). These **secondary tissues** are shown in figure 17.7. They are:

- **Wood**: xylem tissue laid down towards the centre of the stem. The innermost wood (**heartwood**) is denser and drier than the wood further out (**sapwood**). The sapwood is wet because it carries the transpiration stream.
- **Bark**: laid down towards the surface of the stem. The outer part is hard(ish) and consists of dead **corky cells**. The inner part is soft and contains living phloem tissue.

How does gas exchange take place through the bark?

The corky part of the bark is impervious to gases. Gas exchange takes place by diffusion through **lenticels** where the corky cells are loosely packed. This ensures that the living phloem cells underneath get a supply of oxygen.

Annual rings

The secondary xylem of a shrub or tree is laid down every summer, creating a series of **annual rings** in the wood. Each ring consists of a light area (**spring wood**) and a dark area (**autumn wood**). The spring wood contains large vessels which carry the spring flow of the transpiration stream. The autumn wood contains smaller vessels and is therefore denser.

The total number of annual rings in a tree trunk indicates the age of the tree. The width of each annual ring indicates the duration of the summer growing season.

Structure of the root

The internal structure of the root (dicotyledon type) is shown in figure 17.8. Many of the tissues are the same as in the leaf and stem and have the same functions. The packing cells often store large amounts of starch in starch grains. The epidermis has **root hairs** which increase the surface area for the absorption of water and mineral salts. The cambium forms secondary tissues (xylem and phloem) in older roots.

How is water absorbed by the roots?

The concentration of salts in the root hair cells is greater than in the soil water. As a result, water enters the root hair cells by osmosis.

How does water flow from the roots to the leaves?

Two main forces contribute to the flow of water from the roots to the leaves:

- A pushing force from below (**root pressure**).
- A pulling force from above caused by evaporation of water from the leaves (transpiration).

How are mineral salts absorbed by the roots?

Mineral salts are absorbed by the root hair cells by diffusion and/or **active transport** (see page 39). Active transport requires energy from respiration, and oxygen is needed for it. This means that the soil must be well aerated.

Any factor which slows down the respiration of roots will also slow down the uptake of mineral salts.

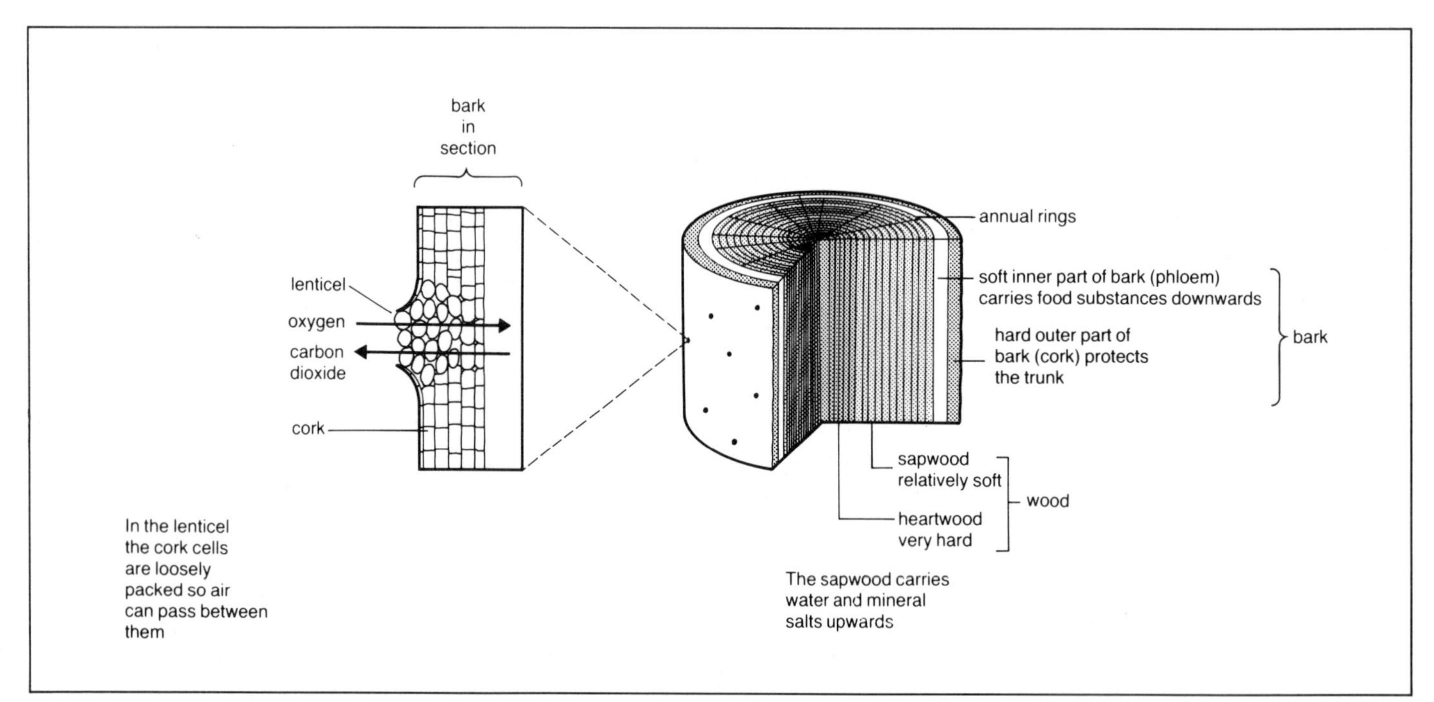

Figure 17.7 *The stem of a tree or shrub showing the secondary tissues (wood and bark).*

How are food substances transported in the phloem?

Phloem transport takes place in the sieve tubes. It is called **translocation**. It is not known for certain how translocation takes places, but it requres energy from respiration and the food substances have to be in solution before they can be transported.

Where are food substances transported to in the phloem?

Depending on the time of year, food substances are transported

- from the leaves to places where growth is taking place (meristems);
- from the leaves to developing storage organs, e.g. tubers;
- from the storage organs to the developing leaves.

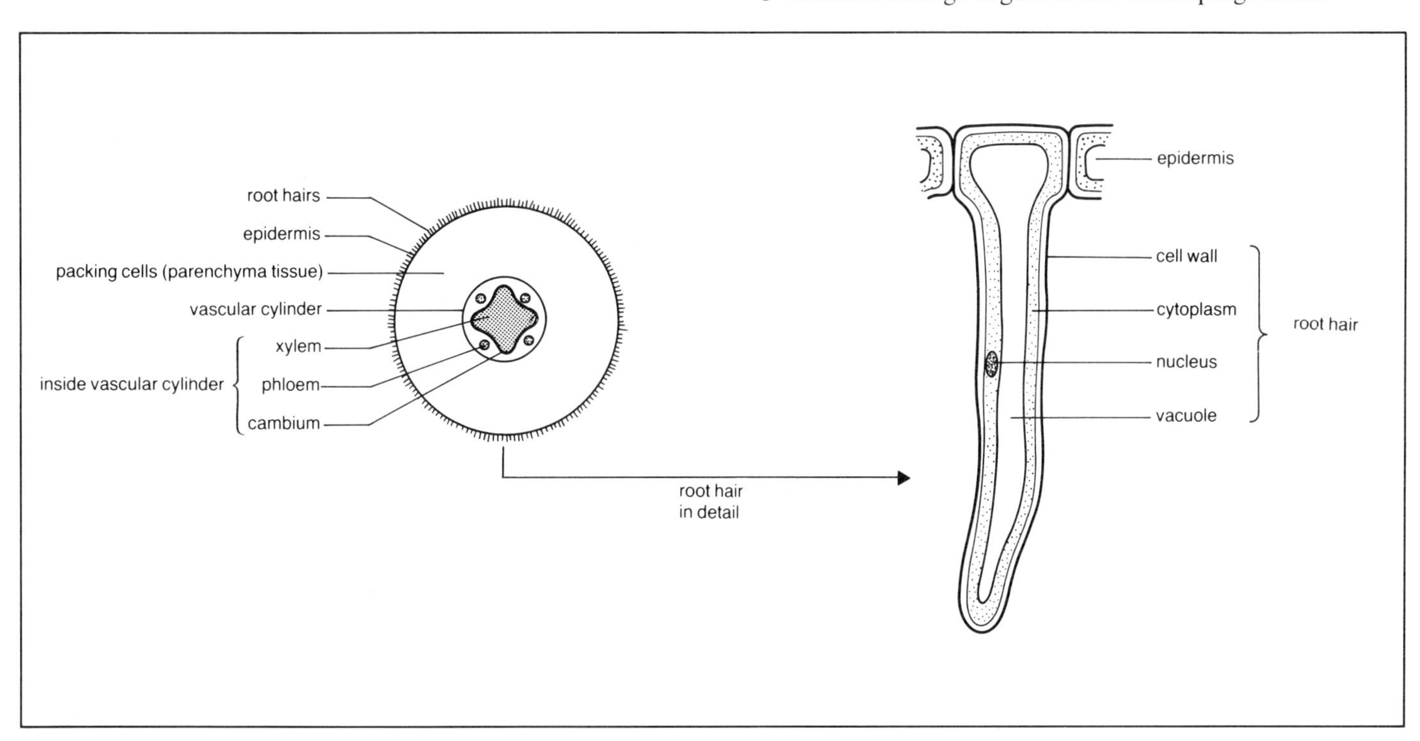

Figure 17.8 *Structure of a young root (dicotyledon type).*

• Questions

1 (a) Why are there fewer stomata on the upper side of a leaf than on the lower side?

(b) What happens to the water content of the guard cells when a stomata closes?

(c) Suggest *one* environmental condition which influences a plant's rate of transpiration other than temperature, humidity and air movement.

(d) At what time of year would you expect food substances to be transported from leaves to tubers in a potato plant?

(e) Why is the cuticle on the upper side of a leaf thicker than on the lower side?

2 Read the list on page 79 of ways in which plants are adapted for living in dry habitats.

(a) How exactly do the following features help a plant to live in a dry habitat?
 (i) Having small leaves.
 (ii) Having a hairy epidermis.
 (iii) Having stomata which are sunk down into pits.

(b) (i) Name *one* succulent plant.
 (ii) The swollen stems of desert succulents are often covered with sharp spines. Suggest why the spines are necessary.

3 (a) (i) Whereabouts in a tree trunk would you find heartwood and sapwood?
 (ii) Write down *two* differences between them in their properties.

(b) (i) Suggest *two* things that should be done to newly sawn-up wood before it is used for house-building.
(ii) Suggest *two* properties which should be possessed by the type of wood which is used for making tennis racquets.

(c) (i) Suggest *one* reason why conifers, rather than oak trees, are grown for timber.
(ii) Suggest *two* reasons why in conifer plantations the trees are grown on steep mountain slopes.

4 The graph below shows the rate of transpiration of a plant growing in moist soil during the course of a sunny day. The time scale is given on a 24 hour clock. The rate of transpiration is expressed as milligrams of water lost per gram of plant per minute.

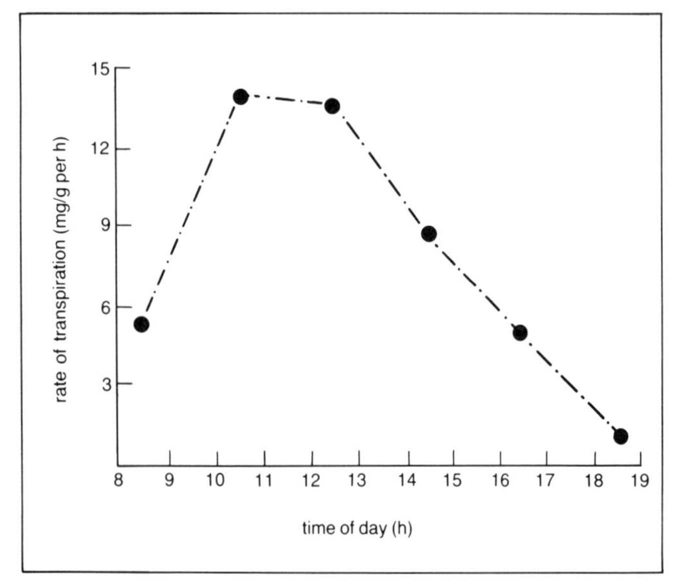

(a) What is meant by the term transpiration?

(b) Suggest reasons why the rate of transpiration
(i) increases between 0830 and 1030 hours, and
(ii) decreases between 1230 and 1830 hours.

(c) Describe the possible appearance of the plant at 1100 hours, and explain the reason for its appearance.

(d) Have you any criticisms of the data on which the graph is based? What could be done to improve the data?

(Data after Bosian)

5 A scientist investigated the effect of oxygen concentration on the rate of respiration and the rate of absorption of bromine ions by the roots of a plant. His results are shown in the graph below.

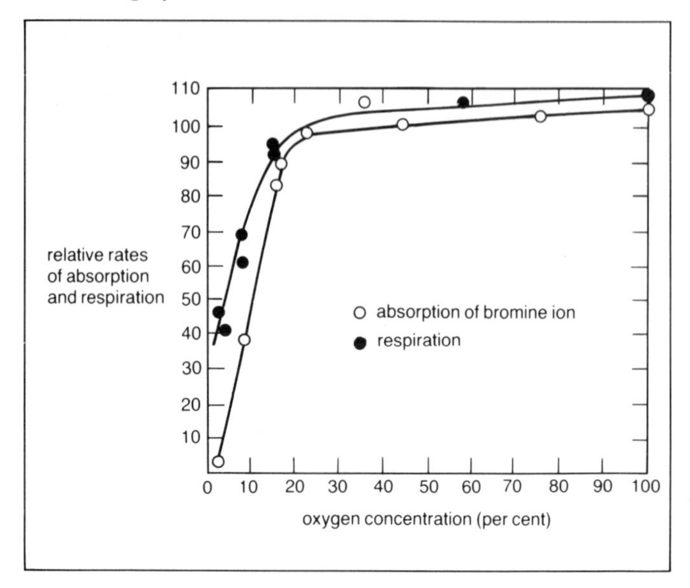

(a) The graph suggests that the absorption of bromine ions is connected with respiration. What feature of the graph suggests this?

(b) Describe in outline an experiment which you would do to support the idea that the absorption of bromine ions is connected with respiration.

(c) Why is respiration necessary for the absorption of bromine ions by roots?

(d) Name one chemical element, other than bromine, which is absorbed by the roots, and say why it is needed by the plant.

(Data after F.C. Steward)

Adjustment and control

Homeostasis

Homeostasis is the keeping of conditions constant inside the body of an organism. It is achieved by **feedback systems**.

How do feedback systems work?

When a particular condition (e.g. body temperature) gets too great, a **corrective mechanism** decreases it and returns it to the normal value.

Conversely, when the condition gets too small, the corrective mechanism increases it and returns it to the normal value (see figure 18.1).

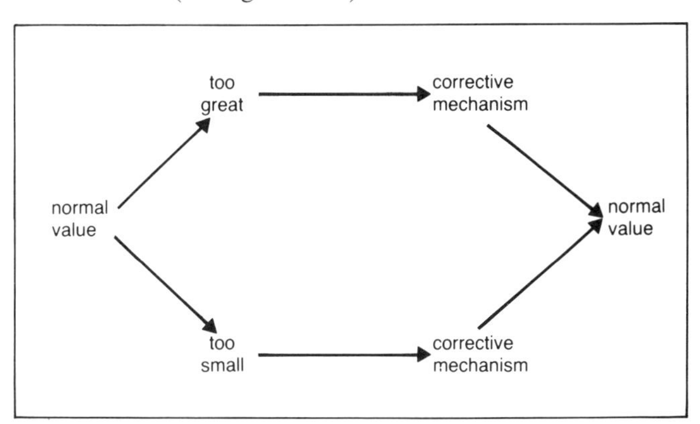

Figure 18.1 *Summary of how homeostasis is achieved by feedback.*

The skin

The skin plays an important part in homeostasis, particularly temperature regulation. Its structure is shown in figure 18.2.

Functions of the skin in a mammal

The skin has six main functions. The main components of the skin which perform each function are added in brackets in the following list.

1. It protects the body from physical damage (epithelium, keratinised hair etc.).
2. It protects the body from harmful ultra-violet rays (melanin).
3. It prevents germs entering the body (keratinised epithelium).
4. It may help to camouflage the animal (hair).
5. It makes the body waterproof (keratinised epithelium, oil).
6. It is sensitive to stimuli (receptors sensitive to touch, pain, temperature and pressure).
7. It insulates the body (hair, subcutaneous fat).

Temperature regulation

The body temperature of a mammal is maintained at approximately 37°C. This is the temperature at which enzymes work best (see page 43).

If the environmental temperature is low:
- The hairs are raised, trapping air which insulates the body.
- Blood is held back from the surface of the skin by constriction (narrowing) of the superficial blood vessels, so less heat is lost.

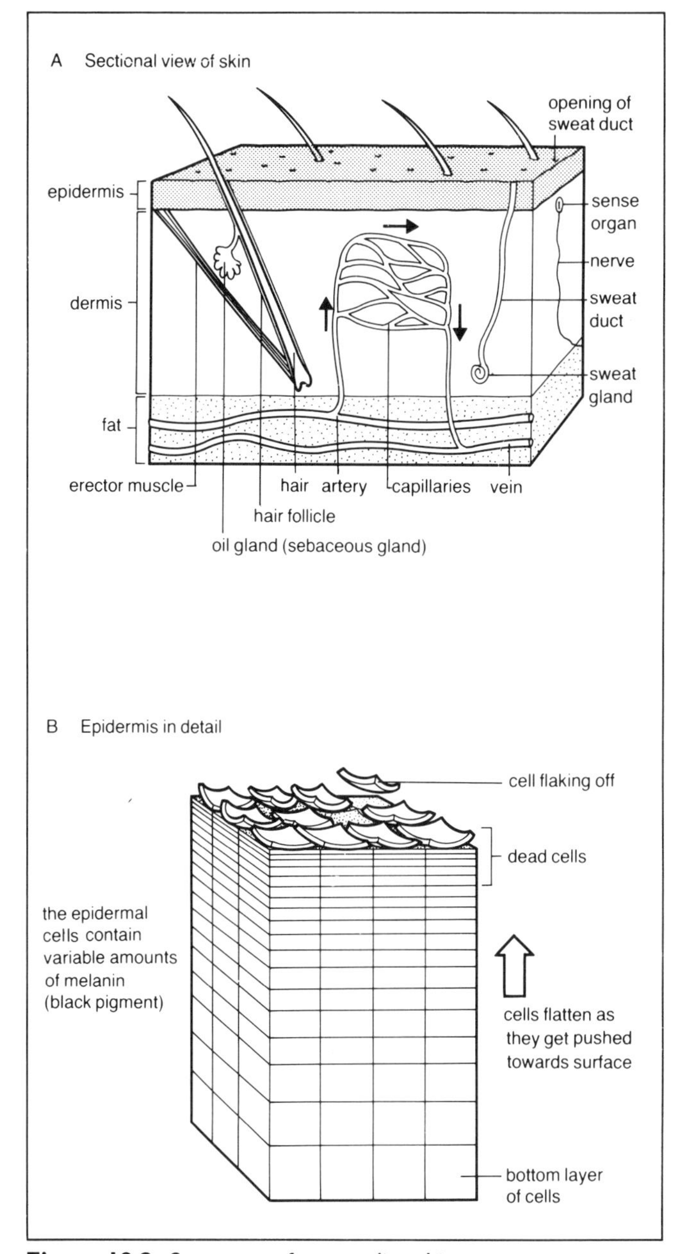

Figure 18.2 *Structure of mammalian skin.*

- Extra heat is generated by the muscles, either voluntarily or involuntarily (shivering).

If the environmental temperature is high:

- The hairs are lowered, thus getting rid of the insulating air.
- Blood flows close to the surface of the skin by dilation (widening) of the superficial blood vessels, so more heat is lost.
- Sweating and/or panting occurs: evaporation of water cools the skin and the blood flowing through it.

What is a warm-blooded animal?

A warm-blooded animal (also called a **homoiothermic** or **endothermic** animal) can keep its body temperature above (or below) that of the environment by *physiological* mechanisms such as those listed in the previous section.

Mammals and birds are warm-blooded.

What is a cold-blooded animal?

A cold-blooded animal (also called a **poikilothermic** or **ectothermic** animal) has a body temperature which is the same as that of its environment. It can only control its body temperature by *behavioural* mechanisms such as basking in the sun or cooling off in the shade.

All animals, apart from mammals and birds, are cold-blooded.

What is hypothermia?

Hypothermia is the fall in body temperature which occurs following prolonged exposure to cold (figure 18.3).

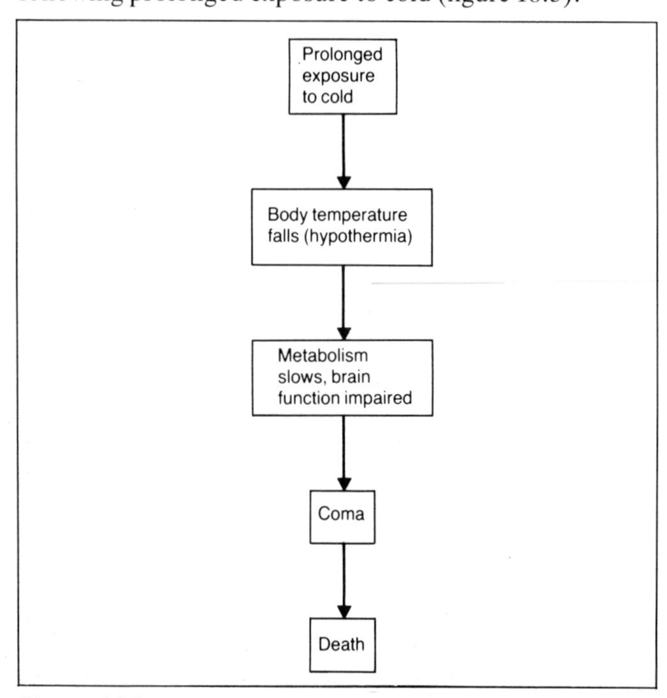

Figure 18.3 *What happens during hypothermia.*

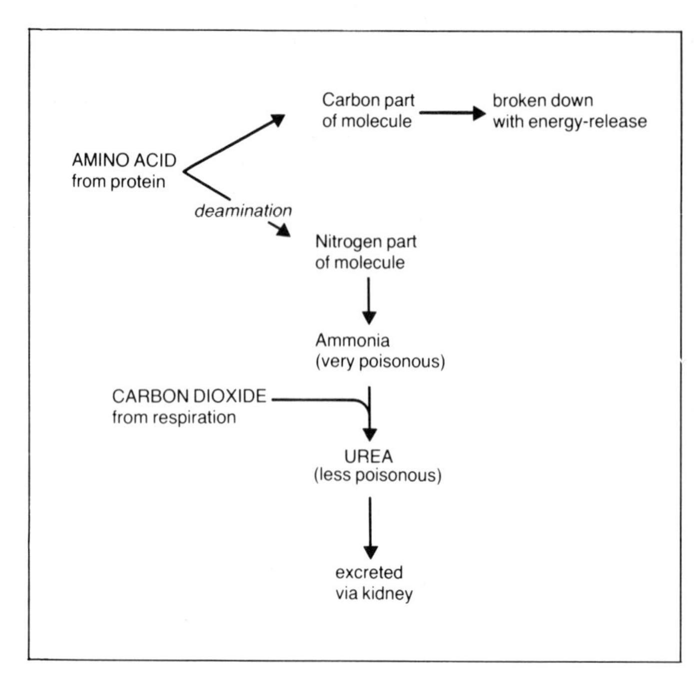

Figure 18.4 *How urea is produced in the liver. Deamination is the process by which the nitrogen part of the amino acid is split off from the rest of the molecule.*

The liver

The liver is the body's largest organ and has many functions, most of them concerned with homeostasis. The connection between liver and gut is shown in figure 12.1, page 55. The blood supply to the liver is shown in figure 14.5, page 67.

Functions of the liver

These are the main functions of the liver:

1. It helps with digestion (by producing bile, see page 57).
2. It produces heat which helps to keep the body warm.
3. It detoxifies poisons.
4. It destroys old red blood cells (the haemoglobin becomes the bile pigments).
5. It stores food substances, particularly glycogen and various vitamins.
6. It gets rid of unwanted amino acids, derived from excess protein.
7. It controls the amount of sugar in the blood.

How does the liver get rid of unwanted amino acids?

The amino acids are **deaminated** with the formation of ammonia which is immediately converted into urea (figure 18.4). The **urea** is eliminated by the kidneys (see page 86).

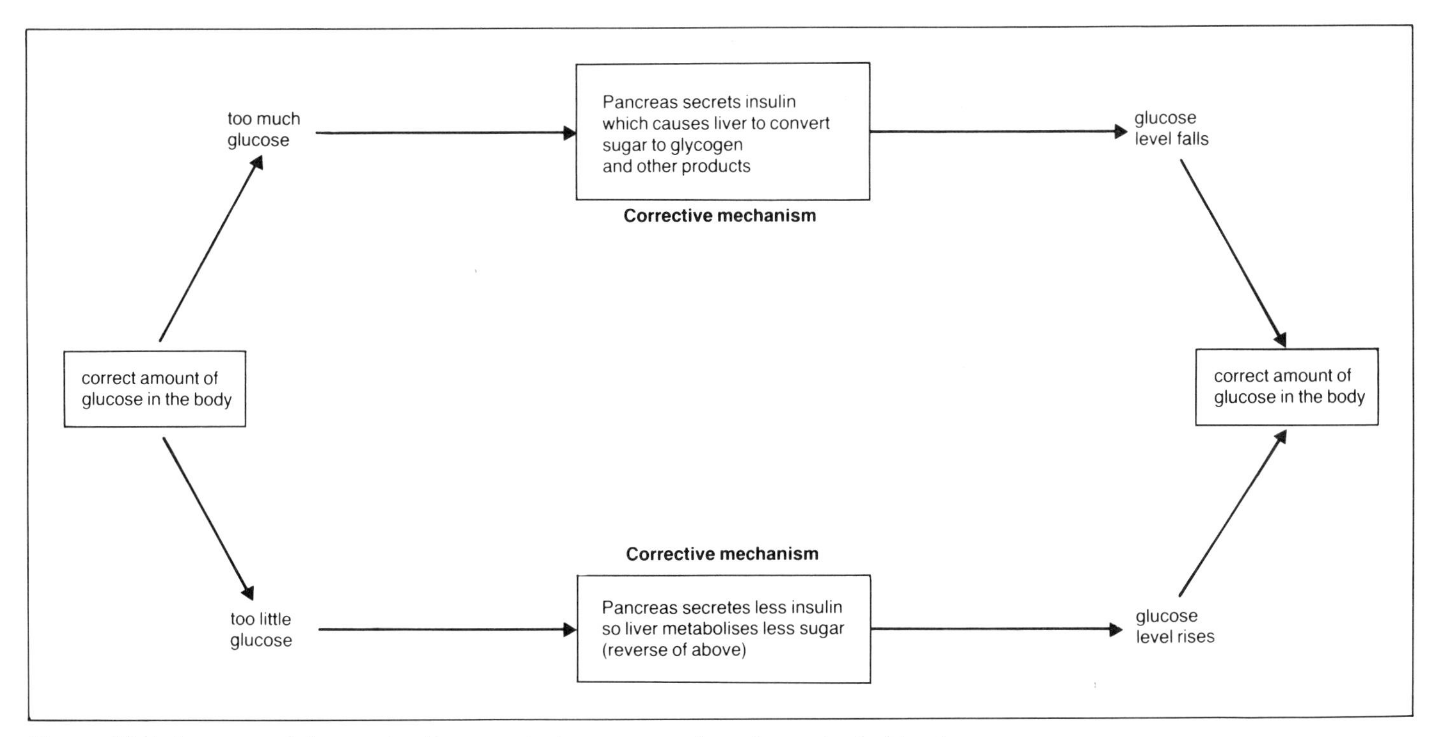

Figure 18.5 *Summary of the way insulin controls the concentration of sugar in the blood.*

How does the liver control blood sugar?

It does so by a feedback mechanism involving the pancreas and the hormone **insulin** (figure 18.5). Insulin is produced by groups of cells in the pancreas called **islets of Langerhans**.

Insulin lowers the blood sugar level by making the liver convert sugar to glycogen. (Another hormone from the pancreas, called **glucagon**, has the opposite effect and raises the blood sugar level.)

What happens if the pancreas produces too little insulin?

The result is **diabetes**, a condition in which there is too much sugar in the blood. It is remedied by injections of insulin.

Excretion and osmo-regulation

Excretion is the elimination from the body of the waste products of metabolism. The main excretory material of mammals is the nitrogenous substance **urea** (see page 84).

Osmo-regulation is the control of the relative concentrations of water and salts in the blood.

Nitrogenous excretion and osmo-regulation are both carried out by the **kidneys**.

The excretory system

The kidneys are the main organs in the **excretory system**. The kidneys produce a watery fluid (**urine**) which is stored in the **bladder** before being passed to the exterior.

The structure of the kidneys

Figure 18.6 shows the internal structure of the kidney. The cortex and medulla are made up of about a million microscopic devices called **nephrons**.

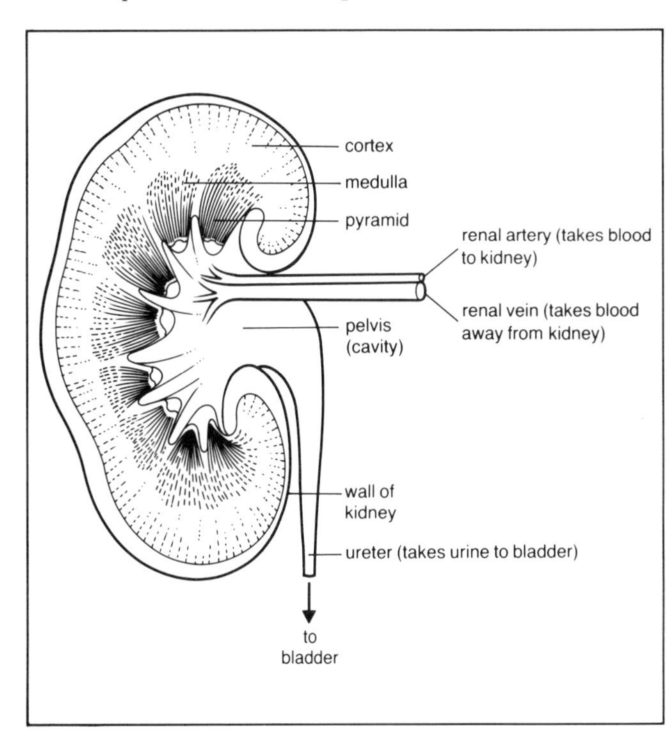

Figure 18.6 *Structure of the mammalian kidney, sectioned to show the inside.*

How do the nephrons work?

The basic structure and function of a nephron is shown in figure 18.7. Blood from the renal artery flows first to the **glomerulus** (in the **capsule**) and then to the **tubule**. As the blood flows past the nephron, it is 'cleaned' by a process of *filtration followed by reabsorption* (figure 18.7).

All the filtration and most of the reabsorption occur in the cortex of the kidney (see figure 18.6). Some water is reabsorbed in the medulla.

The artificial kidney (kidney machine)

This is used for people whose kidneys have failed. Blood from the patient flows over the surface of a selectively permeable **dialysing membrane** on the other side of which is a watery solution. Urea and other unwanted substances pass from the blood to the solution, which is constantly removed and replaced. Meanwhile the 'cleaned' blood returns to the patient.

Kidney transplant

A person who has lost the use of both kidneys permanently may be given a **kidney transplant**. As in other cases of transplant surgery, the donor kidney must come from a person who is genetically similar to the patient. Otherwise the kidney will be rejected.

Water balance in fresh-water animals

In fresh-water animals water enters the body by osmosis, thereby diluting the blood and tissues. *Remedy*: excess water is expelled by an osmo-regulatory device, e.g. contractile vacuole of *Amoeba*, kidney of fish.

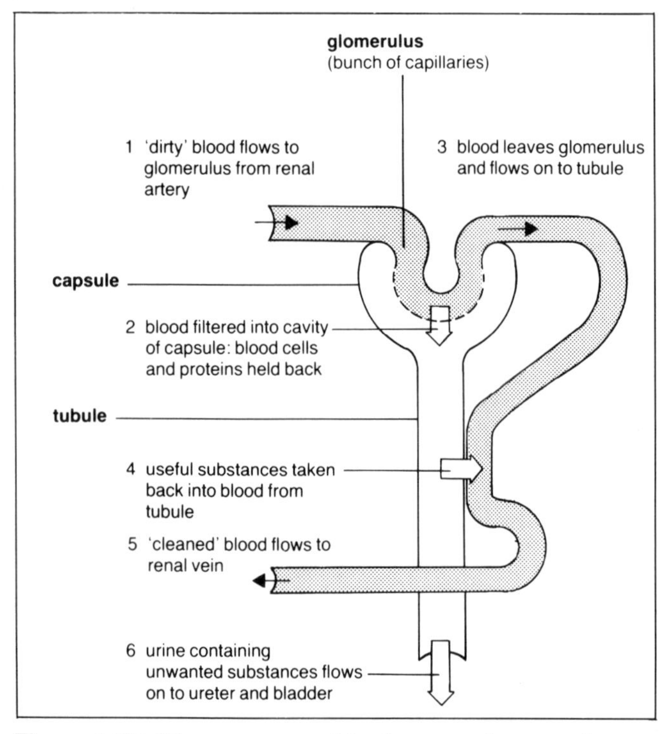

Figure 18.7 *Diagram summarising how a nephron works.*

Water balance in sea-water animals

In sea-water fish water is lost from the body by osmosis, thereby concentrating the blood and tissues. *Remedy*: The kidneys get rid of as little water as possible; excess salt is expelled by the gills.

Water balance in land-dwelling animals

In land animals water is lost from the body by evaporation. Such animals save water by the following means:

1. Having an impermeable skin and/or waxy cuticle (e.g. insects).
2. Excreting nitrogenous waste as solid **uric acid** rather than fluid urea (e.g. insects, birds).

How does the body adjust to high altitudes?

At high altitudes the concentration of atmospheric oxygen is relatively low. To compensate for this the following changes take place.

1. The breathing rate increases (see page 62).
2. The heart rate increases (see page 68).
3. The number of red blood cells increases.

As a result, the oxygen supply to the tissues is increased and the person becomes acclimatised.

How does the body adjust to exercise?

During exercise the metabolic rate of the muscles increases. The muscles' oxygen requirement increases as does the rate at which carbon dioxide is produced.
The body responds to the above changes as follows:

1. The breathing rate increases (see page 62).
2. The heart rate increases (see page 68).
3. Blood is diverted to the muscles from less needful structures. (Blood vessels to the muscles dilate (widen), whereas other blood vessels constrict.)
4. The muscles respire anaerobically, forming lactic acid (see page 48).
5. The body temperature rises (caused by the increased metabolic rate) and this triggers the body's cooling mechanisms (see page 84).

Questions

1 State as precisely as possible where each of the following substances is made.
(a) Bile; (b) Insulin; (c) Sweat; (d) Urea.

2 Study figure 18.2 (page 83).

(a) How would you expect the appearance of the hair and erector muscle to change in very cold weather. Explain the reason for the change.

(b) After running a race, your skin is wet and your face is hot.
(i) Why is your skin wet and how does this help to cool your body?
(ii) Why is your face hot and how does this help to cool your body?

(c) How do you think the sebaceous glands help to keep hair and skin healthy?

(LEAG, modified)

3 The diagram below shows the liver and its connections. The arrows indicate the direction in which blood flows.

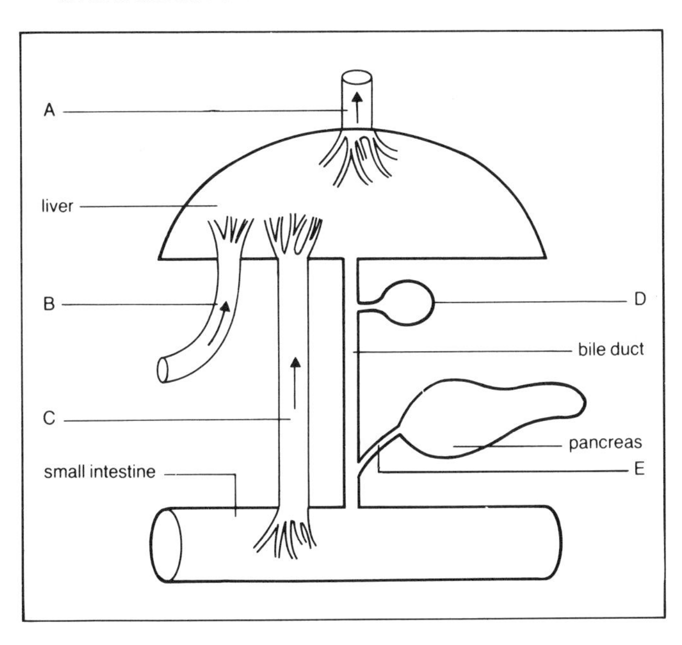

(a) (i) Name structures A to E.
(ii) Which structure supplies the liver with *oxygenated* blood?

(b) (i) Name *one* hormone, produced by the pancreas, which affects the liver.
(ii) How does the hormone get from the pancreas to the liver?
(iii) What effect does the hormone have on the liver?
(iv) What condition arises if insufficient quantities of the hormone are produced?

4 The diagram shows a section through a mammal's kidney.

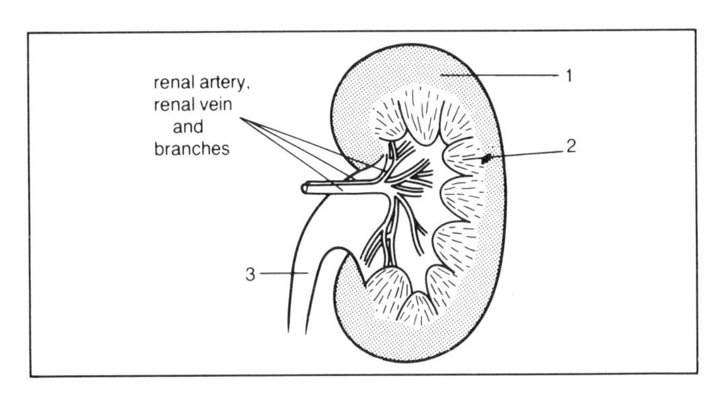

(a) Name the parts 1, 2 and 3.

(b) The table gives information about the human kidney.

Rate of blood flow in kidneys	Rate of filtration into kidney tubules (nephrons)	Rate of urine passing out of kidneys
1.2 dm^3 per minute	0.12 dm^3 per minute	1.5 dm^3 per day

(i) What percentage of blood passing into the kidney is filtered into the kidney tubules?
(ii) Where in the kidney (part 1, 2 or 3 on the diagram) does filtration take place?
(iii) About 172 dm^3 are filtered from the blood into the kidney tubules per day, yet only 1.5 dm^3 of urine are excreted. What happens to the other 170.5 dm^3?

(c) The table shows the average amounts of urine, sweat and salt (sodium chloride) lost on a normal day, a cold day and a hot day. (Assume that food and drink are the same on all days.)

	Urine lost per day (dm^3)	Sweat lost per day (dm^3)	Salt (sodium chloride) lost per day	
			in uring (g)	in sweat (g)
Normal day	1.5	0.5	18.0	1.5
Cold day	2.0	0.0	19.5	0.0
Hot day	0.375	2.0	13.5	6.0

(i) Why is more urine lost on a cold day than on a normal day?
(ii) Why do you think the total amount of salt lost on each of the three days is the same?

(iii) The minimum amount of urine excreted in a day is $0.375\ dm^3$. Why do you think the kidneys always produce some urine?

(iv) What *must* someone losing more than $7\ dm^3$ of sweat in a day do in order to remain healthy?

(SEG)

5 The diagram below illustrates the principle of an artificial kidney machine.

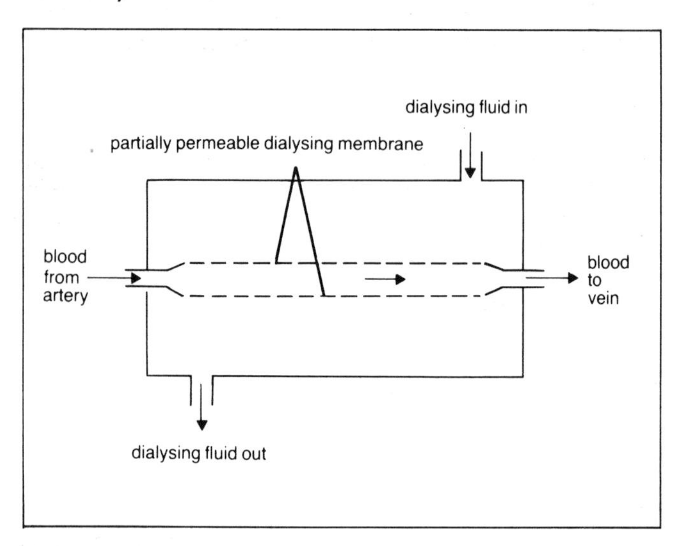

The blood of a patient flows on one side of a partially permeable membrane and dialysing fluid flows on the other side. The dialysing fluid has a composition and concentration equal to that of the plasma of a normal person.

(a) What process causes excess water from the patient to pass into the dialysing fluid?

(b) (i) Name an excretory product, other than water, which will pass out of the blood into the dialysing fluid.

(ii) Name the process by which this occurs.

(c) In reality the dialysing membrane is greatly folded rather than straight. Why is this?

(d) Explain why, in the case of permanent kidney failure, it is more economic and more satisfactory to a patient to be given a kidney transplant rather than treatment with an artificial kidney.

(NEA)

Nervous coordination

Organisation of the nervous system

The main parts of the vertebrate nervous system are summarised in table 19.1.

The central nervous system is surrounded and protected by the skeleton, the brain by the cranium and the spinal cord by the vertebrae.

Table 19.1 *The main parts of the nervous system of a vertebrate such as the human.*

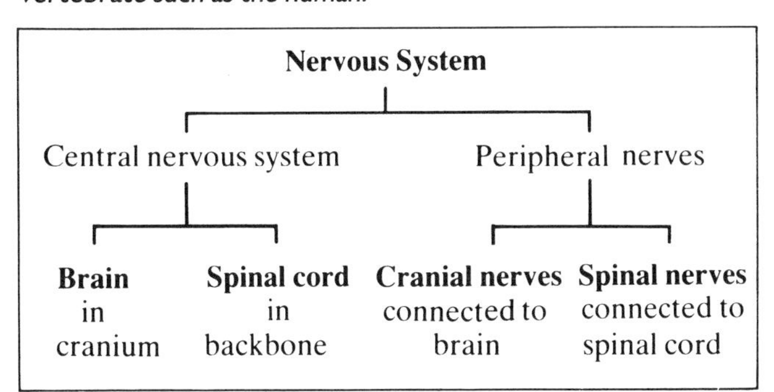

Nervous System			
Central nervous system		Peripheral nerves	
Brain in cranium	**Spinal cord** in backbone	**Cranial nerves** connected to brain	**Spinal nerves** connected to spinal cord

Overall function of the nervous system

The overall function of the nervous system is to transmit messages rapidly from one part of the body to another, and to coordinate the organism's actions.

The nature of the nerve message

A nerve message is a tiny, localised pulse of electricity. It is called a **nerve impulse**. Nerve impulses travel along specific **nerve fibres** within the nervous system. Nerve fibres are long extensions of **nerve cells** (**neurons**).

Reflex action

A **reflex action** is an immediate response (reaction) of the body to a **stimulus**. *Examples*: knee jerk, withdrawing hand from hot object, closing of pupil in response to bright light.

What is a stimulus?

A **stimulus** is a local change in the environment which sets off an impulse (or impulses) in the nervous system. In the knee jerk the stimulus is a tap on the tendon just below the knee cap.

Reflex arc

A **reflex arc** is the route (pathway) along which nerve impulses travel in bringing about a reflex action. A typical vertebrate reflex arc is shown in figure 19.1. Most reflex arcs are connected to the brain, so the brain can control and coordinate reflex actions.

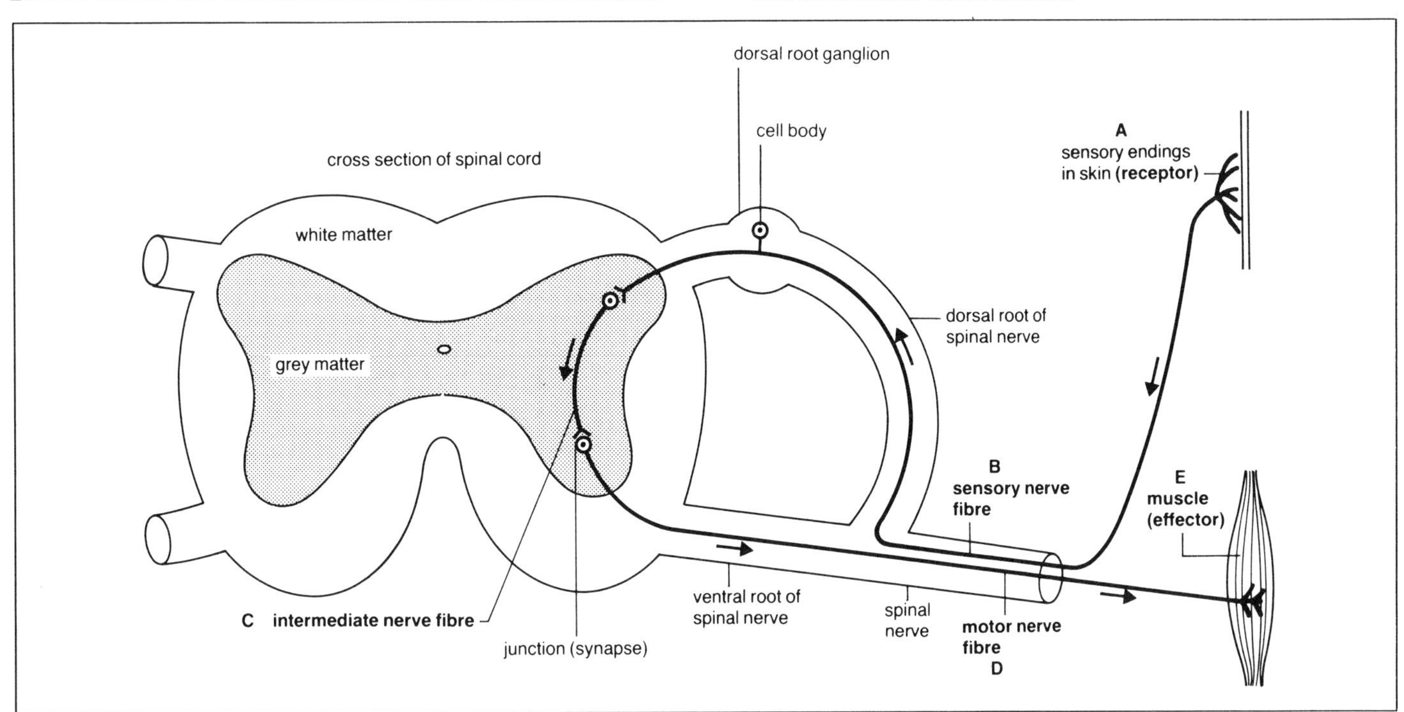

Figure 19.1 *A reflex arc in the nervous system of a vertebrate. The components of the reflex arc are labelled A to E.*

Structure of a nerve cell

A nerve cell is shown in figure 19.2. This particular type of nerve cell is a **motor nerve cell** which is connected to a muscle (see below).

Synapses

A **synapse** is the connection between one nerve cell and the next, or between a nerve cell and a muscle.

At a synapse there is a tiny gap. Nerve impulses cannot get across this gap. When an impulse reaches a synapse a **chemical transmitter** is produced which diffuses across the gap and starts up an impulse on the other side.

Voluntary and involuntary actions

Many actions which are controlled by the nervous system are **voluntary**, e.g. talking, walking etc.

Other actions are **involuntary**, e.g. movements of the gut, speeding up or slowing down the heart beat. These actions are controlled by the **autonomic nervous system**.

Some actions are involuntary early in life but become subject to voluntary control later, e.g. emptying the bladder.

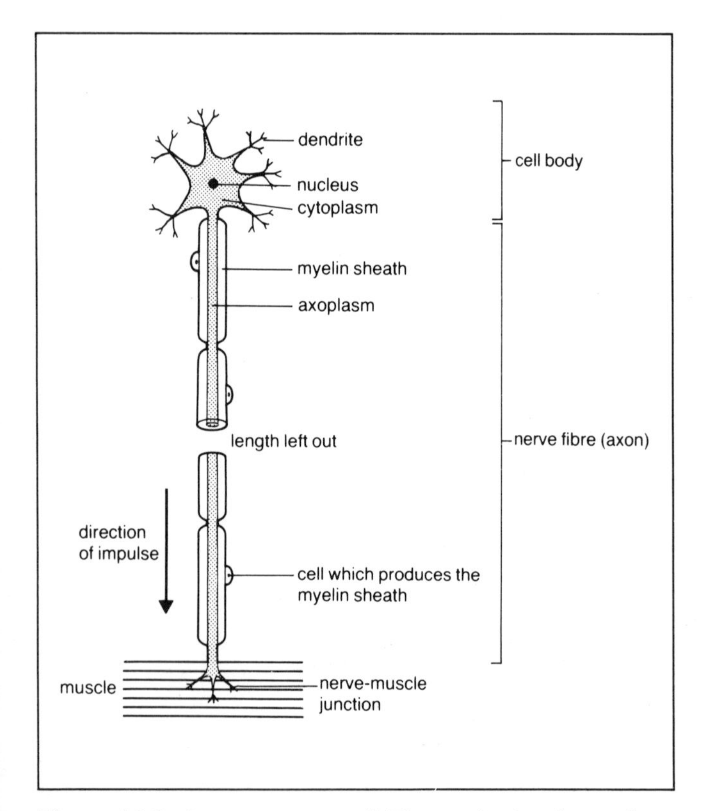

Figure 19.2 *A motor nerve cell. The myelin sheath speeds up the impulse.*

The brain

The main parts of the human brain and their functions are summarised in figure 19.3. In reality the *functional* distinctions are less sharp than the diagram suggests, and there is considerable overlap between one part of the brain and another.

What is a conditioned reflex?

A **conditioned reflex** is a reflex action, triggered by a certain stimulus, which the animal *learns* to associate with a different stimulus.

Ivan Pavlov, a Russian physiologist, introduced the idea of a conditioned reflex. He did experiments with dogs. Dogs salivate when presented with food. In Pavlov's experiments, the dogs learned to associate food with the ringing of a bell, so they would salivate when the bell was rung.

For conditioned reflexes to become established, the brain is necessary. It is one of the simplest types of learning.

Instinct

An instinctive action is inherited from parents (inate) and so does not need to be learned. An example is courtship behaviour (and song) in birds.

Note: the behaviour of animals is rarely completely instinctive. It is usually the result of a combination of learning and instinct.

Drugs that affect the brain

A drug is a substance which alters the way the body works. Drugs which affect the brain include the following:

- **Sedatives**: slow down the brain and make you sleepy, e.g. valium, alcohol.
- **Stimulants**: speed up the brain and make you alert, e.g. caffeine (in coffee and tea), cocaine, nicotine.
- **Painkillers**: suppress the part of the brain responsible for the sense of pain, e.g. morphine, heroin.

Drugs should be taken only under a doctor's supervision. Taken any other way they can be very dangerous.

Why are drugs dangerous?

Drugs are dangerous for three main reasons.

1. They may impair judgment and make you clumsy and inaccurate, e.g. alcohol.
2. You may become addicted to the drug. If you stop taking such a drug, you may suffer from **withdrawal symptoms**, e.g. nicotine, alcohol, heroin.
3. They may injure the body by damaging cells, e.g. alcohol, cannabis. Nicotine can cause heart disease.

Orientation responses

An **orientation response** is a response in which an organism moves into a particular position in relation to a stimulus.

Most motile organisms (i.e. organisms which can move) move towards or away from certain stimuli. Such responses are called **taxis**, e.g. **phototaxis** which is movement towards or away from light.

An organism which moves towards light is **positively phototactic** (e.g. *Euglena*). An organism which moves away from light is **negatively phototactic** (e.g. blowfly larvae).

Three important stimuli to which organisms show orientation responses are light, humidity and gravity. A species' response to, and preferences for, these conditions can be investigated by means of a choice chamber.

What is a choice chamber?

A **choice chamber** is a transparent box, with a perforated floor, which is subdivided into interconnected areas each with different conditions, e.g. light or dark, dry or humid etc. Darkness is created by covering part of the box with opaque paper; humidity by placing water under the perforated floor, dryness by placing a drying agent such as anhydrous calcium chloride or silica gel under the perforated floor.

Small animals such as woodlice are placed randomly in the choice chamber, and you observe which area of the chamber they move into.

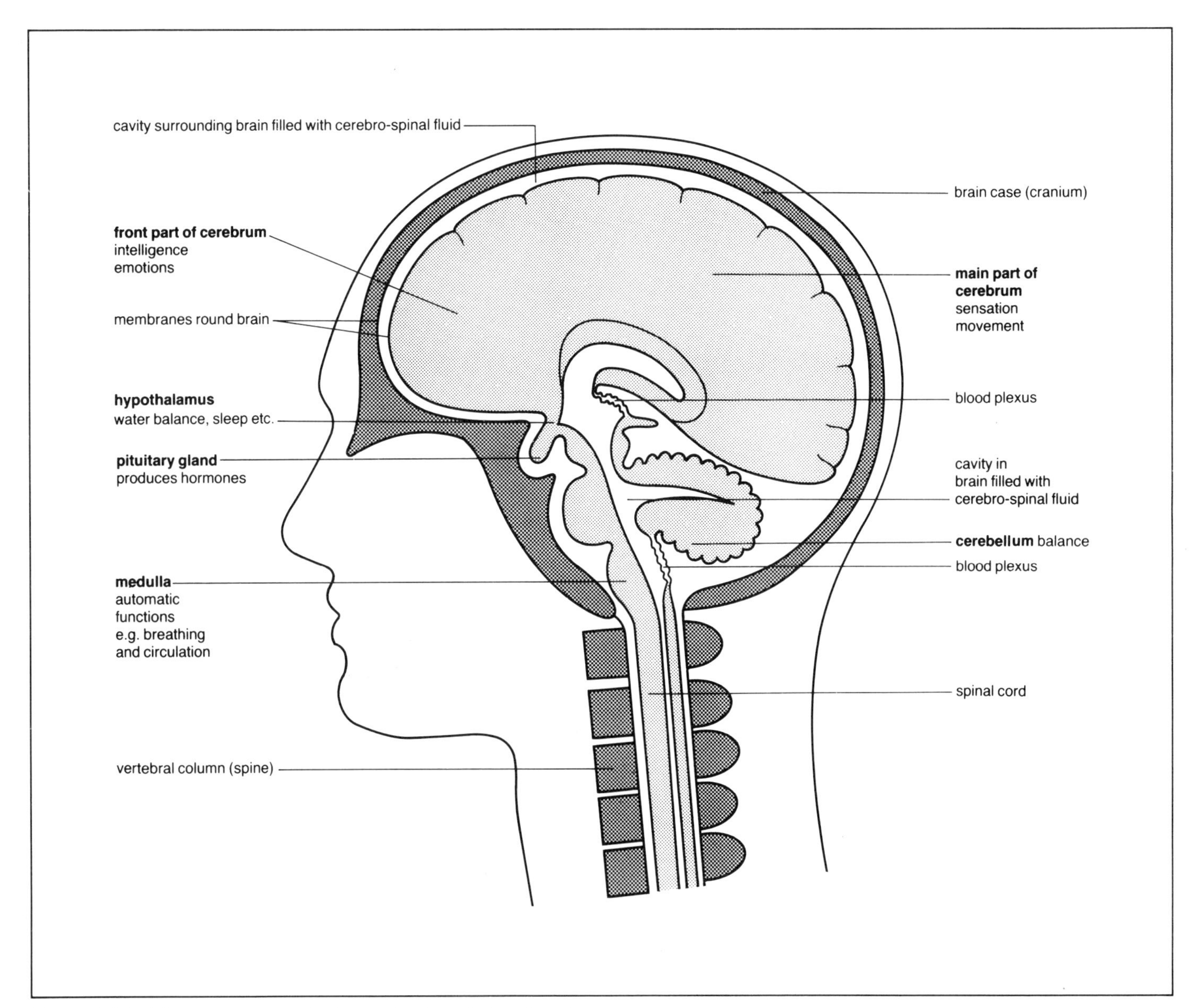

Figure 19.3 *The main parts of the human brain and their functions.*

• Questions

1 (a) The following are all components of a typical reflex arc in a vertebrate. They are listed in alphabetical order.
A Effector
B Intermediate neurone
C Motor neurone
D Receptor
E Sensory neurone
Rearrange them in the order in which an impulse would travel through the reflex.

(b) Explain the terms 'receptor' and 'effector' and give one example of each.

2 (a) What is meant by the term synapse?

(b) How does a message in the nervous system cross a synapse?

(c) Synapses prevent impulses travelling through a reflex arc in the wrong direction. How do you think they achieve this?

(d) Synapses are readily affected by drugs and poisons. Why?

3 The diagram below represents a neurone (nerve cell) in a reflex arc.

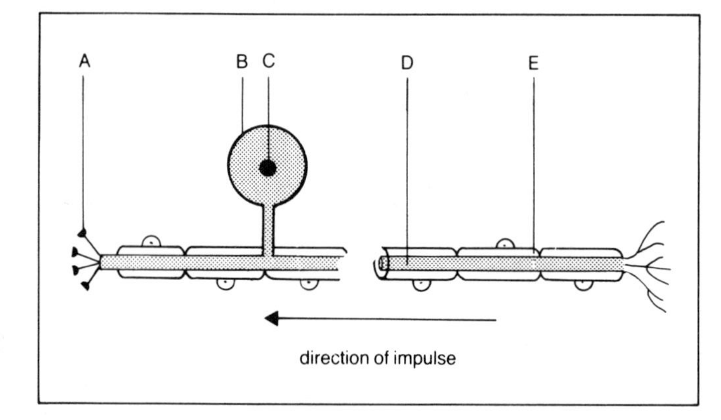

(a) (i) Name the structures labelled A to E.
(ii) Where are structures A and B located?
(iii) Suggest *one* function of structure E?

(b) Suppose the nerve fibres in the human arm transmit impulses at 100 metres per second. What will be the approximate delay between the moment a stimulus is applied to the tip of a finger and the moment the hand starts being pulled away. Explain how you arrive at your answer.

4 A student used a choice chamber to investigate the responses of woodlice to different conditions. The choice chamber was subdivided into four compartments, each with different conditions of light and humidity. Twenty woodlice were placed in the centre of the choice chamber, and their positions after half an hour are shown below.

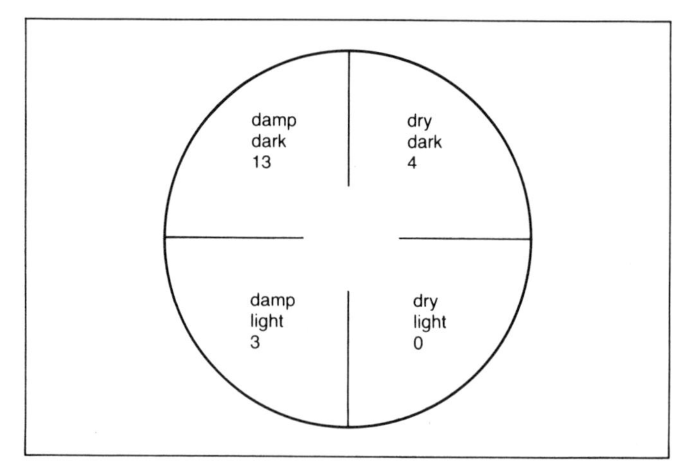

(a) Explain how the student could make conditions damp in one compartment and dry in another.

(b) State *two* conclusions which you would draw from the results.

(c) What should the student do to ensure that the distribution of the woodlice was caused by the conditions in the choice chamber and not by chance?

20

Receptors

What are receptors?

A **receptor** is a structure which receives stimuli and, as a result, sends off impulses in a nerve.

Receptors consist of **sensory cells**. In the case of sense organs such as the eye, other cells and tissues are present to help the sensory cells function efficiently.

What sort of stimuli do receptors receive?

In general receptors are adapted to receive the following types of stimuli:

- **Light**, e.g. the eye.
- **Sound**, e.g. the ear.
- **Touch**, e.g. touch receptors in the skin.
- **Temperature**, e.g. temperature receptors in the skin.
- **Chemicals**, e.g. taste buds in the tongue.

The receptors in the skin are located in the upper part of the dermis (see page 83).

Properties of receptors

Receptors have three important properties:

1. A receptor is normally sensitive to only one kind of stimulus.
2. Only a small stimulus is normally needed to stimulate a receptor.
3. Most receptors cease to respond if stimulated continuously (**sensory adaptation**).

The human eye

The eyes are located in **sockets** in the cranium. Each eye is moved by six **eye muscles** which run from the sides of the eyeball to the back of the socket.

An **optic nerve** connects the eye to the brain. It passes through a hole (foramen) at the back of the socket.

How is the eye protected?

- **Tears**, produced by tear glands, keep the front of the eye clear and moist.
- **Lysozyme** in tears kills bacteria.
- **Blinking**, a rapid reflex, protects the eye from injury.

Internal structure of the eye

The internal structure of the eye is shown in figure 20.1. The two most important parts are:

- the **lens** which focuses light rays on
- the **retina** which contains photoreceptor cells.

Table 20.1 summarises the structure and functions of the parts of the human eye. Study the diagram first and see if you can suggest a function for the various parts *before* you look at the table.

How does the eye work?

Light rays from the object are bent (refracted) inwards by the cornea and lens and meet on the retina. The light stimulates the light-sensitive cells in the retina and nerve impulses pass along the optic nerve to the brain, producing an image of the object.

For an image to be in focus the light rays must meet on the retina.

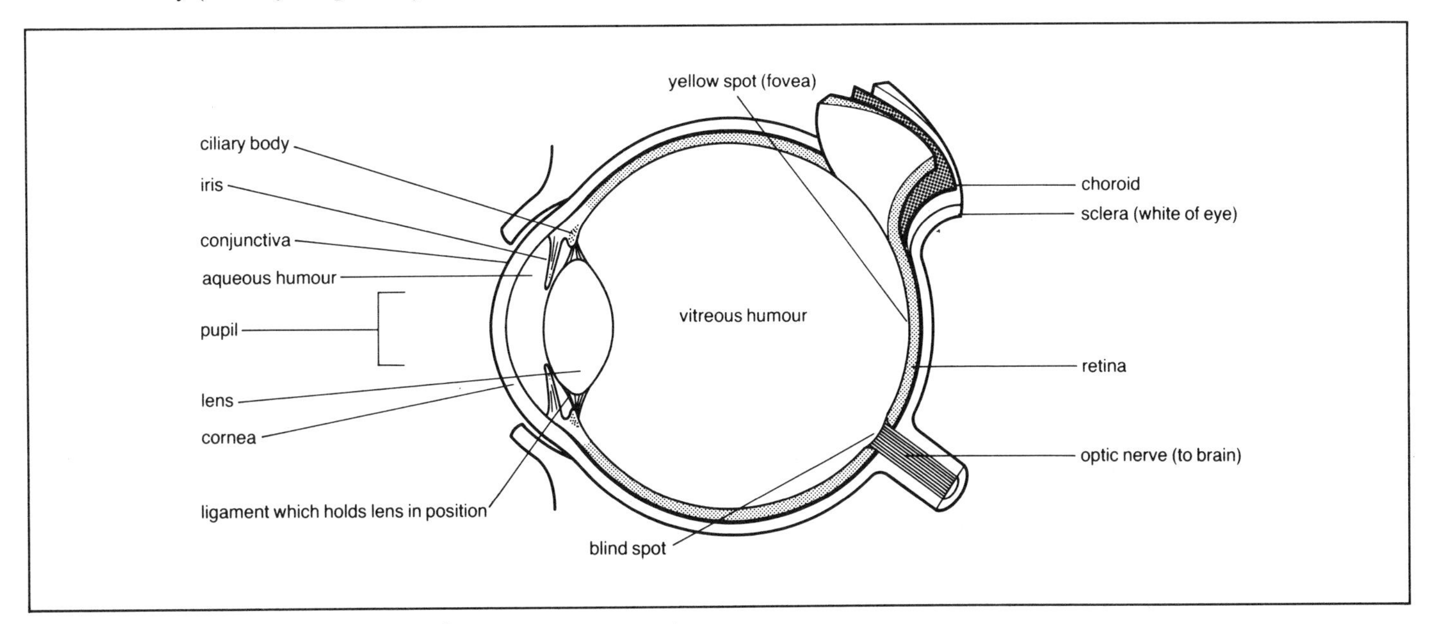

Figure 20.1 *The human eye in sectional view.*

How is the amount of light entering the eye controlled?

The amount of light entering the eye is controlled by the **iris** (figure 20.2).

- In bright light the circular muscle contracts and the radial muscle relaxes: this constricts the iris and the pupil gets smaller. This response is a reflex (the **pupillary reflex**): bright light entering the eye causes impulses to be sent from the retina to the brain; impulses then travel from the brain to the circular muscle of the iris, causing it to contract.
- In dim light the radial muscle contracts and the circular muscle relaxes: this dilates the iris and the pupil gets larger.

Table 20.1 *Structure and functions of the parts of the eye.*

Structure	Notes
Cornea	Transparent, bends (refracts) light rays.
Pupil	Aperture ('hole') through which light rays pass into eye.
Iris	Surrounds pupil, can constrict or dilate thereby regulating amount of light entering eye.
Lens	Transparent, bends (refracts) light rays and brings them to a focus on retina.
Ciliary body	Ring of circular muscle surrounding lens, attached to lens by suspensory ligament.
Aqueous humour	Runny fluid, nourishes cornea and maintains correct pressure in front part of eye.
Vitreous humour	Jelly-like fluid, maintains spherical shape of eyeball.
Retina	Layer of light-sensitive cells lining back and sides of eyeball.
Choroid	Layer of pigmented cells and blood vessels behind retina, absorbs light (preventing it being reflected within eye) and nourishes retina.
Sclera (sclerotic)	Thick wall of eyeball (the 'white' of the eye), protective.
Yellow spot (fovea)	Most central part of retina, responsible for 'precision vision.'
Optic nerve	Carries nerve impulses from retina to brain.
Blind spot	Part of retina where optic nerve is attached to eye, no light sensitive cells are present here.

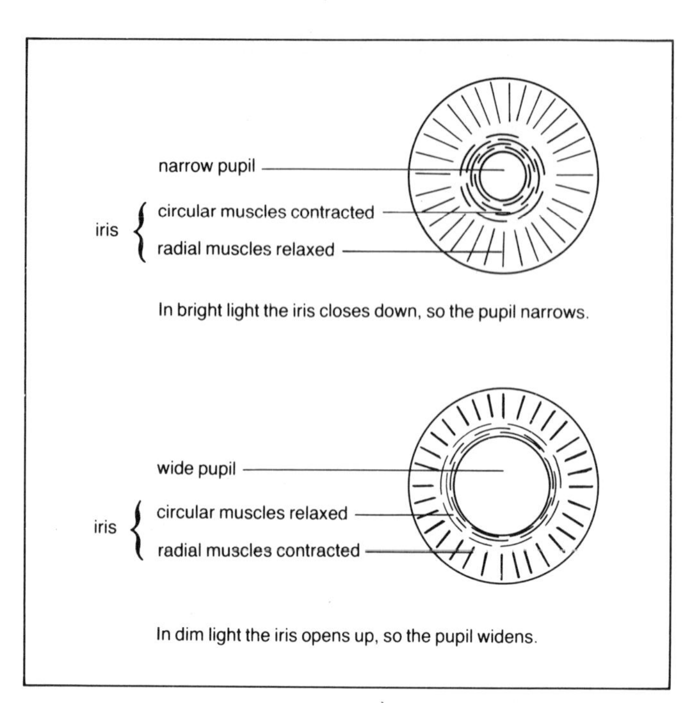

Figure 20.2 *How the iris responds to bright light and dim light.*

How do we see things close at hand?

When we look at an object a long way off and then close at hand, the eye adjusts as follows (figure 20.3):

- The ciliary muscle contracts.
- Tension on the suspensory ligament is released.
- The lens become more spherical.
- Light rays are bent more, so they meet on the retina.

The above process is called **accommodation**. If it did not occur, light rays from a near object would meet behind the retina and the image would be out of focus.

What sort of cells are found in the retina?

The retina contains two types of light-sensitive cells, **cones** and **rods**. They differ in structure and function (table 20.2).

Table 20.2 *Cones and rods compared.*

	Structure	Where found	Function
Cone	Broad	Centre of retina, particularly fovea.	Precision vision in colour in good light: *'Daylight vision'*.
Rod	Narrow	More peripheral parts of retina.	Non-precision vision in black-and-white in poor light: *'Night vision'*.

How do cones and rods work?

Cones and rods contain pigments. When light strikes the cell, the pigment is broken down. This causes the cell to send an impulse into the optic nerve. Meanwhile the pigment is regenerated (re-made) and can be used again.

Why can rods function in poor light?

Rods can function in poor light because their pigment (**visual purple**) is broken down by only very small amounts of light.

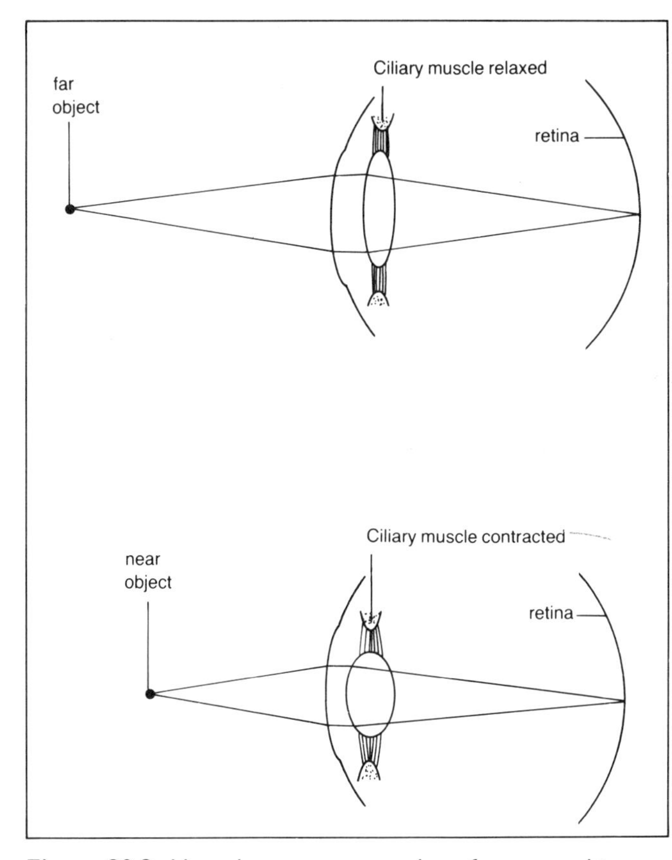

Figure 20.3 *How the eye accommodates for a near object.*

Dark adaptation

This is the gradual process by which you become able to see things in dim light after being in bright light. *Explanation*: the bright light has broken down the visual purple which must be regenerated before the rods can function.

Why are cones able to see clearly?

Cones see things clearly (i.e. give us 'precision vision') because they are very close together and each one has its own nerve fibre to the brain.

Defects of the eye

Here are some of the more common defects of the eye:

- **Short-sight**: the person can focus near objects but not far objects.
- **Long-sight**: the person can focus far objects but not near objects.
- **Colour-blindness**: inability to distinguish colour, particularly red and green.
- **Cataract**: lens becomes opaque and will not let light through.

The ear

The inner ear contains the **cochlea** which is sensitive to sound, and a **balancing apparatus**.

How does the ear work?

This is the sequence of events that occurs when we hear a sound:

1. Sound waves pass down the external ear canal.
2. Ear drum vibrates.
3. Ear ossicles vibrate.
4. Fluid in cochlea moves.
5. Sensory cells stimulated by being distorted.
6. Impulses travel via nerve fibres to brain.

• Questions

1 In figure 20.1 which structure or structures

(a) contains photoreceptors;

(b) can refract light rays;

(c) contain muscles;

(d) alters the shape of the lens;

(e) transmits nerve impulses to the brain?

2 Explain each of these well-known experiences:

(a) A coarse shirt tickles at first but after a time the tickling sensation wears off.

(b) When you enter a dark room from bright sunlight, it takes a few minutes before you can see things.

(c) To see an object clearly and sharply, you must look straight at it.

(d) It is impossible to make out the colour of different objects in dim light.

3 The diagrams below show a mammalian eye as seen from the front.

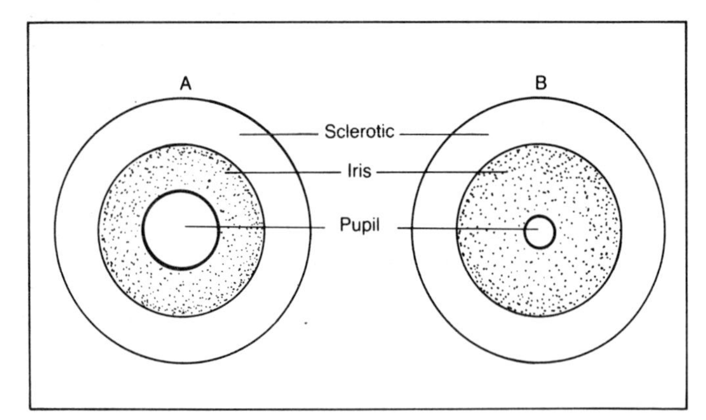

(a) What environmental change would cause the alteration from A to B?

(b) Explain fully how the nervous system and muscles bring about this alteration.

(c) What is the function of the sclerotic?

(d) What structure is found immediately beneath the sclerotic, and what is its function?

(NEA, with addition)

4 The graph below compares the relative numbers of rods and cones in different parts of the retina of the human eye.

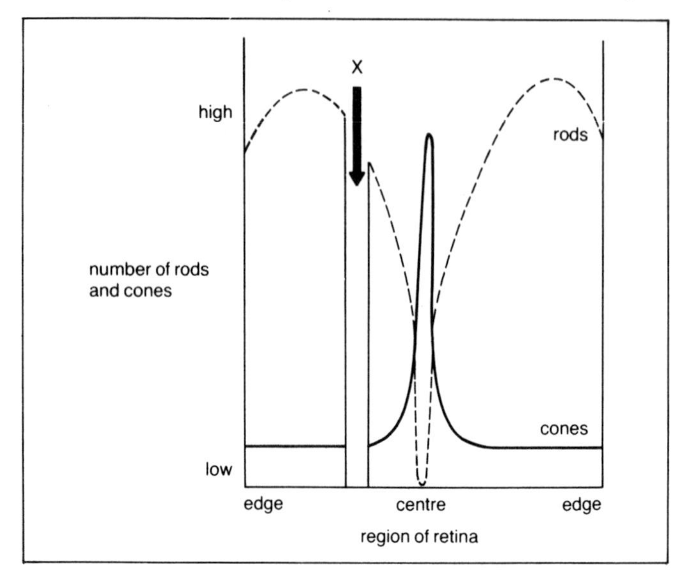

(a) Explain in words what the graph tells us about the distribution of rods and cones in the retina.

(b) What does the region marked X represent?

(c) Criticise the graph and suggest how it might be improved.

21

Glands and hormones

What are glands?

A gland is an organ which produces (**secretes**) a useful substance (**secretion**). Many glands are controlled by the nervous system and are effectors in reflex arcs (see page 89).

Types of gland

There are two types of gland:

① **Ducted glands** shed their secretion into a duct (tube) which leads to, for example, the gut. *Example*: the pancreas which secretes digestive enzymes into the pancreatic duct and thence to the duodenum (see page 55).

② **Ductless glands** shed their secretion into the bloodstream. The secretion is called a **hormone**. Ductless glands are also known as **endocrine glands**.

What are hormones?

A **hormone** is a chemical substance which is produced in one part of the body and has an effect in another part. In most animals hormones travel in the bloodstream. Hormones are known as **chemical messengers**.

Nerves and hormones compared

Hormones provide a means of *communication* within the body. In that respect they are doing the same job as the nervous system, but they work in a different way.

Table 21.1 compares the nervous and hormone systems. The main difference is that the nervous system uses *electrical* messages, whereas the hormone system uses *chemical* messages.

Ductless glands and hormones in the human

Table 21.2 summarises the main glands of the human and the hormones which they produce.

These hormones are particularly important for you to know about: **adrenaline** (also spelled **adrenalin**), **insulin** and **sex hormones**.

Table 21.1 *The main differences between the nervous and hormone systems as methods of communication in the body.*

Nervous system	Hormone system
Message is a nerve impulse (electrical)	Message is a hormone (chemical substance)
Message travels along nerves	Message travels in bloodstream
Messages are carried to specific effectors	Messages are carried all over the body
Responses are localised (e.g. contraction of an individual muscle)	Responses may be widespread (e.g. the effects produced by adrenaline)
Responses usually rapid and short-lived, e.g. movement of arm.	Response often slow and long-lasting, e.g. growth, sexual development.

Table 21.2 *Summary of the human body's main hormoneproducing glands and their secretions.*

Gland	Where it occurs	Hormone	Function
Thyroid	In neck by 'Adam's apple'	Thyroxine	Speeds up the metabolic rate and activities that depend on it, e.g. growth
Adrenals	In abdomen, just above kidneys	Adrenaline	Prepares the body for action and emergency
Pancreas	In abdomen, close to duodenum	Insulin	Decreases the amount of sugar in the blood
Ovaries	In lower part of abdomen	Female sex hormones	Promote sexual development
Testes	In scrotal sacs between legs	Male sex hormones	Promote sexual development
Pituitary	At base of brain	Growth hormone	Speeds up growth
		Thyroid-stimulating hormone	Stimulates the thyroid gland to secrete thyroxine
		Gonad-stimulating hormones	Stimulate the gonads (ovaries and testes) to secrete sex hormones

Adrenaline

Adrenaline has three main effects:

1. It makes the heart beat more quickly.
2. It increases the **metabolic rate**, i.e. the rate at which the cells release energy.
3. It constricts arteries to the less important organs and dilates those to the more important ones (e.g. muscles) so that blood is diverted to where it is most needed.

The overall effect of adrenaline is to prepare the body for emergency. It is nicknamed the **fight or flight hormone**. Its effects are reinforced by the involuntary part of the nervous system. Compared with other hormones the effects of adrenaline are short-lived.

Insulin

The effects of insulin are explained on page 85. Insulin specifically affects the liver: the liver is its target organ. The way insulin controls blood sugar is an important aspect of homeostasis.

Sex hormones

Sex hormones are summarised on page 108. They have long-term effects on the development of the human body.

• Questions

1 Name a hormone directly responsible for each of the following:

(a) causing breasts to develop in the human female;

(b) causing hair to grow on the chest in the human male;

(c) increasing the metabolic rate;

(d) increasing the rate at which the heart beats;

(e) lowering the concentration of blood sugar.

2 What, in the hormone system, is equivalent to each of the following in the nervous system?

(a) Nerve impulse.

(b) Nerve fibre.

3 The pancreas is both a ducted and a ductless gland.

(a) (i) Name *one* substance (other than water) which is produced by the ducted part of the pancreas.
(ii) Where does the substance get to after it leaves the pancreas?

(b) (i) What is the name of the ductless parts of the pancreas?
(ii) Name *one* substance (other than water) which is produced by the ductless part of the pancreas.

(c) Briefly explain the effect on the human body of removing the pancreas.

4 The diagram below illustrates some of the changes which occur in a person's body just before he or she runs a hundred metre sprint.

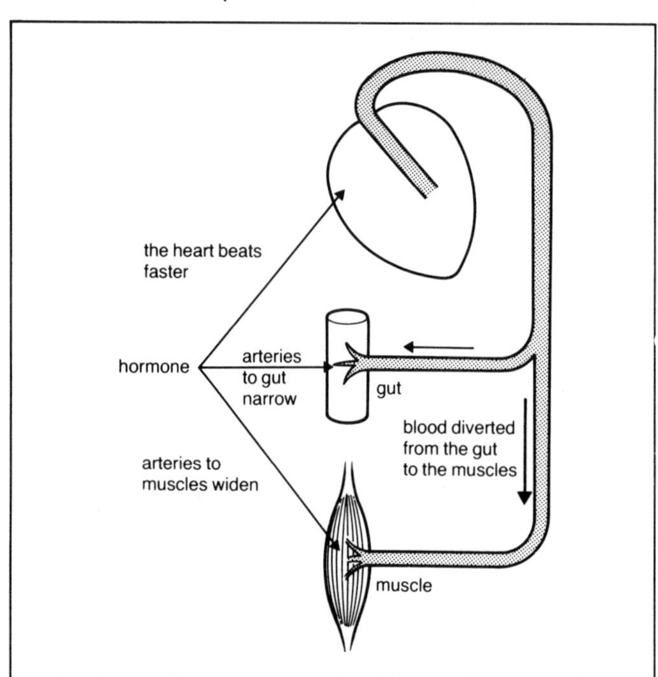

(a) (i) Name the hormone which causes these changes.
(ii) Name the organs which produce the hormone and explain where they occur in the human body.

(b) (i) Why is it helpful for the arteries to the muscles to widen?
(ii) Why is it helpful for the arteries to the gut to get narrower?
(iii) What structure in the walls of an artery enables the artery to get narrower?

22 Muscles and movement

How is movement achieved?

In animals movement is achieved by muscles which contract (shorten), thereby moving some kind of **skeleton**.

Types of skeleton

There are three types of skeleton:

1. **Endoskeleton**: skeleton internal, surrounded by muscles. Found in vertebrates, e.g. human.
2. **Exoskeleton**: skeleton external, muscles inside it. Found in arthropods, e.g. locust.
3. **Hydrostatic skeleton**: skeleton is a fluid-filled cavity surrounded by muscles. Found in soft-bodied invertebrates, e.g. earthworm.

Antagonistic muscles

These are muscles which are arranged in such a way that when they contract they produce opposite effects. *Examples*: the biceps and triceps muscles in the human arm; the circular and longitudinal muscles in the body wall of the earthworm.

Movement is produced by contraction of antagonistic muscles. When one muscle contracts, its antagonist relaxes (or at least does not contract so powerfully).

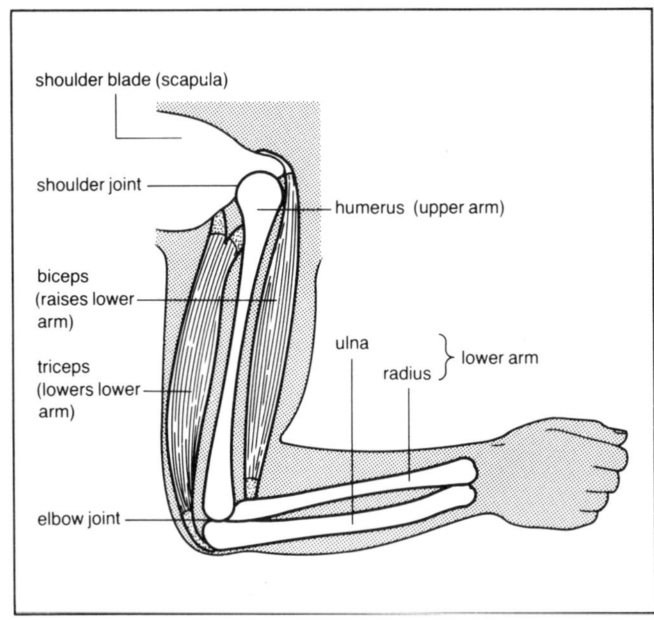

Figure 22.1 *Skeleton and two of the muscles in the human arm. The biceps and triceps muscles are antagonistic in their actions.*

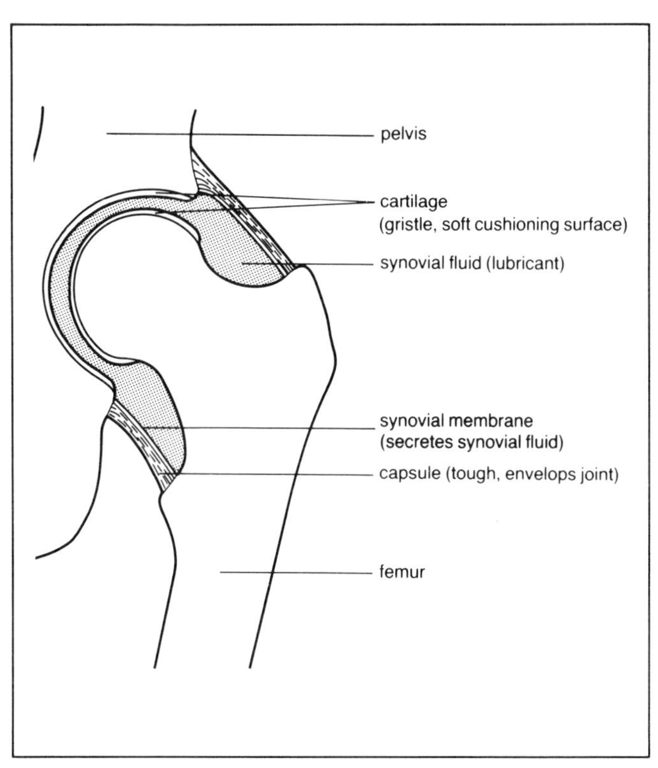

Figure 22.2 *Hip joint of human.*

The human arm

The human arm illustrates most of the important features of the muscles and skeleton of a vertebrate (figure 22.1). Notice, in particular, the relationship between the muscles and skeleton, and the joints.

Joints

A **joint** is the place where two bones move against (i.e. **articulate** with) each other (figure 22.2).
Joints have two important properties:

1. They allow the bones to move easily without friction. This is achieved by a lubricant, **synovial fluid**.
2. They prevent the bones coming apart. This is achieved by the **capsule** surrounding the joint, and by **ligaments** which connect the bones together on either side of the joint.

Types of joint

There are two main types of joint (figure 22.3):

1. **Ball and socket joint** – allows movement in any plane, e.g. hip, shoulder.
2. **Hinge joint** – allows movement in only one plane, e.g. knee, elbow.

Bones as levers

When a muscle moves a bone, the bone functions as a lever. The muscle exerts a **force (effort)** which moves a **load**; the joint serves as the **fulcrum (pivot)**.

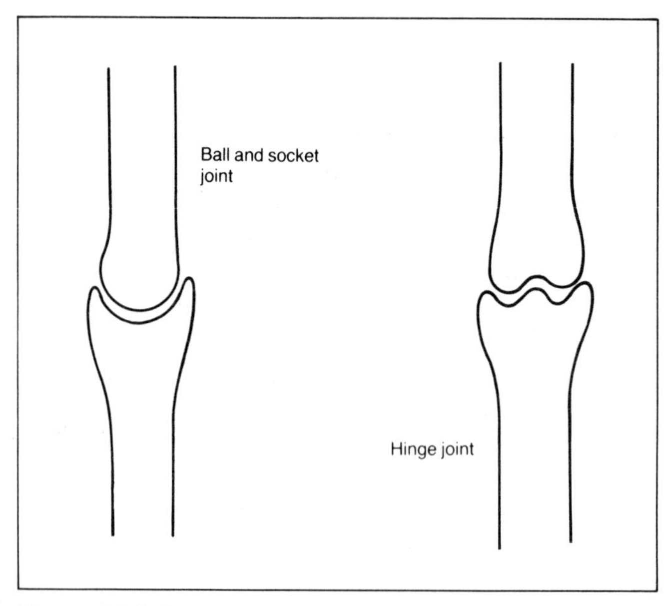

Figure 22.3 *Ball and socket and hinge joints compared.*

The vertebrate skeleton

The main parts of the vertebrate skeleton are summarised in table 22.1.

Details of the skeleton are shown in figure 22.4.

What is the skeleton made of?

The skeleton is made of two types of skeletal tissue:

① **Bone**: very hard because it contains minerals, particularly calcium.
② **Cartilage (gristle)**: soft, serves as a shock-absorber between bones.

Tendons and ligaments

Tendons and ligaments are types of **connective tissue**:

- **Tendons** connect muscles to bone. They have little elasticity and so are virtually unstretchable.
- **Ligaments** connect bones to bones. They are elastic and can be stretched. They hold the skeleton together while allowing muscles to move the individual bones.

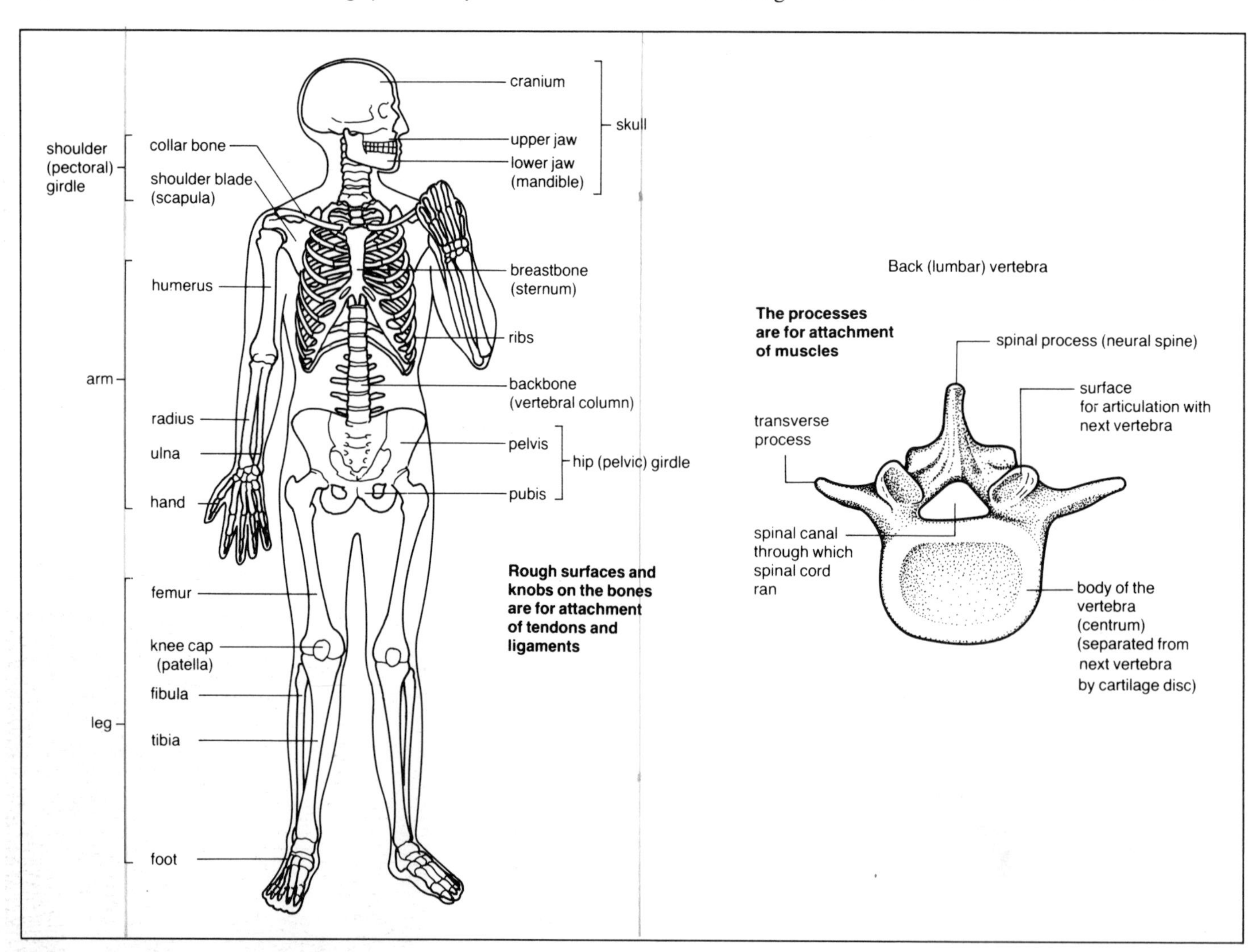

Figure 22.4 *Main parts of the human skeleton with detail of one of the vertebrae.*

Table 22.1 *The main parts of the vertebrate skeleton.*

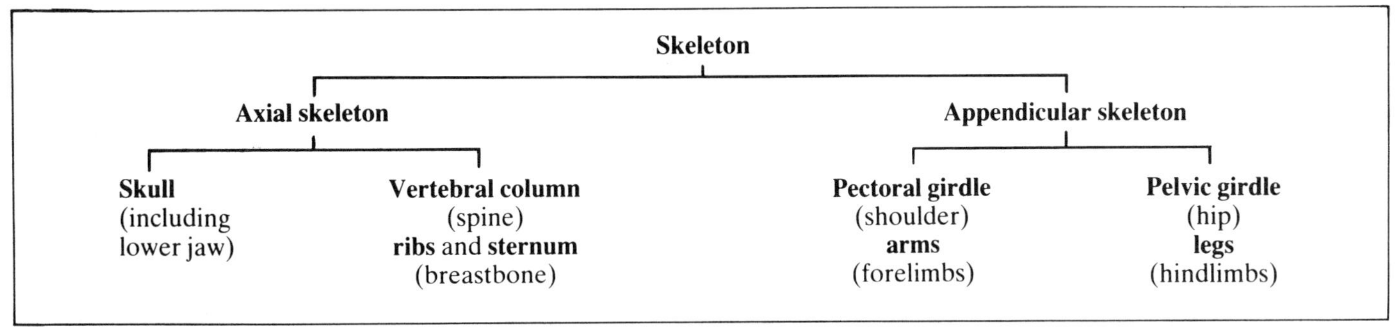

Internal structure of a bone

Figure 22.5 shows the internal structure of the femur:

- **Spongy bone** at the end is a lattice (criss-cross) of bone tissue which can carry a heavy load at an angle to the main axis.
- **Compact bone** in the shaft is dense bone tissue which can carry a heavy load acting along the main axis.

What is bone marrow?

Bone marrow is a soft material found in the centre of bones.

- **Yellow marrow** is found in limb bones, e.g. femur, and consists mainly of fat.
- **Red marrow** is found in, e.g. ribs and pelvis and is where blood cells are made.

Summary of functions of the skeleton

The skeleton has four important functions:

① It supports the body, maintaining its shape and form.

② It protects the soft organs, e.g. brain, heart, lungs etc.

③ It makes blood cells (in the red bone marrow).

④ It brings about movement, along with muscles.

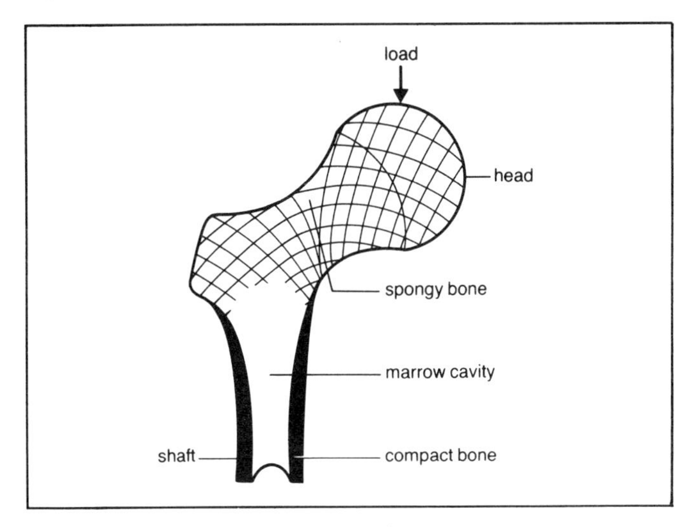

Figure 22.5 *Internal structure of femur.*

• Questions

1 (a) Name two bones of the human skeleton which occur in pairs (left and right) and two which occur singly.

(b) Name one place in the skeleton where cartilage occurs. What is the function of the cartilage in this place?

2 The photograph here shows the internal structure of a human limb bone. Look at the photograph carefully and describe *three* ways in which the structure of this bone helps in either support or movement of the body.

(SEG, modified)

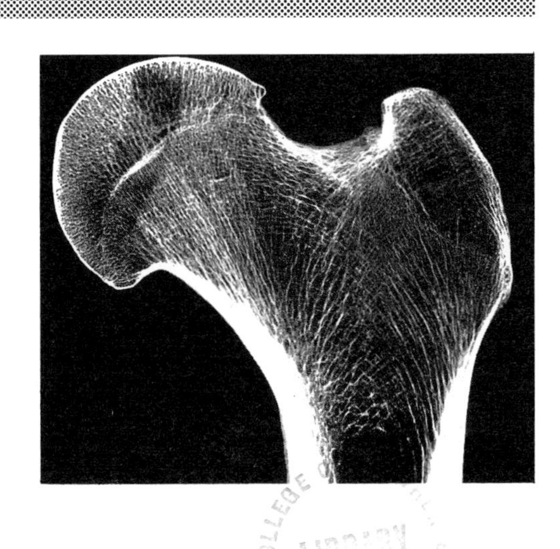

COLLEGE LIBRARY

3 The diagram below shows a section through the elbow joint of a human.

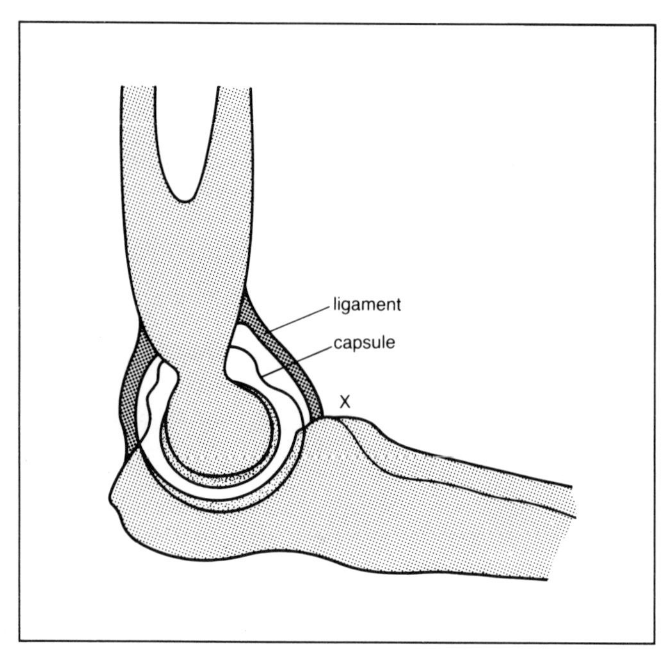

(a) (i) State *two* functions of the capsule.
(ii) State *one* function of the ligaments.

(b) State *two* properties of ligaments, and say why each is important.

(c) What would be attached to the lower arm bone at the position marked X in the diagram. Be as precise as possible.

4 The diagram below shows the positions of the bones and main muscles of the legs of a human when running. (*For clarity, each muscle is shown on one leg only.*)

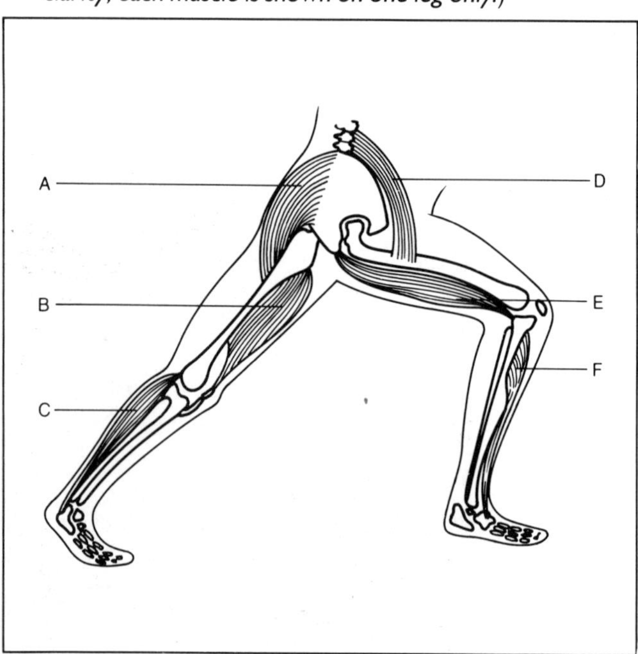

(a) Explain the term 'antagonistic muscles', illustrating your answer by reference to *two* sets of antagonistic muscles from the muscles labelled A to F on the diagram.

(b) Suggest why muscle C is more powerfully developed than muscle F.

(NEA)

5 The diagram below shows the cross-section of an earthworm.

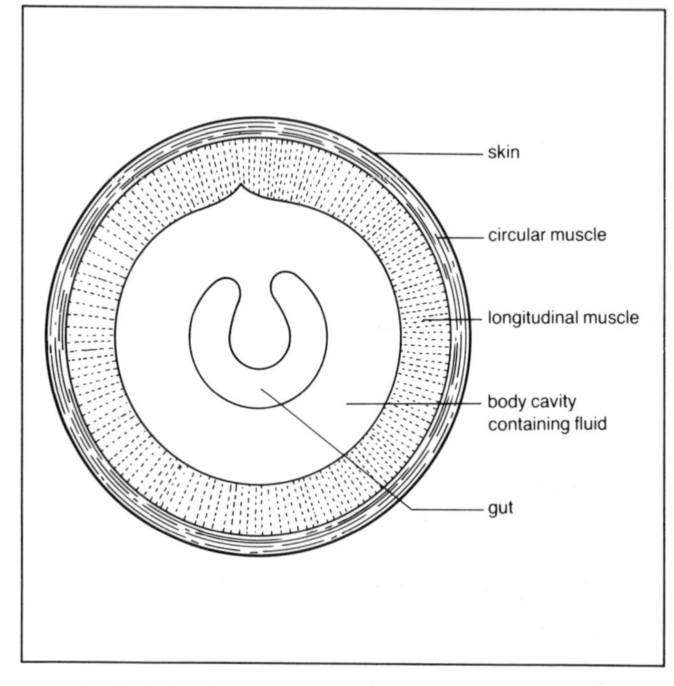

(a) (i) What happens to the shape of the body when the circular muscle contracts?
(ii) What happens to the shape of the body when the longitudinal muscle contracts?

(b) (i) Where is the skeleton in the diagram, and what does the skeleton consist of?
(ii) Explain briefly how the muscles and skeleton work together to enable the earthworm to burrow through soil.

23

Producing offspring

How do organisms produce offspring?

Offspring are produced by **reproduction**. There are two types of reproduction: **asexual reproduction** and **sexual reproduction**.

What is asexual reproduction?

Asexual reproduction is the production of offspring by an individual without the use of a partner.

Methods of asexual reproduction

- **Fission**: Used by single-celled organisms. The cell splits into two 'daughter cells' which grow to full size then split again – and so on. *Examples*: bacteria, *Amoeba*.
- **Budding**: An outgrowth (bud) grows out from the parent. Eventually it breaks away and becomes a self-supporting individual. *Examples*: yeast, *Hydra*.
- **Spore-formation**: A spore is a single cell which may be surrounded by a thick, resistant coat. Spores are produced in large numbers, usually inside a spore case or capsule from which they are released. *Examples*: pin mould (*Mucor*), moss, fern.
- **Vegetative reproduction**: Usually involves the production of a perennating organ (e.g. bulb, corm) which survives the winter and gives rise to new individuals the following year. Characteristic of flowering plants.

What is sexual reproduction?

Sexual reproduction is the production of offspring by two individuals, normally a **male** and a **female**. The male and female produce **gametes**.

What are gametes?

A gamete is a cell which cannot develop any further until it fuses with (i.e. unites with) another gamete.

The male gamete is called a **sperm (spermatozoon)**, the female gamete is called an **egg (egg cell, ovum)** (figure 23.1).

The fusion of a sperm with an egg is called **fertilisation**. The fertilised egg is called a **zygote**. The zygote develops into an **embryo** which in turn develops into the **adult**.

Why is the egg much larger than the sperm?

The egg contains a food store to nourish the embryo during the early stages of its development. The sperm has only to fertilise the egg, after which its function is complete.

What happens during fertilisation?

The sperms swim by lashing their tails from side to side. They bump into the egg. A sperm penetrates the egg membrane. The tail is discarded. The egg membrane thickens to prevent other sperms entering. The head is drawn towards the egg nucleus. The sperm nucleus combines with the egg nucleus.

What is external fertilisation?

Sperms and eggs are released into the surrounding water where fertilisation takes place. External fertilisation is characteristic of aquatic animals, or land animals which breed in water. *Examples*: bony fish (e.g. cod), amphibians (e.g. frog, toad).

External fertilisation is chancy, as is the subsequent development of the zygote and embryo. Animals with external fertilisation produce large numbers of eggs to overcome the slender chances of survival.

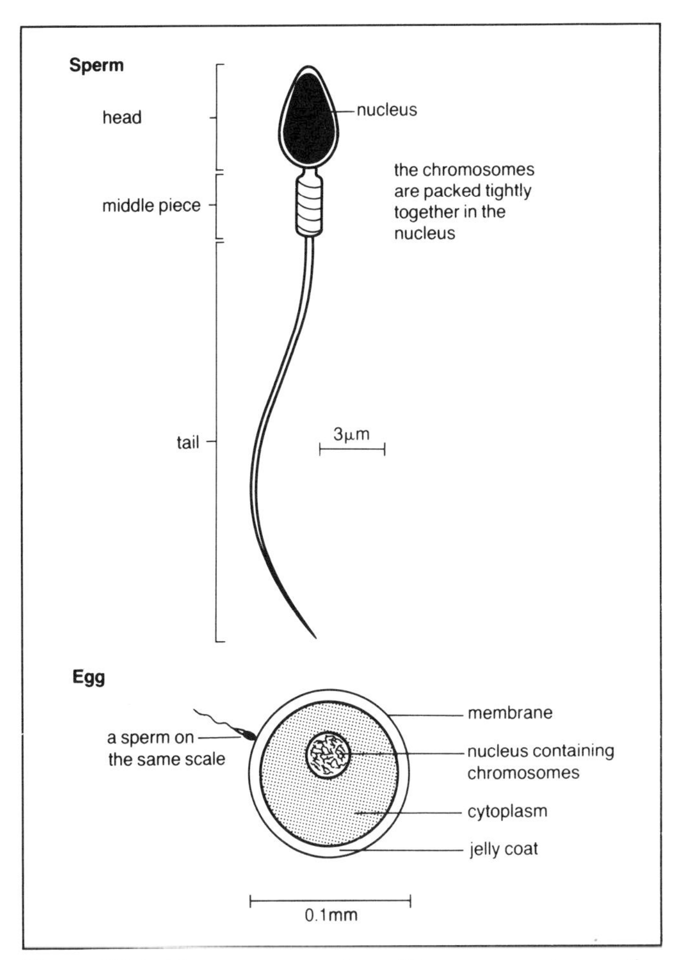

Figure 23.1 *Human sperm and egg. The sperms and eggs of other species are basically similar although there may be differences in detail.*

What is internal fertilisation?

Sperms are put inside the female's body where they fertilise the eggs. *Examples:* insects, cartilaginous fish (e.g. sharks), reptiles, birds, mammals. Internal fertilisation is characteristic of land animals, though some aquatic animals have it too.

What are the advantages of internal fertilisation?

1. It is less chancy than external fertilisation, so fewer eggs need to be produced.
2. The fertilised egg may be coated with a protective case or shell before it is laid, e.g. shell of birds' eggs.
3. The fertilised egg may develop into the embryo within the female's body where it is protected and nourished, e.g. mammals.
4. The result of 2 and 3 is that the young are born at a relatively advanced stage of development when they are less vulnerable and more likely to survive. Giving birth to 'live' young is called **viviparity**. Viviparity is characteristic of mammals but is also seen in other vertebrates, e.g. certain reptiles.

Flowering plants

Flowering plants differ from other organisms in their reproduction. Their sexual method involves the production of **egg cells** and **pollen grains**. The pollen grain contains the equivalent of a sperm. It is transferred to the female by wind or insects. (Details on pages 114–5.)

How do animals care for their young?

Amongst animals in general, care of the young by the parent(s) may be shown in the following ways:

- **Protection**: e.g. male stickleback fans eggs (to oxygenate them) and protect young from attack by predators.
- **Incubation**: e.g. bird sits on eggs, snake coils its body round eggs. (Incubation keeps the eggs warm, at a constant temperature.)
- **Feeding**: e.g. bird feeds nestlings, worker bee feeds larvae, human female feeds her baby on milk.
- **Training**: e.g. humans train and educate their children, as do most 'higher' animals.

Sexual reproduction in lower organisms

Some lower organisms reproduce by the union of cells, or parts of cells, which do not take the form of eggs and sperm. This is usually referred to as **conjugation**.
The following types of conjugation are recognised:

1. **Union of whole cells**, e.g. *Spirogyra* (the contents of two neighbouring cells fuse to form a zygote).
2. **Union of nuclei**, e.g. *Paramecium* (a nucleus from one individual moves into, and fuses with, the nucleus of another individual).
3. **Union of chromosomes**, e.g. bacteria (part of a chromosome from one individual moves into, and links up with, the chromosome of another individual.)

What is a hermaphrodite?

A **hermaphrodite** is an individual which can produce both eggs and sperms. Hermaphrodite species are known as **monoecious**. *Examples*: *Hydra*, earthworm. (Species with separate sexes are known as **dioecious**.)

A hermaphrodite organism may achieve fertilisation in one of two ways:

1. The sperms may fertilise the eggs of the same individual (**self-fertilisation**).
2. The sperms may fertilise the eggs of a different individual (**cross-fertilisation**).

Self-fertilisation, if carried out repeatedly, is bad for the species: harmful genes are more likely to be passed to the offspring than is the case with cross-fertilisation. For this reason, most hermaphrodite organisms have ways of preventing self-fertilisation and promoting cross-fertilisation. For example, when two earthworms copulate, an ingenious mechanism ensures that the sperms of each individual are transferred to the other.

How can sexual reproduction be speeded up?

The rate of production of offspring by sexual reproduction can be increased in two main ways:

1. Sperm from a single mating may fertilise numerous eggs. *Example*: bees and other insects. (The sperms are stored in a sperm receptacle in the female, and they fertilise the eggs as they leave the body.)
2. Offspring may be formed from unfertilised eggs, thus avoiding the need for males. This is called **parthenogenesis**. *Example*: aphids.

Advantages and disadvantages of asexual and sexual reproduction:

Asexual reproduction

Advantages:

- Often rapid and prolific (e.g. spores).
- No need for a partner.

Disadvantages:

- No variety in the offspring (they are genetically identical with the parent).
- Harmful genes in the parent will be handed on to the offspring.

Sexual reproduction

Advantages:
- There is variety in the offspring (they are not genetically identical with the parents or with each other).
- Harmful genes in the parents will not necessarily be handed on to the offspring.

Disadvantages:
- Not normally so rapid and prolific (but there are exceptions).
- Normally a partner is needed. (Some humans might regard this as an advantage!)

How does reproduction help organisms to survive unfavourable periods?

Reproduction often involves the production of dormant, resistant structures which can survive unfavourable periods such as winter. *Examples*:
- **Spores** produced by fungi and bacteria.
- **Perennating organs** (e.g. bulbs) produced by flowering plants.
- **Seeds** produced by flowering plants.
- **Zygospores** produced by, e.g. *Spirogyra* and pin mould (Mucor). (A zygospore is a zygote with a thick wall round it.)
- **Pupae** produced by certain insects, e.g. butterflies (see page 122).

How are offspring dispersed?

① By **self-motility**: the offspring themselves are capable of moving to new localities, e.g. birds and most other animals.
② By **wind and air currents**: many spores, seeds and fruits are dispersed this way, e.g. hairy parachute-like fruits of dandelion, and winged fruits of sycamore.
③ By **animals and other moving objects**: certain fruits have hooks for clinging to animal fur, e.g. goose-grass, burdock.
④ By **water**: some fruits can float, e.g. coconut, and larvae may swim or be wafted by currents.
⑤ By **being eaten**: fleshy fruits are eaten by birds and seeds pass out unharmed with faeces, e.g. cherry, blackberry.
⑥ By an **explosive mechanism**: spore- or seed-case may split open violently, throwing contents over a wide area, e.g. spores of fern, seeds of pea and bean plants and wallflower.
⑦ By a **pepper-pot mechanism**: open spore- or seed-case may be shaken by the wind, scattering contents, e.g. spores of moss, seeds of poppy.

Why is it important for offspring to be widely dispersed?

If the offspring are not widely dispersed, the population may get too dense, leading to overcrowding and competition (see page 23).

When do organisms reproduce?

Amongst lower organisms such as *Amoeba*, *Spirogyra* and *Hydra*, asexual reproduction tends to occur in the spring and early summer when food is plentiful and growing conditions are good. Sexual reproduction (with the production of dormant, resistant structures) tends to occur in the autumn when food is getting scarce and growing conditions are not so good.

In higher animals such as insects and vertebrates, sexual reproduction tends to occur in the spring or early summer when there is plenty of food and good growing conditions for the offspring.

In flowering plants sexual reproduction, and the formation of perennating organs such as bulbs and corms, occurs mainly in the spring and summer when conditions are good for photosynthesis and the plants are making food.

• Questions

1 For animals with external fertilisation it is important that the males and females should release their gametes at the same time and in the same place.

(a) (i) What is the meaning of the term gamete?
(ii) What is the meaning of the term fertilisation?

(b) Suggest *one* way which would help to ensure that the males and females release their gametes at the same time.

(c) Suggest *one* way which would help to ensure that the males and females release their gametes in the same place.

(d) The male and female gametes of an animal are known as sperms and eggs. State *four* observable differences between a typical sperm and a typical egg.

2 *Hydra* can reproduce asexually by budding, or sexually by producing eggs and sperms. After fertilisation, the zygote acquires a thick wall and becomes a zygospore.

(a) Asexual reproduction takes place in the spring whereas sexual reproduction takes place in the autumn. Account for this difference.

(b) (i) Suggest *one* advantage to *Hydra* of its asexual method of reproduction.
(ii) Suggest *one* advantage to *Hydra* of its sexual method of reproduction.

(c) *Hydra* has remarkable powers of regeneration. If the animal is cut into small pieces, each can grow into a new individual. Suggest *one* reason why this should, and *one* reason why it should not, be regarded as one of *Hydra's* methods of reproduction.

3 The diagram below shows spores being released from a moss plant.

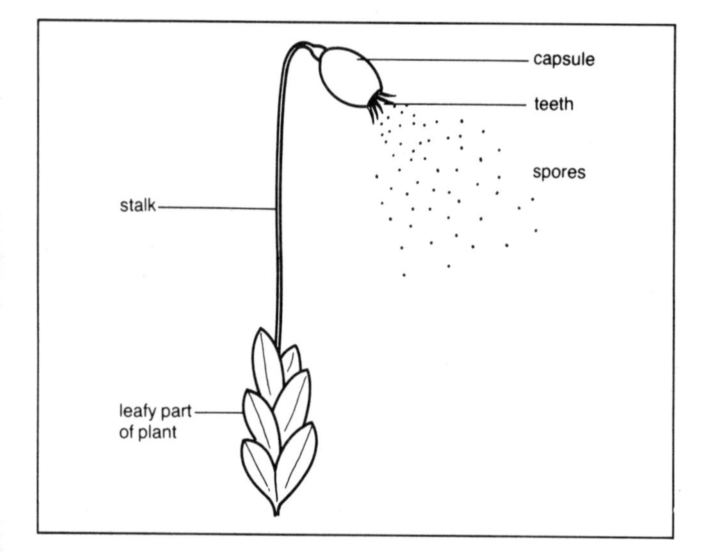

(a) The spores are very light. Why is this an advantage?

(b) Suggest *two* ways in which the stalk may help to disperse the spores over a wide area.

(c) The spores are released in dry conditions. In moist conditions the teeth close over the opening of the capsule and the spores stop being released. Suggest *one* reason why it is advantageous for the spores to be released in dry conditions.

(d) Although the spores are released in dry conditions, they will only germinate in damp conditions. Suggest *one* reason why moisture is needed for the spores to germinate.

(e) In ferns, the spore capsule opens by suddenly splitting down one side. In what way might the splitting process help in dispersing the spores?

24 Human reproduction

Structures involved in reproduction

The structures involved in reproduction make up the **reproductive system.** They are shown in figures 24.1 and 24.2.

The functions of the various structures are summarised in tables 24.1 and 24.2.

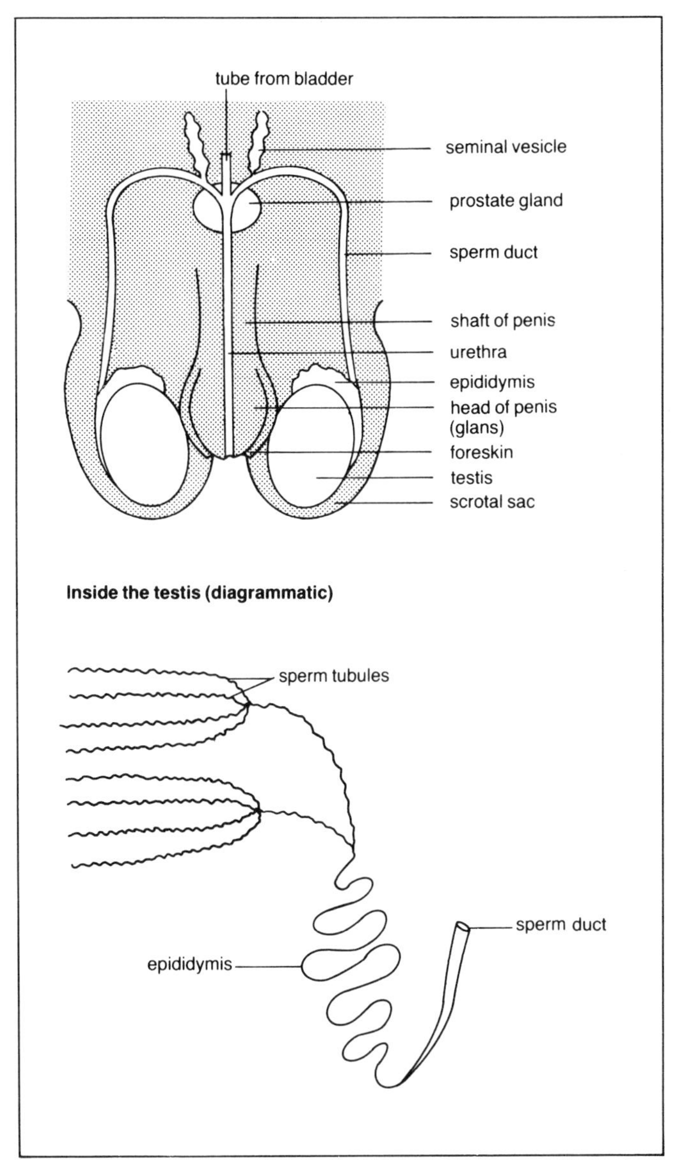

Figure 24.1 *The male reproductive system viewed from the front.*

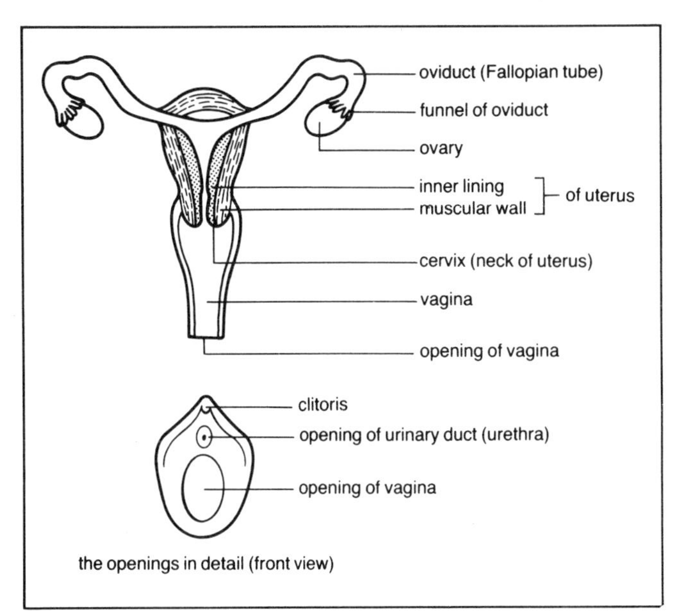

Figure 24.2 *The female reproductive system viewed from the front.*

Sexual development

Table 24.3 summarises the way our sex organs and related structures develop as we get older.

In the male sperms are produced continuously. In the female an egg is shed from one of the two ovaries approximately once every 28 days.

How do sperms reach the egg?

This is normally achieved by **intercourse** (**copulation**). The erect penis is inserted into the vagina and moved repeatedly until **ejaculation** occurs. The sperms then swim in the watery mucus lining the cervix and uterus to the Fallopian tubes. **Fertilisation** takes place in the upper part of one of the Fallopian tubes.

What happens after fertilisation?

The **fertilised egg** (**zygote**) divides up into a ball of cells (**embryo**). The embryo becomes **implanted** in the inner lining of the uterus. **Conception** has now taken place and the woman is **pregnant**.

When can conception occur?

Conception is only possible when an egg is present in a Fallopian tube, ready to be fertilised. An egg is present every 28 days, in the middle of the menstrual cycle.

What is the menstrual cycle?

The **menstrual cycle** is the sequence of changes which occurs in the ovary and uterus in the course of 28 days. The changes are coordinated by **hormones** secreted by the pituitary gland and ovary.

Table 24.1 *Functions of the reproductive organs in the human male.*

Structure	Function
Testis	
Sperm tubules	Manufacture sperms
Epididymis	Stores sperms in a viable but immobile state.
Sperm duct (vas deferens)	Contractions sweep sperms into urethra during ejaculation. Otherwise, sperms degenerate.
Seminal vesicles and **prostate gland**	Secrete fluid which activates and nourishes sperms (secretions + sperms = semen).
Urethra	Contractions expel semen from penis during ejaculation. (Urethra also carries urine to exterior).
Penis	
Shaft	Contains spongy tissue which, when full of blood, causes erection.
Glans	Sensitive part of penis: repeated stimulation results in ejaculation and orgasm (reflex).
Foreskin	Protects glans. Removed by circumcision.

Table 24.2 *Functions of the reproductive organs in the human female.*

Structure	Function
Ovary	Manufactures eggs.
Oviduct (Fallopian tube)	Site of fertilisation. Conveys fertilised egg/embryo to uterus.
Uterus	Inner lining receives, protects and nourishes embryo. Contractions of muscular wall expel baby during birth.
Cervix (neck of uterus)	Produces watery mucus which serves as lubricant for penis and medium in which sperms swim after ejaculation.
Vagina	Receives penis during intercourse. Baby passes along it during birth.
Clitoris	Equivalent to male penis: can become erect, and repeated stimulation results in orgasm.

Table 24.3 *Summary of sexual developments of the human male and female.*

	Male (♂)	**Female** (♀)
At birth	Testes have descended into scrotal sac but they do not make sperms yet.	Ovaries containing immature eggs are present in abdomen but they do not produce eggs yet.
12–14 years Puberty	Pituitary gland ↓ gonad-stimulating hormones ↓ Testes ↓ male sex hormones (androgens) ↓ Testes start producing sperms. Secondary sexual characteristics develop, e.g. growth of body hair and breaking of voice.	Pituitary gland ↓ gonad-stimulating hormones ↓ Ovaries ↓ female sex hormones (oestrogens) ↓ Ovaries start producing eggs. Secondary sexual characters develop, e.g. growth of breasts and laying down of fat in thighs.
45–50		Menopause ('change of life') Ovaries stop producing eggs

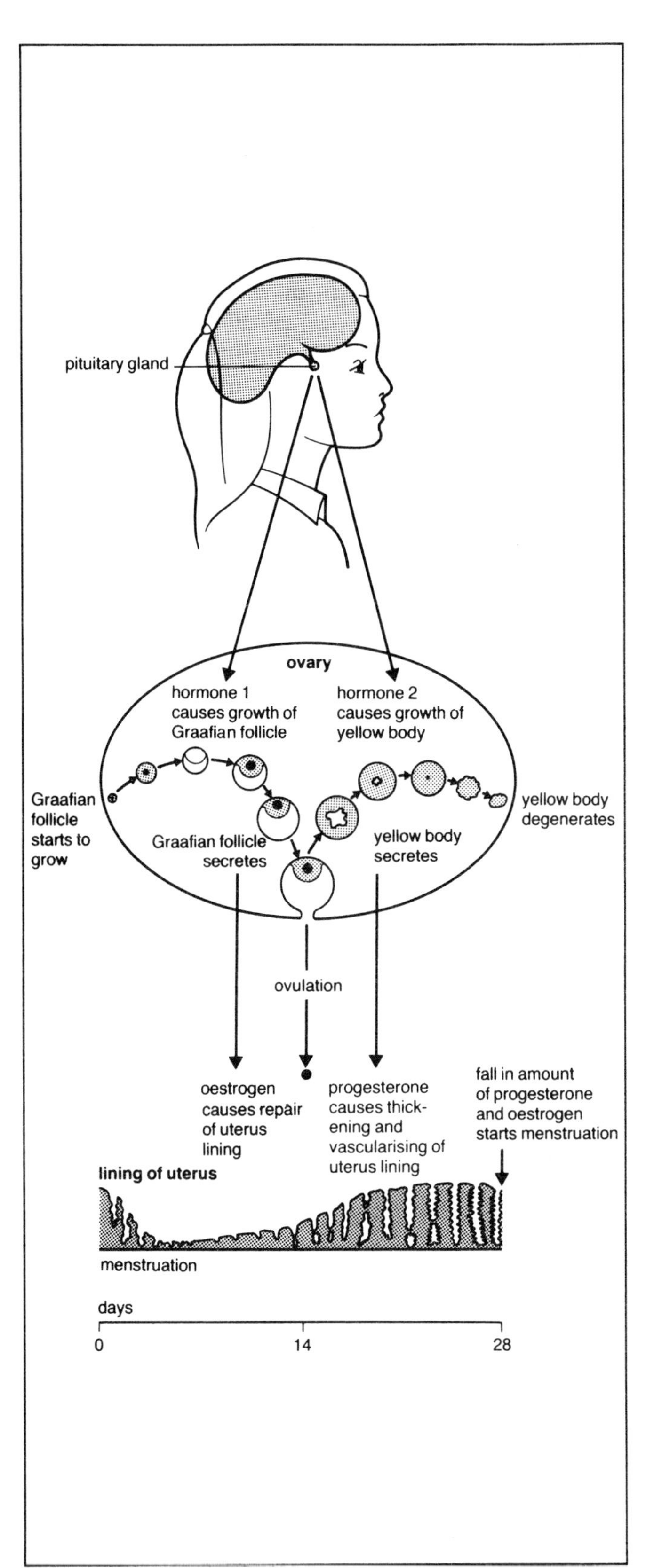

Figure 24.3 *Summary of the events which take place during the menstrual cycle and how they are coordinated by hormones.*

The main events in the menstrual cycle

Three main events occur in the course of the menstrual cycle:

- **Menstruation** (figure 24.3): this is the disintegration, with bleeding, of the inner lining of the uterus.
- **Preparation of the uterus**: this is the repair, thickening and vascularisation of the inner lining of the uterus so that it is ready to receive an embryo if fertilisation occurs.
- **Ovulation**: this is the shedding of a mature egg from the ovary.

What happens to the menstrual cycle if a woman conceives?

If a woman conceives, menstruation (and ovulation) cease for as long as she is pregnant. This is because progesterone goes on being produced, first by the yellow body (which persists in the ovary) and later by the placenta.

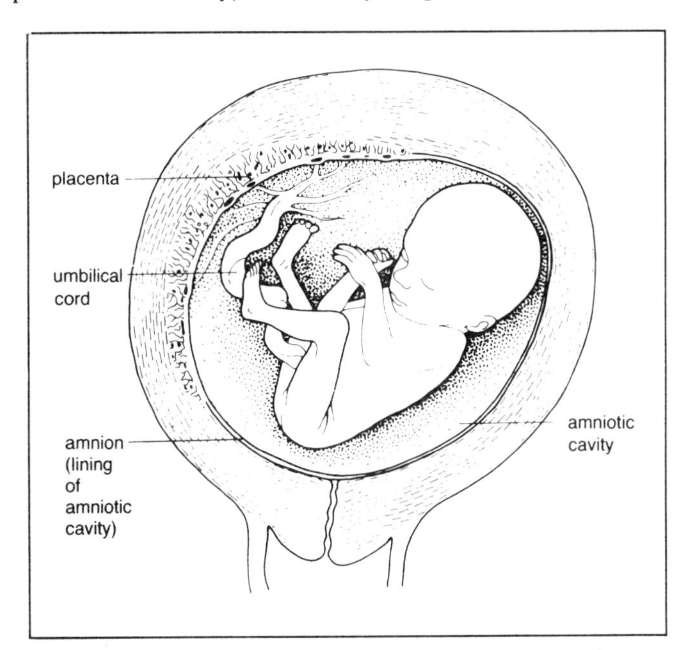

Figure 24.4 *The foetus 14 weeks after the beginning of pregnancy.*

What happens in the uterus during pregnancy?

1. The embryo grows into a miniature human being (**foetus**).
2. The foetus becomes surrounded by a fluid-filled **amniotic cavity**.
3. The **placenta** develops: the foetus is attached to it by the **umbilical cord**.
4. The uterus expands to accommodate the growing foetus.

Figure 24.4 shows a 14-week old human foetus in the uterus.

Structure of the placenta

The placenta contains finger-like **villi** which project into a large blood space containing the mother's blood. The villi contain capillaries which belong to the foetus's circulation (figure 24.5).

The bloodstreams of the foetus and mother are separated by a thin barrier. This prevents the bloodstreams mixing but allows exchange of chemical substances.

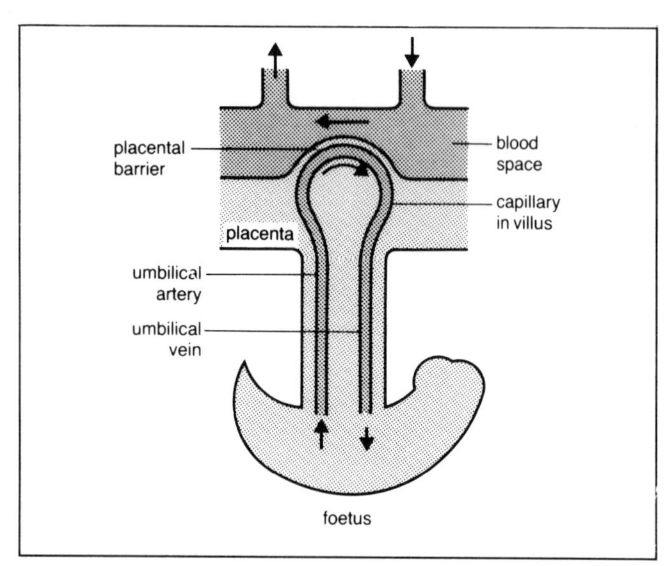

Figure 24.5 *Schematic diagram showing the relationship between the blood of the foetus and mother. Although the two bloodstreams come very close to each other, they never mix. The arrows indicate the flow of blood.*

Why must the foetal and maternal bloodstreams not mix?

There are two reasons:

1. The high blood pressure of the mother would burst the foetal blood vessels.
2. If the two bloods belonged to different blood groups, agglutination would occur.

What substances pass across the placenta?

- **From mother to foetus**: oxygen, nutrients, antibodies.
- **From foetus to mother**: carbon dioxide, excretory waste (urea).

Note: Drugs, poisons, bacteria and viruses can also pass from mother to foetus. This is a further reason why they are dangerous to a pregnant woman.

The foetal circulation

The foetus's blood is oxygenated by the placenta. Its lungs are therefore non-functional. In the foetus blood bypasses the lungs by means of:

- a hole in the heart connecting the right and left atria,
- a vessel which connects the pulmonary artery with the aorta.

What happens during birth?

1. The uterus undergoes occasional contractions ('labour').
2. The amnion bursts and the amniotic fluid is discharged.
3. The uterus contracts powerfully, expelling the baby.
4. The lungs start functioning and the baby takes its first breath.
5. The umbilical cord is tied and cut.
6. The lung-bypasses close so that blood flows to the lungs.
7. The placenta is discharged ('afterbirth').
8. The breasts produce milk when the nipples are sucked.

What causes birth?

Birth is brought about by:

- a change in the amounts of hormones (oestrogen and progesterone) produced by the placenta towards the end of pregnancy.
- the secretion of **birth hormone** (**oxytocin**) by the pituitary gland.

The combined effect of these hormonal changes is to make the uterus muscles contract.

What causes the breasts to produce milk?

- Before birth, the hormones from the placenta cause the mammary glands in the breasts to make milk.
- At birth, **milk-stimulating hormone** (**prolactin**) from the pituitary gland causes milk to flow when the nipples are sucked.

Breast-feeding or bottle-feeding?

There are four main advantages to breast-feeding:

- Mother's milk contains all the necessary nutrients in the right proportions.
- Mother's milk is delivered at exactly the right temperature.
- Mother's milk contains antibodies which help to protect the baby from disease.
- Breast-feeding allows close contact between the mother and her baby.

However, some mothers are not able to breast-feed their babies and it is certainly possible to bottle-feed a baby satisfactorily.

How are twins produced?

There are two types of twins:

1. **Identical twins** arise when the fertilised egg splits into two cells, each of which develops into a new individual. The two individuals are genetically identical and the same sex.

② **Non-identical** twins arise when the ovary produces two eggs, each of which develops into a new individual. The two individuals are genetically different and may be different sexes.

What is contraception?

Contraception is any procedure which prevents conception. Some methods involve the use of an artificial device (**contraceptive**). Table 24.4 summarises the various methods of contraception and their reliability.

Contraception results in **birth control** and is the basis of **family planning**.

Note: Frequent sexual intercourse with multiple partners before puberty is complete carries health risks.

What is abortion?

Abortion is the destruction of a foetus in the uterus. In the United Kingdom the law says that an abortion may be carried out only if two or more doctors consider that by continuing the pregnancy a woman's physical or mental health is at risk. An abortion is not normally carried out after the 24th week of pregnancy, although at the time of writing 28 weeks is the legal limit.

What is infertility?

An infertile person is one who cannot procreate offspring (i.e. produce children).

What causes infertility?

Possible causes include the following:

In the male:

- No, or too few, sperms produced by the testes (sterility).
- Inability to get an erection and/or ejaculate.

In the female:

- No eggs produced by the ovaries (sterility).
- Sperms die before they reach the egg.
- Fallopian tube(s) blocked.
- Embryo fails to implant in the uterus.

How can infertile couples be helped?

Infertile couples can be helped in four main ways:

① **Artificial insemination**: semen from the husband or a donor is injected into the woman's vagina.
② **Fertility drug**: the woman is given a hormone preparation which stimulates her ovaries to produce eggs.
③ **In vitro fertilisation**: an egg is taken from the ovary, fertilised outside the body and then put into the uterus to develop ('test tube baby').
④ **Adoption**: the couple acquire, usually through an adoption society, a baby whose natural parents cannot look after it themselves.

Table 24.4 *The main methods of contraception.*

Method	What is involved	How it works	Reliability
Condom (sheath)	Rubber sheath placed over penis before intercourse.	Catches semen during ejaculation.	Good if used properly.
Cap (diaphragm)	Rubber diaphragm placed over cervix before intercourse.	Prevents sperms entering uterus.	Good if used with a spermicide.
Spermicide	Chemical substance placed in vagina before intercourse.	Kills sperms.	Good but only if used with sheath or cap.
IUD (intra-uterine device)	Plastic or metal device placed in uterus by doctor.	Stops embryo implanting.	Good.
Oral contraceptive ('pill')	Tablet (hormone preparation) taken daily.	Stops ovulation.	Very good, but health risks if overused.
Injectable contraceptive	Hormone preparation injected into bloodstream by doctor every three months.	Stops ovulation.	Very good, but suspected health risks.
Rhythm method	Intercourse restricted to 'safe period' when egg is not available for fertilisation.	Avoids fertilisation.	Poor except under expert guidance.
Male sterilisation (vasectomy)	Sperm ducts cut and tied.	Stops sperms getting into semen.	Excellent, but irreversible.
Female sterilisation (tubal ligation)	Fallopian tubes cut and tied.	Stops sperms reaching egg.	Excellent, but irreversible.

What are sexually transmitted diseases?

A **sexually transmitted disease (STD)** is a disease which can only be spread by sexual intercourse or very close sexual contact.

In a wider sense, STDs include all diseases which are treated regularly in STD clinics.

Table 24.5 summarises the main STDs and their causes. One of the most serious is **acquired immune deficiency syndrome (AIDS)**. It is particularly serious because it is deadly and incurable.

How is AIDS spread?

AIDS is spread in three main ways:

① By close sexual contact, particularly intercourse.

② By drug-users sharing the same needle.

③ By contaminated blood being given in a transfusion.

(3 is now prevented in the U.K. by testing ('screening') all blood which is to be used in transfusions.)

How can the spread of AIDS be prevented or at least slowed down?

① By using a condom when having intercourse.

② By practising safer sex. Safer sex is sexual contact without transfer of body fluids (blood, semen, saliva, mucus).

③ By having as few sexual partners as possible.

④ By getting to know your partner and avoiding casual sex.

⑤ By drug-users not sharing needles.

The above precautions can help to cut down the spread of other STDs as well.

How does the AIDS virus work?

Helper cells in the bloodstream help our lymphocytes to destroy germs. The AIDS virus attacks these cells.

The incubation period for AIDS can run into years. This is because the virus may remain dormant in the body for a long time before destroying the helper cells.

Table 24.5 *Some important sexually transmitted diseases. The agents responsible can pass across the placenta and infect babies in the womb.*

Disease	Organism which causes it	Main symptoms	Treatment
Syphilis	Bacterium	Sores, fever, insanity. Fatal if untreated.	Antibiotics.
Gonorrhoea	Bacterium	Discharge from genital opening. Painful joints.	Antibiotics.
AIDS	Virus	Reduced resistance to disease. Fatal.	No cure. Drugs help. Vaccine not yet developed.
Non-specific urethritis	Virus	Inflamed urethra.	Drugs.
Genital herpes	Virus	Painful blisters.	Drugs.

• Questions

1 Look at figures 24.1 and 24.2. Which structure or structures in the diagram:

(a) manufacture sperms;

(b) manufacture eggs;

(c) carry urine and sperms to the exterior;

(d) contain an egg immediately after it has been fertilised;

(e) expand greatly during birth?

2 Using table 24.3 to help you, write a short account of the sequence of changes which occurs at puberty in (a) the human male and (b) the human female.

3 Read through table 24.4. Which of the contraceptive methods listed in the table:

(a) provide a barrier which prevents sperms reaching the egg;

(b) prevent eggs being produced;

(c) prevent the embryo implanting;

(d) involve the use of chemical substances;

(e) cannot be reversed?

4 The diagram below shows the circulation of blood in the wall of the uterus of a pregnant woman. The arrows indicate the direction of flow of blood.

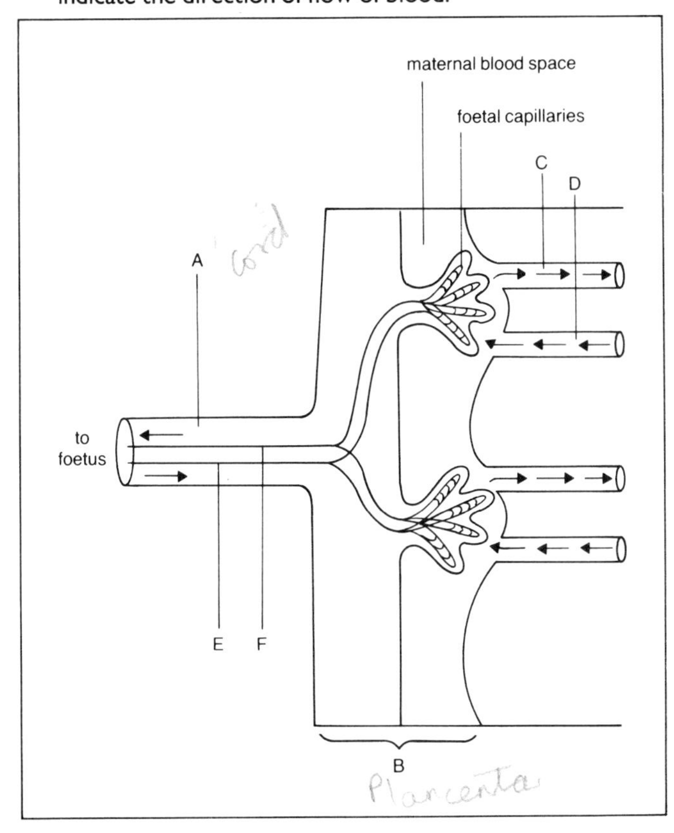

(a) Name structures A and B.

(b) Structures C, D, E and F are blood vessels. State the function of each one.

(c) Name *three* substances, useful to the embryo, which pass from the maternal blood space to the blood in the foetal capillaries.

(d) Name *two* waste substances which the foetus produces and explain how they are got rid of from the foetus.

(e) The foetal capillaries are located inside numerous finger-like villi. Why is this a useful arrangement?

5 In answering the following questions you are advised to refer to figure 24.2.

(a) Explain why a woman with blocked Fallopian tubes cannot get pregnant in the normal way.

(b) Read the following passage and, with the help of the diagram, answer the questions below.

> A woman with blocked Fallopian tubes can now have a 'test-tube baby'. She is given hormones to increase the number of eggs maturing in each ovary. A doctor, using a fine tube through the body wall, searches for and sucks up several eggs from the surface of the ovary just before they are released naturally. The eggs are put in a culture solution in a dish. Semen containing sperm is added and fertilisation usually occurs. Three days after fertilisation, embryos of between eight and sixteen cells have formed. Two or three of these embryos are gently transferred by a fine tube via the cervix into the uterus. If the process is successful, at least one of the embryos develops into a baby.

(i) Suggest why the doctor wants to increase the number of eggs maturing in each ovary (lines 3–4).
(ii) Suggest why the doctor wants to collect eggs just before they are released naturally and at what stage of the menstrual/oestrous cycle the doctor would do this (lines 6–7).
(iii) Suggest features of the culture solution (line 7–8) which are essential for a successful test-tube baby.
(iv) What is 'fertilisation' (line 9)?
(v) Suggest why the doctor waits three days (lines 9–10) before transferring the embryos to the uterus.
(vi) Suggest why, when embryos are put in the uterus, the tube is passed through the cervix (lines 12–13) and not through the body wall (line 5).

(c) Do you think the term 'test-tube baby' (line 2) is a good one? Give reasons for your answer.

(LEAG)

25 Reproduction of the flowering plant

How do flowering plants reproduce?

Flowering plants reproduce sexually and asexually. The organs for sexual reproduction are contained in the **flower**.

Structure of the flower

The flower consists of a series of rings (**whorls**) of structures. They are, from the outside inwards:

- **Sepals**: usually small and leaf-like.
- **Petals**: often large and showy, may have nectaries at base which produce nectar.
- **Stamens**: male part of the flower, produce pollen grains; each pollen grain contains a **male gamete**.
- **Carpel**: female part of flower, contains one or more **egg cells**.

A generalised flower showing details of the individual parts is shown in figure 25.1. Nectar is a sugary liquid which serves as a bait for insects (see below).

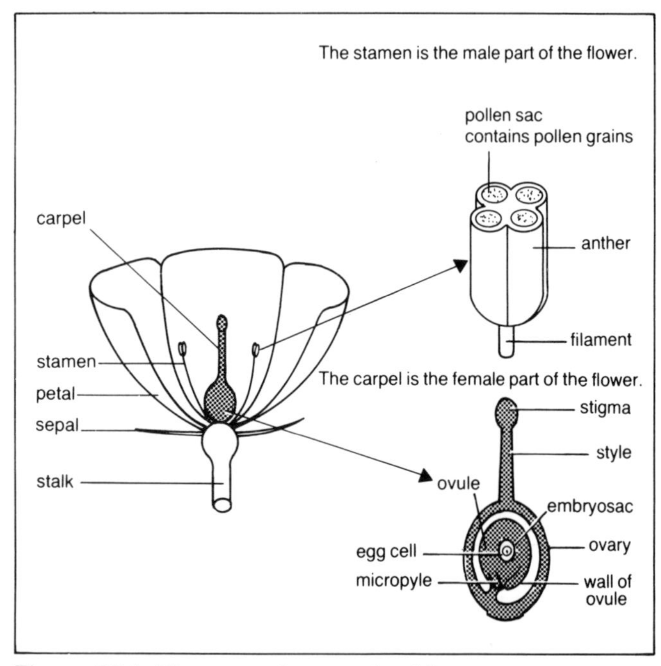

Figure 25.1 *The parts of a generalised flower.*

Variation in flower structure

Flowers of different species vary in the following ways:

① The number of sepals, petals, stamens and carpels. (Some may be absent altogether.)

② The number of ovules in a carpel.

③ The structures may be joined or fused together to varying extents.

④ The flower may be **radially symmetrical** or **bilaterally symmetrical** (see page 34).

Arrangement of flowers on the stem

Flowers may be single (solitary) or numerous. When numerous, the flowers together constitute an **inflorescence**. Within an inflorescence the flowers may be arranged in various different ways, depending on the species.

Pollination

Pollination is the transferring of pollen grains from an anther to a stigma. There are two types of pollination:

① **Self-pollination**: pollen grains transferred to a stigma in the same flower or to a stigma in another flower on the same plant.

② **Cross-pollination**: pollen grains transferred to a stigma in a flower on a *different plant*.

In general cross-pollination is preferable to self-pollination because it creates variety in the species (see page 105). Most plants have features which favour cross-pollination, and/or reduce the chance of self-pollination.

How is cross-pollination brought about?

In cross-pollination the pollen grains are carried by **wind** or by **small animals**, usually insects. Flowers are adapted for one or other type of pollination.

Table 25.1 *Summary of the main differences between typical wind-pollinated and insect-pollinated flowers.*

Wind-pollinated flowers	Insect-pollinated flowers
1 Generally small	Generally larger
2 Petals green or dull coloured	Petals often brightly coloured
3 Do not produce nectar	Petals have nectaries which produce nectar
4 Flower hangs down for easy shaking	Flower faces upwards
5 Stamens and stigma hang out of the ring of petals	Stamens and stigma inside the ring of petals
6 Large number of pollen grains produced	Smaller number of pollen grains produced
7 Pollen grains very light with smooth surface	Pollen grains heavier with spikes for sticking to insect
8 Stigma has feathery branches for catching pollen	Stigma is like pinhead and lacks branches

How are flowers adapted for wind and insect pollination?

The adaptations are listed in table 25.1. This table also summarises the differences between wind and insect-pollinated flowers.

How is fertilisation brought about?

This is the sequence of events leading to fertilisation:

1. A **pollen tube** grows out of the pollen grain.
2. The pollen tube grows into the stigma and down the style to the ovary.
3. The pollen tube grows into the ovule (usually through the micropyle) and releases a **male nucleus** into the embryosac.
4. The male nucleus fuses with the egg cell (**fertilisation**), thus forming a fertilised egg (zygote).

What happens after fertilisation?

After fertilisation the following changes take place:

1. The zygote develops into an **embryo**.
2. The embryo becomes surrounded by **endosperm tissue** which nourishes it.
3. The ovule develops into the seed, the wall of the ovule becomes the seed coat.
4. The ovary develops into the **fruit**.
5. The seed dries out, becoming **dormant**.
6. The sepals, petals and stamens wither away and/or drop off.

Structure of seeds

Figure 25.2 shows the structure of a broad bean seed.

The **cotyledons** store food (starch) for use when the seed germinates (see below).

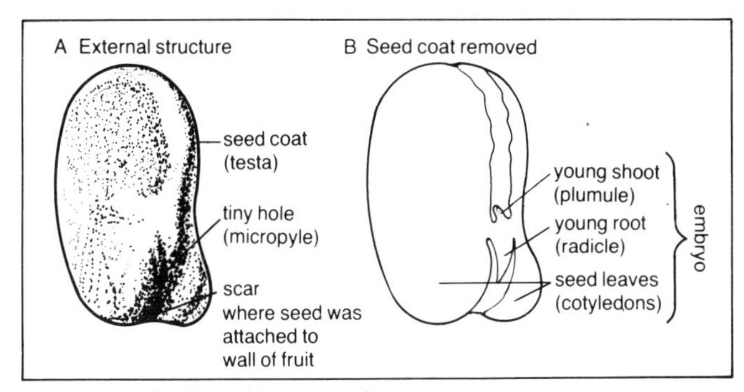

Figure 25.2 *Broad bean seed.*

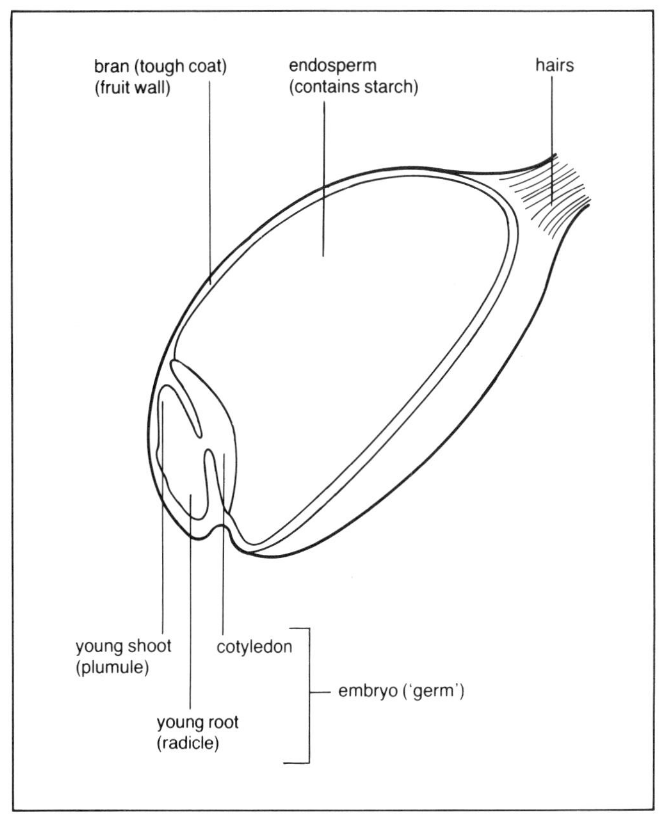

Figure 25.3 *Wheat seed ('grain') sliced down the middle.*

Some seeds contain **endosperm** tissue which provides food when the seed germinates. Such seeds tend to have small cotyledons. An example is the wheat seed (figure 25.3).

Variations in seed structure

Seeds of different species vary in the following main ways:

- Size and shape (related to the method of dispersal – see below).
- The number of cotyledons (dicotyledons have two cotyledons, monocotyledons have one).
- The size of the cotyledon(s).
- The presence or absence of endosperm tissue.

What is germination?

Germination is the sprouting of a seed to produce a new plant. The young plant is called a **seedling**.

What happens when a seed germinates?

The following changes occur during germination:

1. The dry seed takes in (imbibes) water through the micropyle, and expands.
2. The seed coat ruptures.
3. The young root emerges and grows downwards.
4. The young shoot emerges and grows upwards.

⑤ Starch in the cotyledons is broken down into soluble sugar which is transported to growing points at the tip of the shoot and root.

⑥ Root hairs increase the surface area for absorption of water and mineral salts from the soil.

⑦ On emergence from the soil, the shoot develops green leaves which start photosynthesising.

⑧ The cotyledons shrink and wither away.

Structure of seedlings

Figure 25.4 shows the structure of a broad bean seedling. In broad bean and its relatives, the shoot is bent back as it pushes up through the soil: this protects the plumule.

In grasses (which include wheat, barley and other cereals), the plumule is protected by a sheath called the **coleoptile**. Coleoptiles are much used for experiments on auxin (see page 121).

What conditions are needed for germination?

These conditions are needed for successful germination:

① **Water**: This is needed for the swelling and bursting of the seed, movement of food reserves, and for growth of the shoot and root.

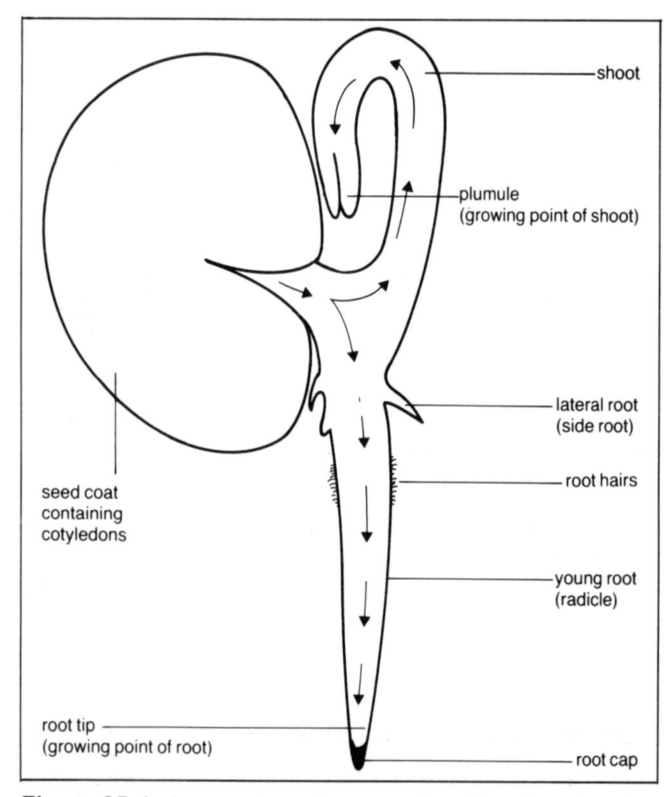

Figure 25.4 *A young broad bean seedling. The plumule is the youngest part of the shoot, where growth takes place. The root tip is the youngest part of the root, where growth takes place. The arrows indicate the transport of food from the cotyledons to the growing points.*

② **Oxygen**: This is needed for respiration (energy release) by the growing seedling.

③ **Suitable temperature**: This is needed for efficient functioning of enzymes in the embryo.

Note: Light is not normally needed for germination, but it is needed for the formation of chlorophyll and for photosynthesis by the seedling.

How can you find out which conditions are needed for germination?

Place some seeds of, e.g. cress on filter paper or cotton wool in a series of dishes or test tubes. Provide one of them (the control) with all the conditions needed for germination. Provide each of the others with all the necessary conditions except the one you wish to investigate.

Dispersal of seeds

Seed dispersal is the process by which seeds are moved from the parent plant to another location. Fruits play an important part in the dispersal of seeds.

For methods of dispersal, and the reasons why dispersal is important, see page 105.

How do humans make use of seeds?

- Peas and beans contain starch and are used as a general food.
- Wheat grain is used as a source of starch for making flour. (The seed coat and fruit wall (bran) provide fibre in wholemeal flour; white flour is made from the endosperm (starch).
- Barley grain is used as a source of malt sugar (maltose) for making beer.
- Rape seeds are used as a source of oil for making margarine.

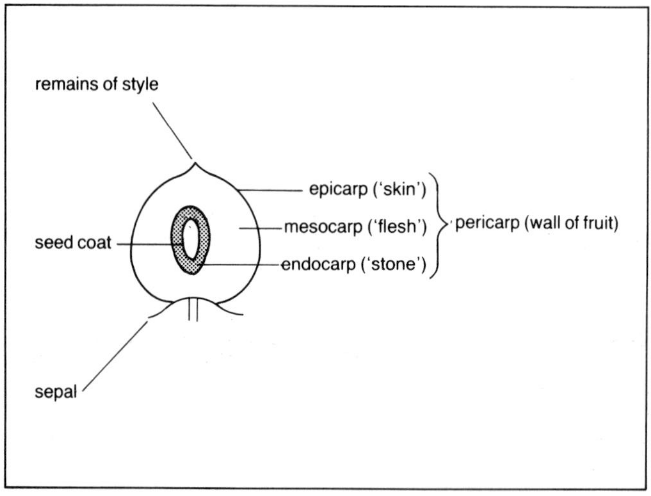

Figure 25.5 *Section through a generalised fruit based on plum.*

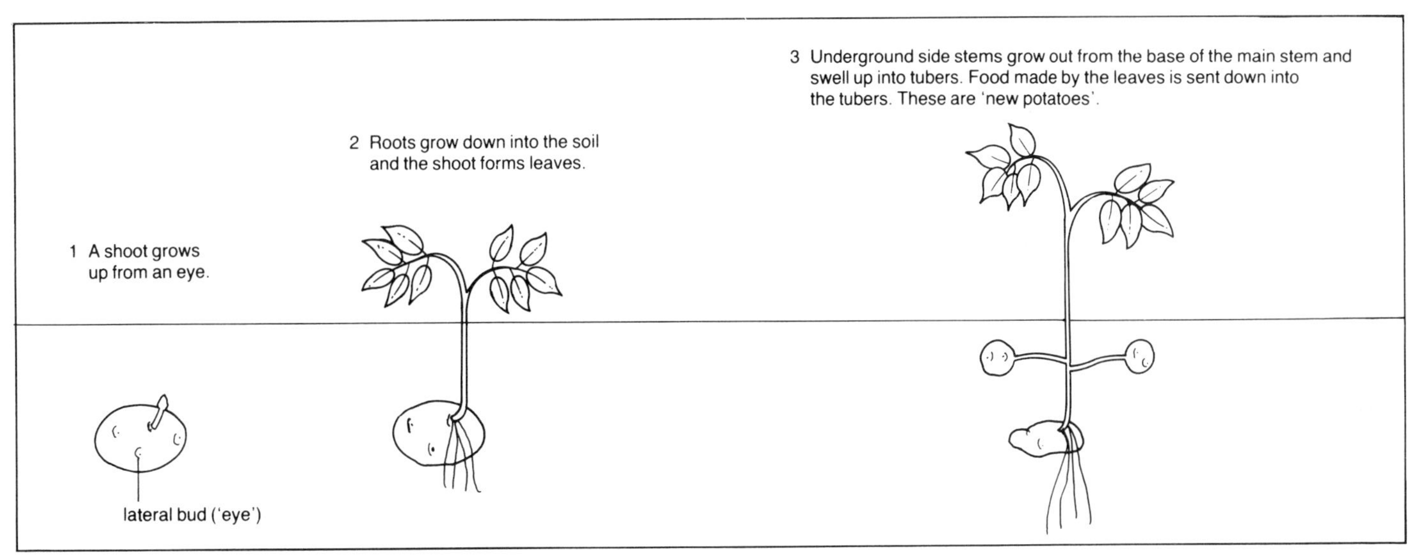

Figure 25.6 *How a potato plant reproduces by means of stem tubers.*

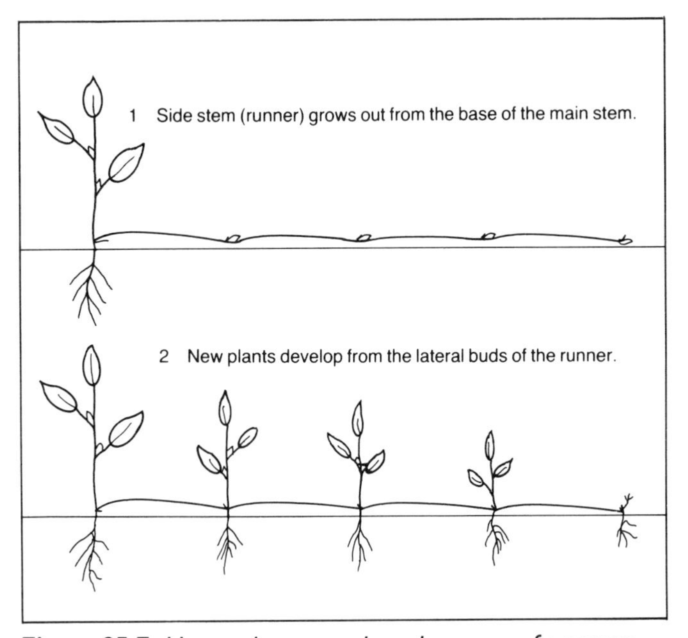

Figure 25.7 *How a plant reproduces by means of a runner.*

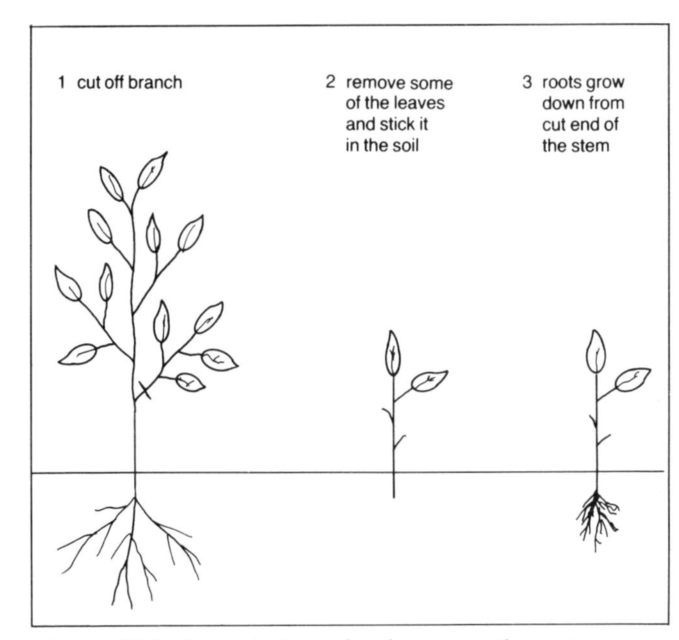

Figure 25.8 *Reproducing a plant by means of a cutting.*

Structure of fruits

The basic structure of a generalised fruit is shown in figure 25.5. All fruits have the parts shown in this diagram. However, the details vary, depending on how the fruits and/or seeds are dispersed.

Asexual reproduction in flowering plants

Methods of asexual (vegetative) reproduction in flowering plants fall into two groups:

① By **perennating organs** (see page 105), e.g. potato tubers (figure 25.6).

② By **growth from part of the plant**, e.g. strawberry runners (figure 25.7).

As these are *asexual* methods of reproduction, all the offspring have the same genetic constitution as the parent plant. They belong to the same **clone**.

Human's use of vegetative reproduction

Horticulturalists reproduce (propagate) plants by asexual means, e.g. grafting and taking cuttings (figure 25.8). As all the offspring are genetically, identical, useful features can be perpetuated (see page 104). The process of producing a population of genetically identical individuals is called **cloning**.

• Questions

1 In the following list, which structures belong to the male part of a flower, which structures belong to the female part, and which structures belong to neither the male nor the female parts?
(a) filament; (b) petal; (c) receptacle;
(d) sepal; (e) stigma.

2 (a) The following are features of an insect-pollinated flower. Explain the reason for each feature:
(i) coloured petals;
(ii) produce nectar;
(iii) wall of pollen grain has spikes.

(b) The following are features of a wind-pollinated flower. Explain the reason for each feature:
(i) long pendulous stamens;
(ii) feathery stigma;
(iii) very small pollen grains.

3 Look at figure 25.5 and note that the wall of a fruit consists of three layers: epicarp, mesocarp and endocarp. What can you say about the *possible* features of:

(a) the epicarp of a fruit which is eaten by birds;

(b) the mesocarp of a fruit which floats in water;

(c) the epicarp of a fruit which is blown about by the wind?

4 The diagram below shows the external appearance of a mung-bean seed.

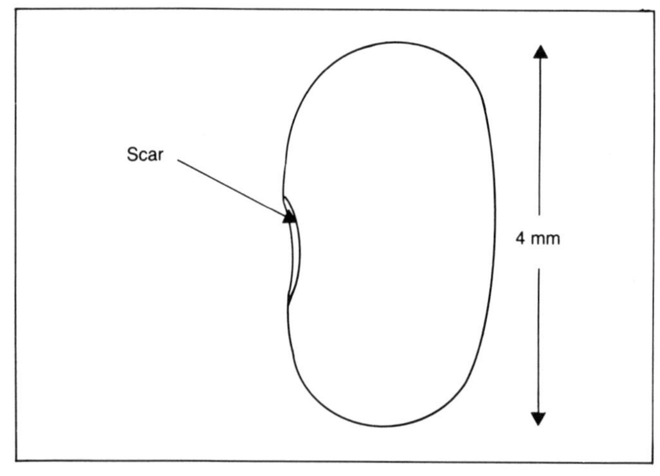

(a) (i) From what part of the flower has the seed developed?
(ii) What does the scar represent?

(b) You are asked to find the best temperature for germinating mung-bean seeds. You are given 100 seeds, 10 test tubes, cotton wood, 10 thermometers, a clock and a supply of water.
(i) Draw and label one of the test tubes you would set up for the experiment.
(ii) Explain how you would carry out the experiment.
(iii) Describe *one* difficulty you might have in carrying out the experiment.

(LEAG)

5 Two batches of vegetable seeds were stored in two separate rooms A and B. The temperature in each room was kept constant at 10°C.

The humidity in room A was kept at 25% whereas the humidity in room B was kept at 75%.

At the time intervals given in the table below, 200 seeds from each batch were removed from the rooms and placed in conditions ideal for germination. The percentage of seeds which germinated was recorded. The results are summarised in the table.

Years of storage	Percentage germination of seeds stored in room A	Percentage germination of seeds stored in room B
0.5	56.0	52.0
1	52.0	45.0
2	49.0	34.0
3	47.0	23.0
4	46.0	15.0
5	45.5	9.0

(a) Plot these results on graph paper, joining the points with straight lines.

(b) From your graph find
(i) the percentage germination of seeds stored for 3.5 years at 75% humidity.
(ii) the maximum length of time seeds could be stored at 25% humidity to ensure 50% germination.

(c) Suggest why seeds survived better in room A than in room B.

(d) Suggest how a manufacturer might best package these seeds for sale in shops.

(NEA)

26 Growth and development

What is growth?

Growth is the permanent increase in size which takes place as an organism develops.

How can growth be measured?

There are three main ways of measuring the growth of an organism. Each has certain disadvantages.

1. Measure a **linear dimension**, e.g. height, at regular intervals of time.
   Disadvantage: takes no account of changes in shape.
2. Measure **fresh mass** (by weighing the organism) at regular intervals of time.
   Disadvantage: takes no account of changes in the fluid content of the body.
3. Measure **dry mass** at regular intervals of time. The dry mass is the mass of the organism after all moisture has been driven off by heating.
   Disadvantage: the organism is killed by the drying process, so a different individual has to be used for each measurement.

How is the dry mass method carried out?

This method is mainly used with plants. Sow a large number of seeds of a particular species at the same time and in the same conditions. Ideally the plants should all be genetically identical or as similar as possible (see page 104). Take samples (say five plants) at regular intervals of time and measure their dry mass as explained in the previous section.

Why can measuring the growth of the whole organism by misleading?

It takes no account of the fact that different parts of the organism may grow at different rates and may stop growing at different times.

What is a growth curve?

A **growth curve** is the line obtained on a graph when size measurements such as height or mass are plotted against time.

What do most growth curves look like?

Most organisms have growth curves which look like the one in figure 26.1. The main thing to notice is that the curve is *smooth*. This shows that the organism increases in size gradually and steadily.

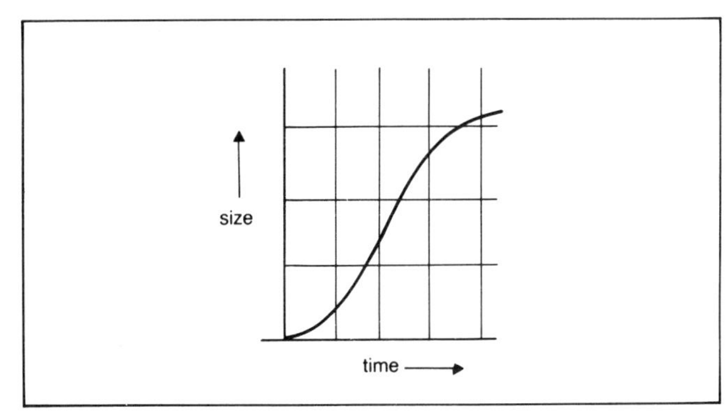

Figure 26.1 *Growth curve typical of most organisms.*

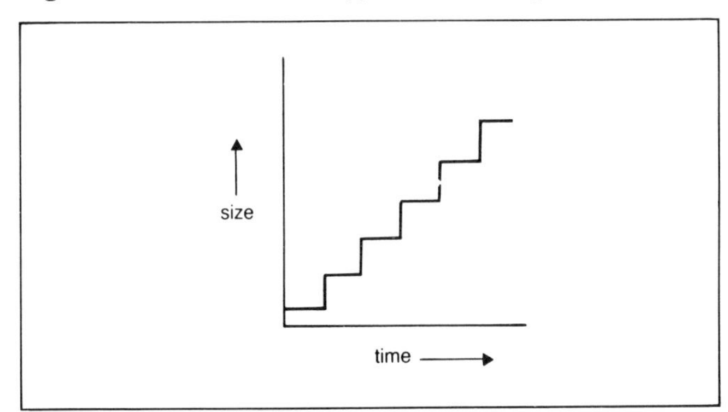

Figure 26.2 *Growth curve typical of an arthropod.*

The growth curve of arthropods

Arthropods (e.g. insects) have a growth curve which is *stepped* (figure 26.2). This shows that the animal increases in size intermittently, i.e. in spurts.

Why do arthropods have a stepped growth curve?

Arthropods have a hard cuticle (exoskeleton) which cannot stretch. The animal can only increase in size immediately after it has moulted, i.e. shed its cuticle. At this stage the new cuticle is soft and stretchable. After being stretched, the new cuticle hardens and no further increase in size takes place until the next time the animal moults.

What is meant by an organism's growth rate?

An organism's **growth rate** is its increase in size (i.e. height, mass etc.) per unit time.

What is percentage growth rate?

This is the organism's increase in size over a given period of time, expressed as a percentage of the size at the beginning of the period.

$$\text{percentage increase in size} = \frac{\text{final size} - \text{initial size}}{\text{initial size}} \times 100$$

The percentage growth rate can give useful information on the way an organism's growth rate changes as it gets older.

What factors affect the growth rate?

An organism's growth rate (and its final size) may be affected by the following factors:

1. **Oxygen supply**. (Oxygen is needed for respiration on which growth depends.)
2. **Supply of nutrients** (see pages 50 and 71).
3. **Temperature**. (The growth rate is increased by raising the temperature so long as it does not exceed about 40°C, the temperature at which enzymes are destroyed.)
4. **Genetic constitution** (i.e. the genes which the organism inherits from its parents.)
5. **Hormones** (see below).
6. **Light**. (This is only relevant to plants.)

What is the effect of lack of light on plant growth?

A growing plant deprived of light becomes **etiolated**. An etiolated plant has these features:

- It is tall and spindly. Darkness causes the plant to grow more rapidly than usual.
- Its leaves and stem are yellow instead of green. Light is needed for the plant to make chlorophyll.
- The leaves fail to expand.

How does growth take place?

In organisms generally, growth takes place by:

- **cell division**: this increases the number of cells in the organism;
- **cell expansion**: this increases the size of the cells in the organism.

Differences between animal and plant growth

Table 26.1 summarises the main differences between the growth of a typical animal and plant. Although animals stop growing at a certain age, cell division still continues in the bone marrow and epidermis, and in places where damaged tissues are healing.

Table 26.1 *The main differences between animal and plant growth.*

Animals, e.g. human	Plants, e.g. geranium
Growth occurs mainly by cell division.	Growth occurs by cell division and cell expansion.
Growth occurs all over the body.	Growth is restricted to certain regions (meristems).
Growth ceases when the organism reaches a certain age.	Growth continues throughout life.

How does cell expansion take place in plants?

The young cell acquires a vacuole into which water passes by osmosis. At this stage the cell wall is still thin; as a result the cell expands. Subsequent thickening of the cell wall prevents further expansion.

Different plant cells acquire different shapes as they expand. Some become spherical (forming packing cells), others become elongated (forming, e.g. xylem vessels).

Where does growth occur in a flowering plant?

In all flowering plants, **primary growth** takes place at the tips of shoots and roots. Here cell division and expansion produce **primary tissues** which increase the length of the stem or root.

In woody perennials (trees and shrubs), **secondary growth** takes place within the stems and roots. Here cell division and expansion produce **secondary tissues** (mostly vascular) which increase the width of the stem or root. At the same time **corky cells** are formed at the surface of the stem and root: they form the hard part of the bark.

Meristems

A **meristem** is a region of the plant where active cell division, and therefore growth, is taking place. Flowering plants have the following meristems:

1. **Primary meristem** in the tip of the shoot and root where primary growth takes place.
2. **Secondary meristems** in the stem and root where secondary growth takes place. These contain the:
   (a) **Vascular cambium**: the layer of cells which produces secondary vascular tissues.
   (b) **Cork cambium**: the layer of cells which produces cork.

Growth of the shoot and root

If you look at a longitudinal section of a shoot or root under a microscope, you will be able to recognise three zones. These zones are, from the tip backwards:

- **Zone of cell division**: the cells are actively dividing.
- **Zone of cell expansion**: the cells are expanding.
- **Zone of cell differentiation**: the cells are becoming specialised.

How is growth controlled?

Growth is controlled by a combination of external and internal factors:

- **External factors** include light.
- **Internal factors** include hormones.
  (Hormones are defined on page 97.)

Hormones controlling growth in plants

Several hormones control growth in plants. One of the most important is **auxin**. Auxin is produced at the tip of the shoot. It then passes down the shoot, causing it to grow.

What is the evidence that a hormone from the tip causes the shoot to grow?

One of the most conclusive experiments is illustrated in figure 26.3.

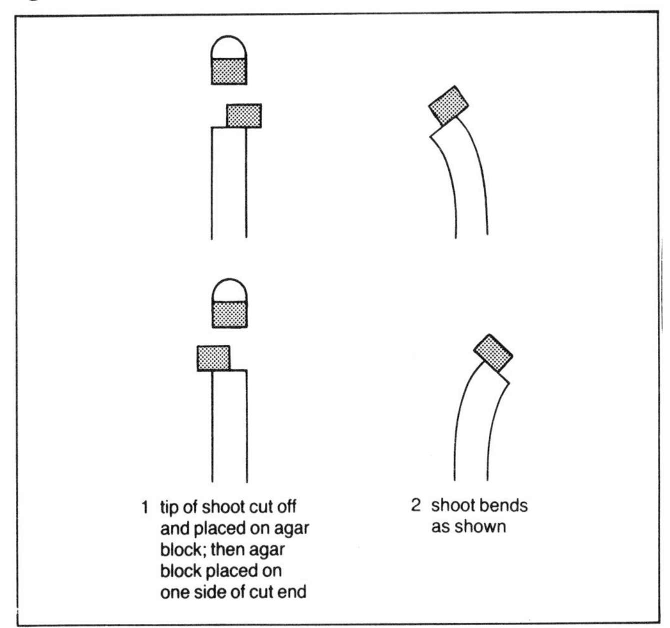

Figure 26.3 *Experiment providing evidence that a hormone from the tip of a shoot causes the shoot to grow. Experiments of this sort are often carried out with coleoptiles (see page 116).*

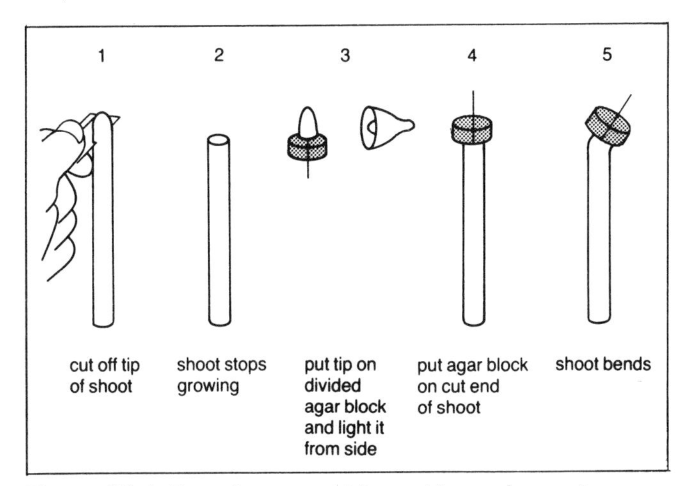

Figure 26.4 *Experiment providing evidence that auxin accumulates on the darker side of a shoot which is illuminated from one side.*

What else does auxin do?

- Auxin prevents lateral buds developing into side branches. (In this case auxin is inhibiting growth.)
- Auxin stimulates the growth of adventitious roots (roots which grow out of stems.)
- Auxin enables shoots and roots to respond to stimuli such as light and gravity. A growth response is called a **tropism**.

What happens if you illuminate a shoot from one side?

The shoot bends towards the light, i.e. it is **positively phototropic**.

How can we explain the shoot's response to light in terms of auxin?

Light causes auxin to accumulate on the darker side of the shoot, with the result that the shoot grows faster on that side.

What is the evidence that auxin accumulates on the darker side of the shoot?

One of the most conclusive experiments is illustrated in figure 26.4.

How do the shoot and root respond to gravity?

Suppose you lay a young bean seedling on its side so that the shoot and root are horizontal. What happens? The shoot bends upwards away from gravity (i.e. it is **negatively gravitropic**), whereas the root bends downwards towards gravity (i.e. it is **positively gravitropic**).[1]

[1] Another word for gravitropic is *geotropic*. Gravitropic is becoming the more widely used term nowadays.

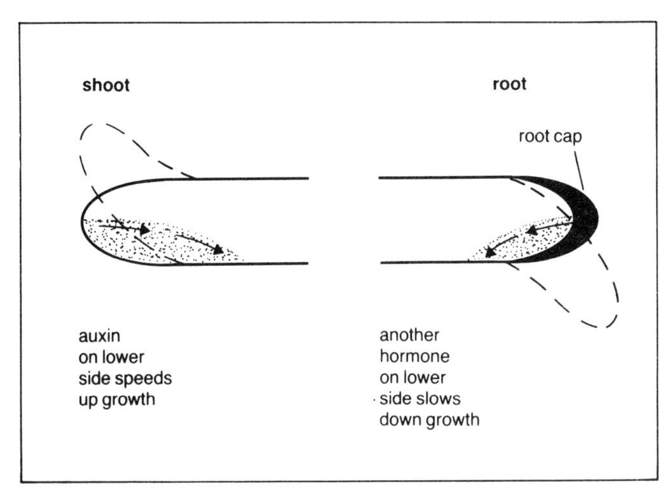

Figure 26.5 *The mechanism by which the shoot and root are thought to respond to gravity.*

How can we explain the response to gravity?

An explanation is given in figure 26.5. It is not the only possible explanation but it is the one for which there is most evidence.

What use can humans make of auxin?

Auxin-like substances are manufactured commercially. They are called plant growth substances. Here are a few of their uses:

- They help to make cuttings take. (The cut end of the stem is dipped into the growth substance which stimulates the growth of roots.)
- They are used as hormone weedkillers. (If applied to a lawn, the weedkiller causes excessive growth of broad-leaved plants, resulting in death.)
- They can cause fruits to develop without fertilisation.

Hormones controlling growth in humans

In humans growth is controlled by two main hormones:

- **Growth hormone** produced by the **pituitary gland**. Under-secretion of growth hormone during the growing period causes **dwarfism**. Over-secretion causes **gigantism** (the person becomes a 'giant'.)
- **Thyroxine** produced by the **thyroid gland**. Under-secretion of thyroxine during the growing period causes **cretinism** (stunted growth, mental retardation).

What is differentiation?

Differentiation is the process in which cells acquire certain structural features which enable them to carry out specific functions. Differentiation takes place early in development and it results in **cell specialisation** (see page 32).

What is dormancy?

A dormant organism is in an inactive state and has a very low metabolic rate. Dormancy may occur in reproductive products such as spores and seeds (see page 105), or in adult organisms. It enables the organism to survive an unfavourable period.

Hibernation is an example of dormancy. It enables animals in temperate regions to get through the winter, e.g. dormouse, hedgehog, amphibians and reptiles.

What is a larva?

A **larva** is a free-living, self-supporting organism into which an embryo may develop. The larva then develops into the adult. *Examples*: amphibians (the larva is the tadpole), butterflies (the larva is the caterpillar).

What are the functions of larvae?

Larvae have two main functions:

- To help disperse the species. (This is particularly important in sedentary or slow-moving animals such as mussels and clams.)
- To build up a store of food to enable development to continue to the adult. (This is particularly true of insect larvae such as caterpillars.)

How does a larva develop into the adult?

A larva develops into the adult by undergoing a total change of form: this is called **metamorphosis**. The change from tadpole to adult frog is an example of metamorphosis.

Metamorphosis also occurs in insects. Insects have two types of metamorphosis:

① **Complete metamorphosis**
egg → larva → pupa (chrysalis) → adult

② **Incomplete metamorphosis**
egg → nymph (several moults) → adult

The moults which take place in the nymph give rise to the stepped growth described on page 119. The nymphal stages are called **instars**.

• Questions

1 Suggest an explanation for each of the following:

(a) In estimating a plant's growth rate it is better to measure dry mass than wet (fresh) mass.

(b) Plant roots show poor growth in waterlogged soil.

(c) Cutting off the apex of a young cypress tree makes the tree bushier.

(d) If you plant a broad bean seed upside down, the shoot still grows upwards.

(e) Seeds will only germinate after they have absorbed water.

2 The graph below shows the increase in body mass of an aquatic insect over a period of nearly 70 days. The sudden increases in mass, indicated by the arrows, correspond to moulting of the cuticle. When moulting occurs the insect swallows water, thereby expanding the new cuticle before it hardens.

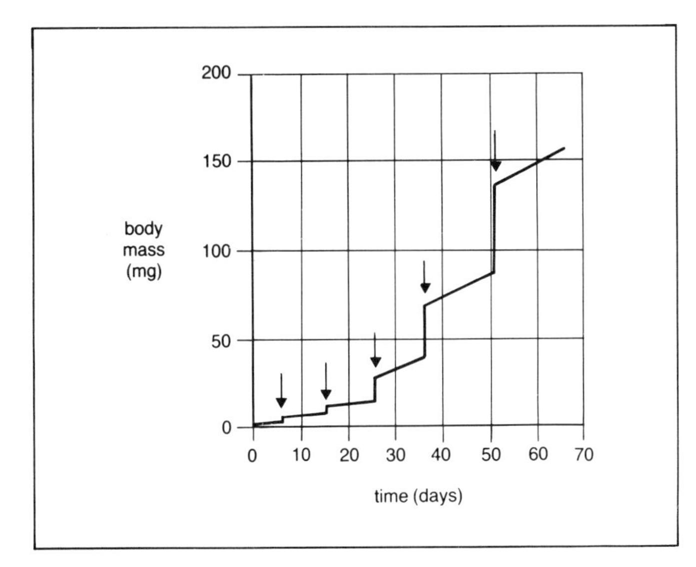

(a) What causes the mass to increase rapidly when the insect moults?

(b) What causes the mass to increase slowly between one moulting and the next?

(c) (i) If the insect's increase in length, rather than mass, was plotted, how would the graph differ in appearance from the one shown here?
(ii) Account for the difference.

(d) How would the growth curve for a human differ in appearance from the one shown here?

3 The graph below shows the average heights of English boys of different ages in 1874 and 1958.

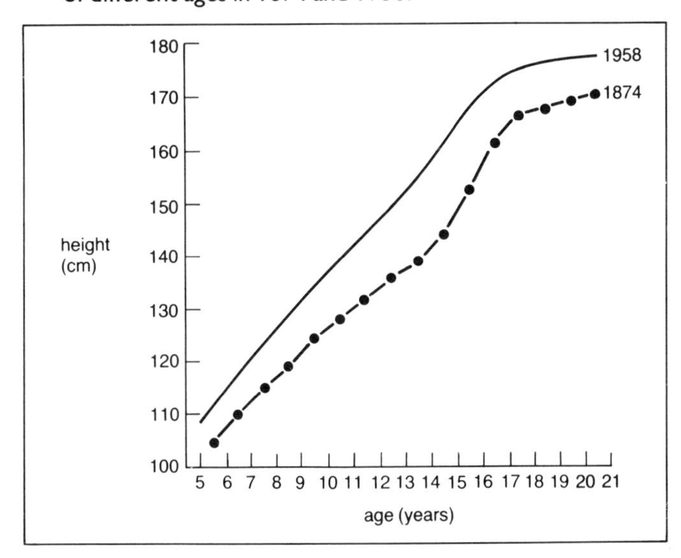

(a) What is the difference in height between the two groups at the ages of 8, 13½ and 17 years?

(b) Suggest *one* reason why boys of a given age should be taller in 1958 than in 1874.

(c) Suggest *one* reason why the height difference is particularly great between the ages of 13 and 15.

(d) Suggest *two* factors, besides the one mentioned in your answer to (b), which might influence the growth rate of an *individual* boy or girl today.

4 The illustration below shows an experiment which was performed to find out where growth takes place in the root of a broad bean seedling.

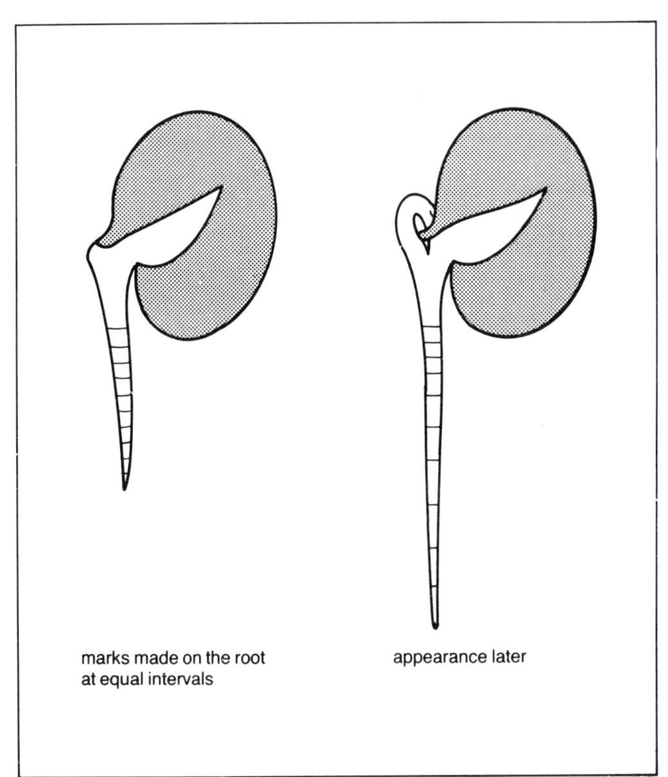

(a) What conclusion do you draw as to where growth takes place?

(b) What is the percentage increase in the length of the root between the first and second diagrams?

(c) How does growth take place within the root?

(d) From what structures does the root derive nourishment during its growth?

(e) The right hand diagram shows the shoot beginning to emerge.
(i) Why does the shoot emerge after the root has done so?
(ii) Why is the shoot bent back on itself?

5 The diagrams here show details of an experiment which was carried out by a student on oat coleoptiles. The student was trying to test the hypothesis that the stimulus of light is perceived by the tip of the coleoptile rather than by the part just behind the tip (the part that bends).

(a) Does the result of the experiment support the student's hypothesis?

(b) The experiment should be accompanied by a control. What should the control be, and why is it needed?

(c) Suggest a further experiment which the student could do to show that the light stimulus is not perceived by the part of the coleoptile behind the tip.

(d) Explain briefly how the bending response to light takes place.

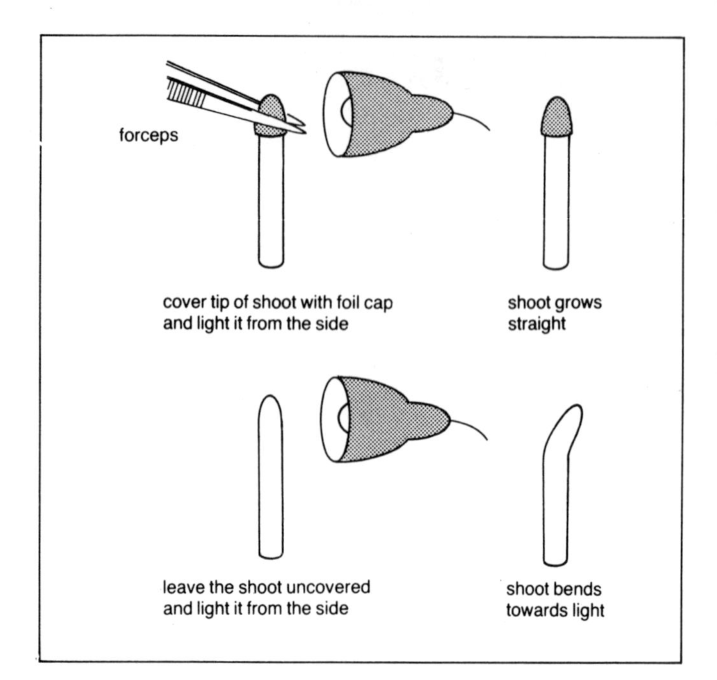

Chromosomes and genes

What are chromosomes?

Chromosomes are thread-like bodies which occur in the nucleus of a cell. They contain **genes**.

A gene is a structure within a chromosome which determines an individual's characteristics. Characteristics are transmitted from parents to offspring via the genes.

Chromosomes exist in pairs. The chromosomes belonging to a pair are called **homologous chromosomes**. Homologous chromosomes look exactly alike.

When a cell divides the chromosomes and their genes are shared out between the daughter cells in a precise and organised way.

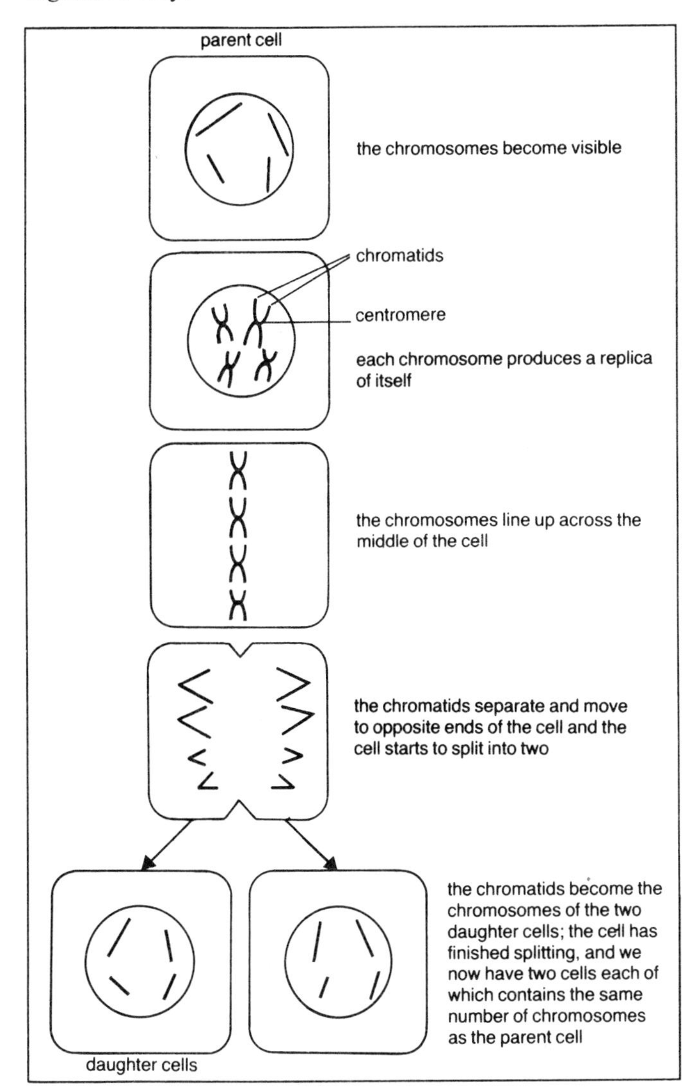

Figure 27.1 *What the chromosomes do during mitosis.*

Cell division

There are two types of cell division:

① **Mitosis**: The chromosomes behave in such a way that the daughter cells contain the same number and types of chromosomes as the parent cell (figure 27.1).

② **Meiosis**: The cell divides in such a way that the daughter cells contain *half* the number of chromosomes as the parent cell, i.e. one chromosome of each homologous pair (figure 27.2).

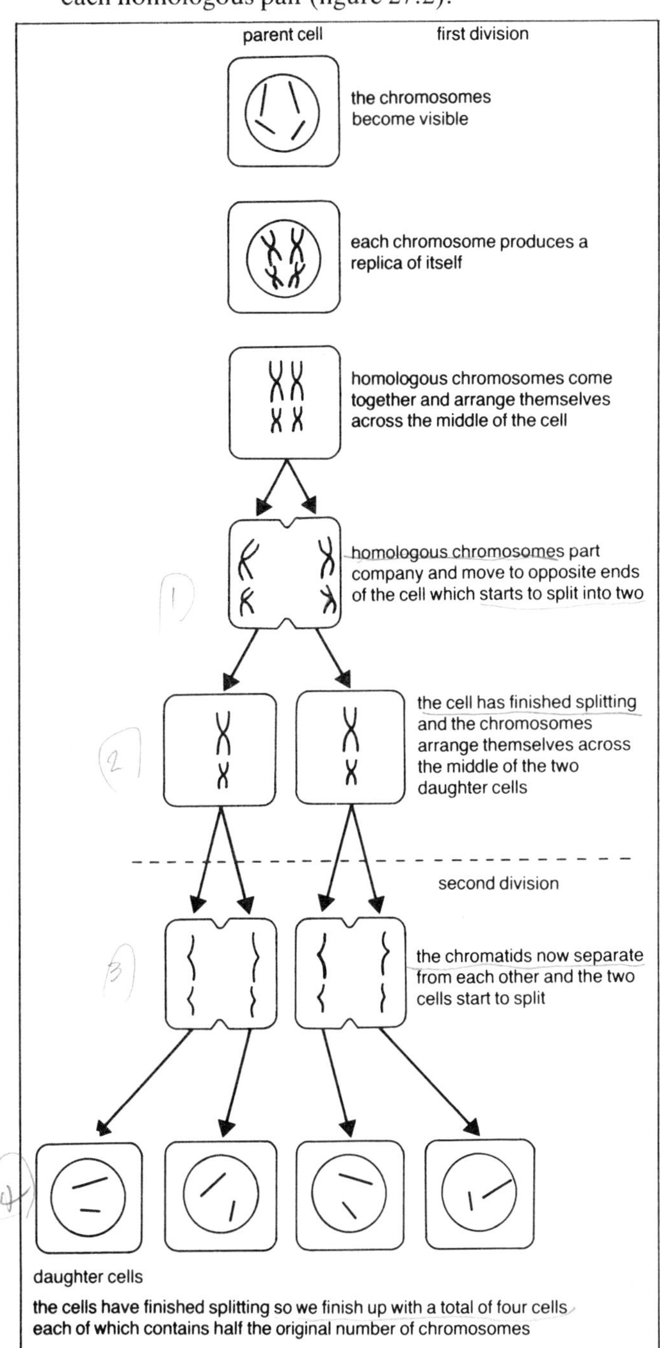

Figure 27.2 *What the chromosomes do during meiosis.*

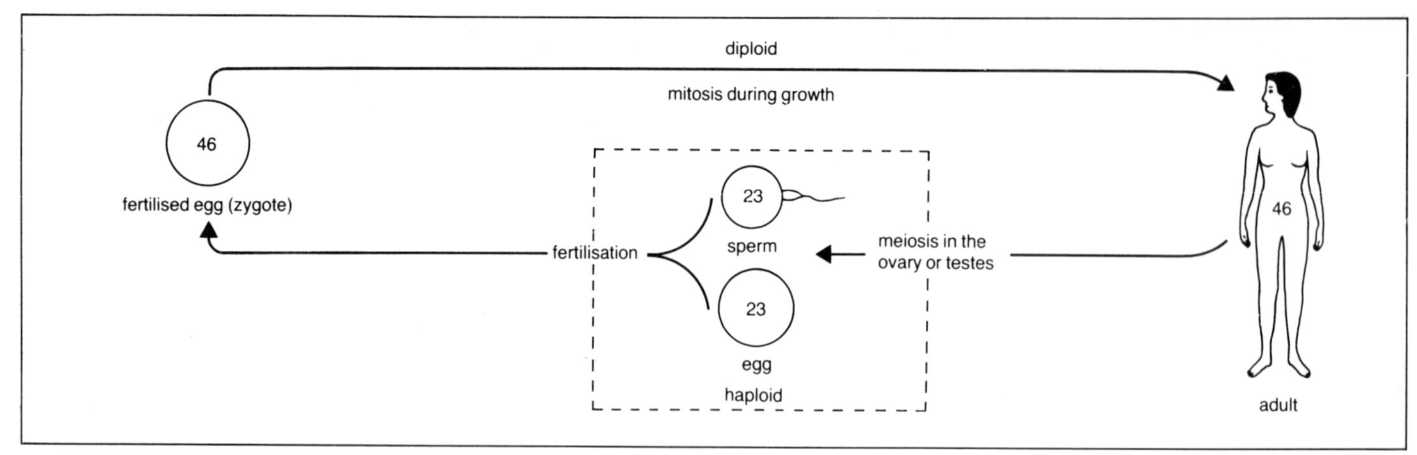

Figure 27.3 *The human life cycle.*

Diploid and haploid

A **diploid** cell contains the full number of chromosomes, i.e. both chromosomes of each homologous pair.

A **haploid** cell contains half the number of chromosomes, i.e. one of each homologous pair.

Meiosis in the human life cycle

In organisms such as the human, meiosis occurs during the formation of eggs and sperms. The eggs and sperms are therefore haploid. When fertilisation occurs the diploid condition is restored. The fertilised egg then divides by mitosis so the cells of the adult are diploid (figure 27.3).

How is an individual's sex determined?

Sex is determined by the sex chromosomes: **X** and **Y**. The **X** and **Y** chromosomes are homologous, but unlike other homologous chromosomes they differ in size: the **X** chromosome is much longer than the **Y** chromosome.

- A female contains two **X** chromosomes in her cells.
- A male contains an **X** and a **Y** chromosome in his cells.

Figure 27.4 shows how the sex chromosomes are transmitted from parents to offspring.

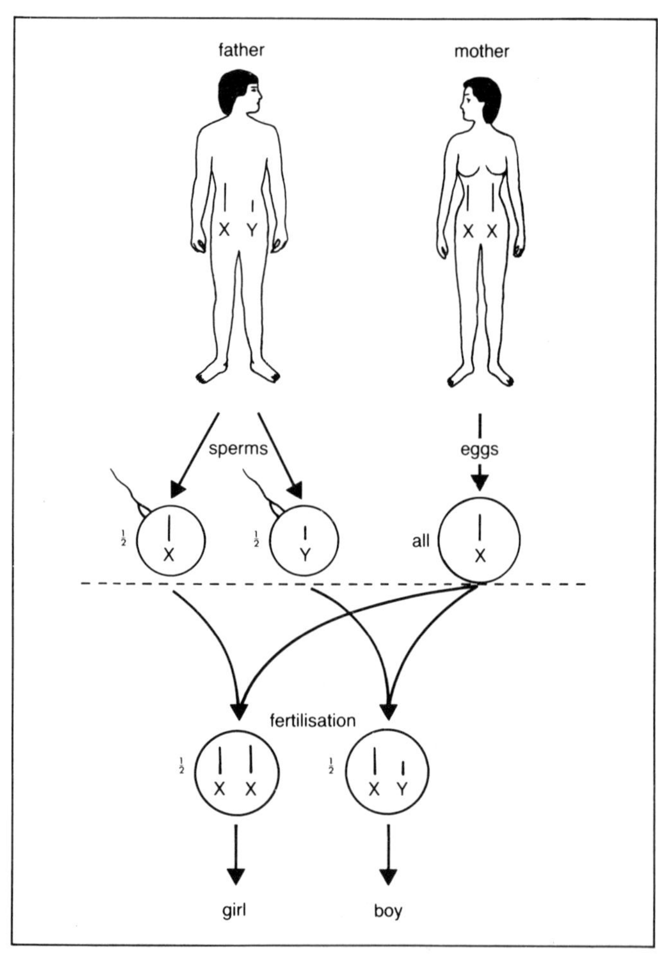

Figure 27.4 *A boy or a girl? It all depends on the sex chromosomes.*

What is DNA?

DNA stands for **deoxyribonucleic acid**. This is the chemical substance which genes are made of.

The DNA molecule is shaped like a twisted ladder (**a double helix**). The rungs of the ladder consist of **organic bases**. There are four organic bases altogether and they are arranged in pairs, one pair for each rung of the ladder.

The genetic code

The DNA molecule contains the **genetic code**. The genetic code is a set of instructions, contained within each gene, which tells the cell what proteins to make. The proteins include enzymes. The enzymes control the cell's chemical reactions and thus determine the organism's characteristics.

So genes determine an organism's characteristics by controlling what proteins, particularly enzymes, they make.

How does DNA control protein synthesis?

A protein is made by joining amino acids together in a particular order (see page 42). This occurs on the **ribosomes** in the cytoplasm (see pages 31–2).

The order in which the amino acids are joined together is determined by the order of base-pairs in the DNA.

Instructions for telling the cell how to assemble the protein is carried from the DNA in the nucleus to the ribosomes in cytoplasm by **ribonucleic acid** (RNA).

This is similar to DNA and carries the same message. The type of RNA involved in this process is called **messenger RNA**.

Genetic engineering

Genetic engineering is the process in which an organism's gene's are altered (by humans) so that its characteristics are changed.

A commonly used type of genetic engineering is to take a specific piece of DNA out of mammalian cells and transfer it to bacteria. The bacteria then produce substances, characteristics of mammalian cells, which may be useful to humans.

Examples of useful substances produced by genetic engineering include **growth hormone** (needed to counteract dwarfism) and **insulin** (needed to counteract diabetes).

• Questions

1 Explain the meaning of the following terms:
(a) Gene (b) Chromosome
(c) Homologous chromosomes.

2 (a) When and where does meiosis occur in the human life cycle?
(b) Why is it essential for meiosis to occur in the life cycle?

3 The diagrams below show the chromosomes in two different cells. In answering the following questions refer to each cell as A or B.
(a) Which cell was formed by mitosis? How can you tell?
(b) (i) Which cell is haploid?
(ii) Explain the meaning of the term haploid.
(c) Which cell is *not* a gamete? How do you know?
(d) Briefly explain the function of chromosomes in a cell.

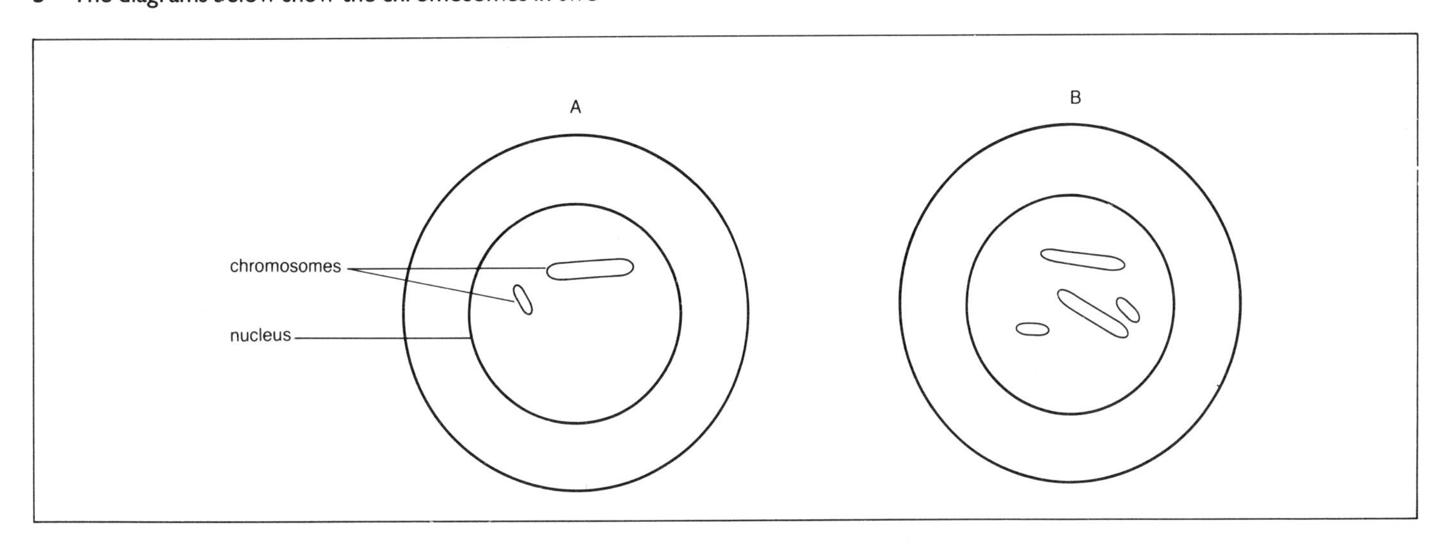

4 The picture below shows a set of chromosomes of a human male. The chromosomes have been arranged in homologous pairs.

(a) How many chromosomes are present altogether?

(b) How many chromosomes would be present in a sperm produced by this person?

(c) How can you tell from the picture that this person is male and not female?

(d) In what way do chromosomes A and B differ in where they came from?

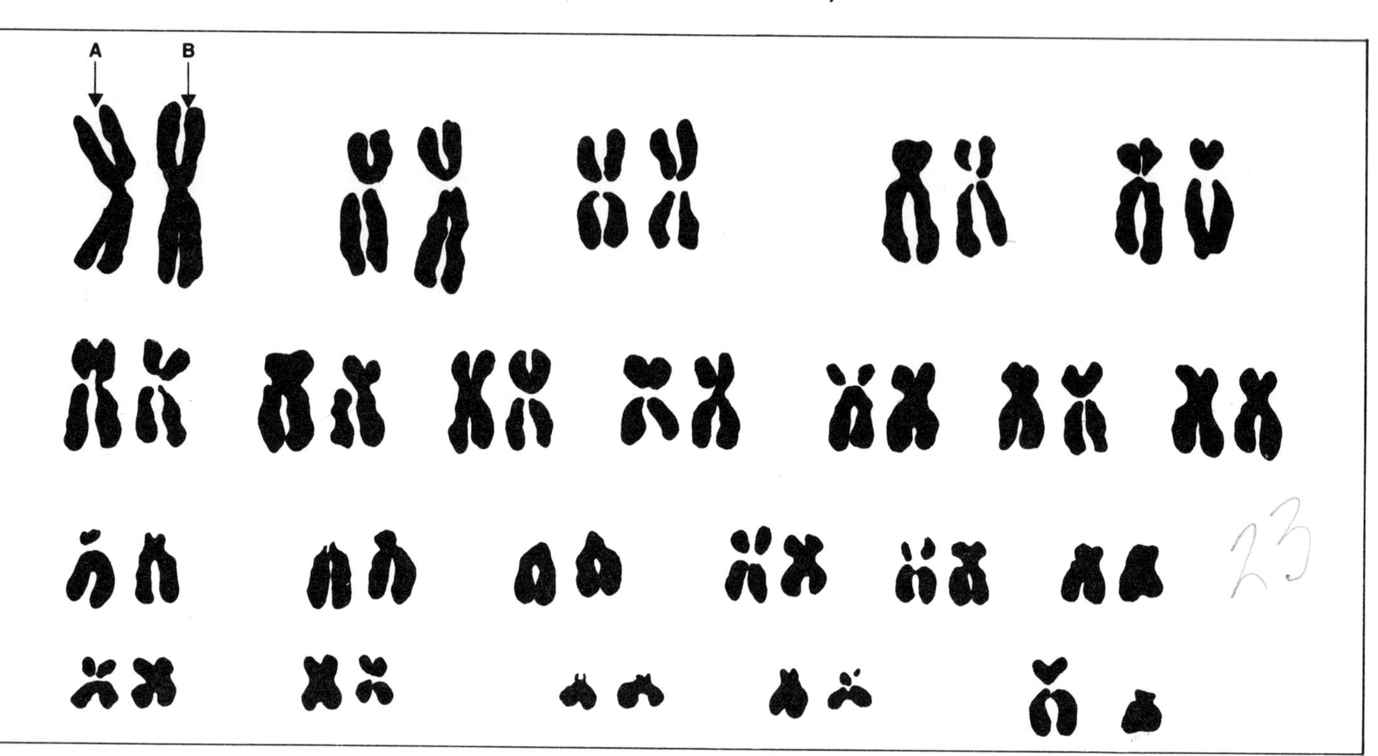

5 The diagram below shows an example of genetic engineering.

(a) What is DNA, and whereabouts does it occur in the cells of higher organisms?

(b) What is meant by genetic engineering?

(c) Where would cell A have come from?

(d) What kind of organism does cell B belong to?

(e) To ensure that large quantities of growth hormone are produced, cell B must do something as soon as it has received the DNA from cell A. What must it do?

(f) Give *one* other example of a useful substance which is produced by genetic engineering.

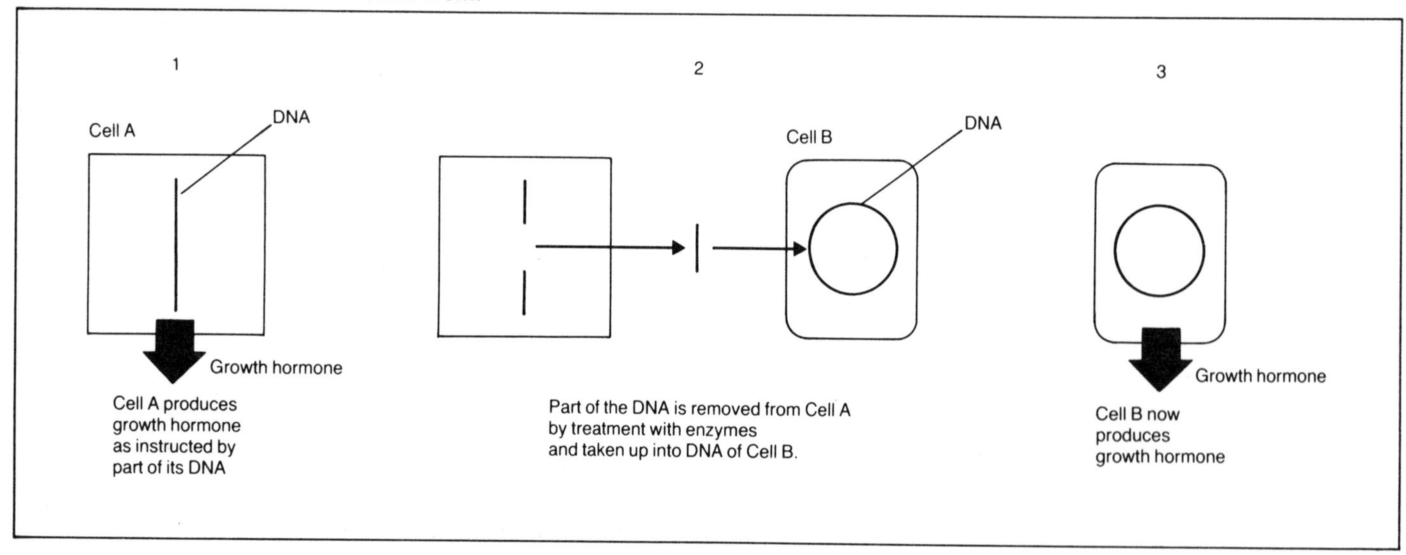

28

Heredity

What is heredity?

Heredity (also known as **genetics**) is the study of how characteristics are passed from parents to offspring.

The principles of heredity were discovered by Gregor Mendel (1822–1884).

The principles of heredity

You should have no difficulty understanding heredity if you know the basic principles and technical terms. Here they are:

1. Characteristics are passed (i.e. transmitted) from parents to offspring via genes which are located in **chromosomes** in the nuclei of cells (see page 125).
2. The gene controlling a particular characteristic can exist in two forms called **alleles**. One of the alleles may be **dominant**, the other **recessive**.
3. Alleles occur in pairs which are located in the same relative position (**locus**) on homologous chromosomes (figure 28.1).
4. An individual may possess two identical alleles for a given characteristic. Such an individual is **homozygous**. If both alleles are dominant, the individual is **homozygous dominant**. If they are recessive, the individual is **homozygous recessive**.
5. An individual may possess two non-identical alleles, one dominant and the other recessive, for a given characteristic. Such an individual is **heterozygous**.
6. The alleles which an individual possesses for a given characteristic (i.e. its genetic constitution) is called the **genotype**. The genotype is shown by letters: a capital letter for the dominant allele, and the corresponding small letter for the recessive allele.
7. The observable characteristics of an individual (i.e. the way the alleles express themselves) is called the **phenotype**.
8. When a dominant and recessive allele are present together (i.e. in a heterozygous individual), only the dominant allele produces an effect (i.e. expresses itself) in the phenotype.
9. A recessive allele will only express itself when it is in the homozygous state, i.e. when the dominant allele is absent.
10. In a gamete only one of a pair of alleles is present. This is because gametes are formed by meiosis in which homologous chromosomes become separated (see page 126).

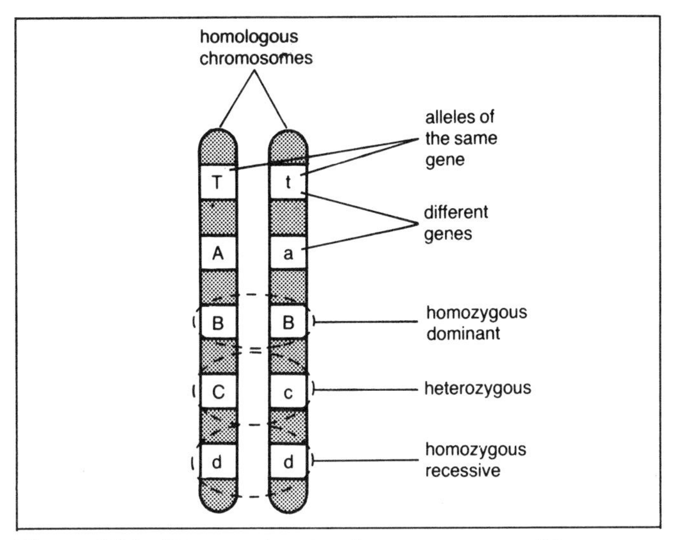

Figure 28.1 *Diagram showing the arrangement of five imaginary genes on a pair of homologous chromosomes.*

Crosses showing complete dominance

Figure 28.2 shows two successive crosses between plants in which red flower is dominant to white flower. Follow

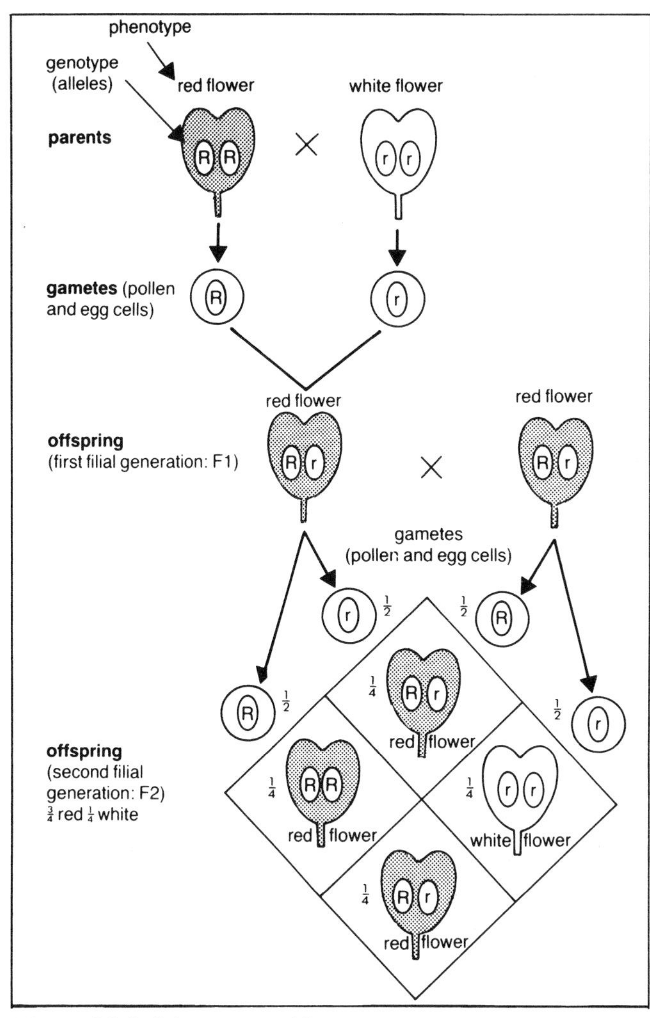

Figure 28.2 *Inheritance of flower colour in a plant.*

the sequence of events step by step. Make sure you understand how the alleles are transmitted from one generation to the next, and how they are expressed in the phenotypes.

Can you deduce an individual's genotype from its phenotype?

In the second generation offspring in figure 28.2, a white flowered plant *must* be homozygous recessive (rr); there is no other possibility (note why).

However, a red flowered plant could be either homozygous dominant (RR) or heterozygous (Rr).

How can you find out if an individual is homozygous dominant or heterozygous?

You must cross it with a homozygous recessive individual (**test cross**). If it is homozygous dominant you will get only homozygous dominant offspring (red flowered in the example above). If it is heterozygous you will get red flowered *and* white flowered offspring.
Note: If the test cross involves crossing the offspring with one of the parents, it is called a **back cross**.

Incomplete dominance

Sometimes an allele is only *partially* dominant over the recessive allele. So, in heterozygous individuals the recessive allele can express itself, at least to a slight extent.

In figure 28.2, if the allele for red flower was only partially dominant, the first generation offspring would be pink instead of red.

Human genetics

Various human characteristics, including certain inherited diseases, are transmitted in a Mendelian manner.

An example is the disease cystic fibrosis. This is caused by a recessive allele:

- Homozygous recessive individuals have the disease.
- Homozygous dominant individuals are completely normal.
- Heterozygous individuals do not have the disease but they can transmit the recessive allele to their children, i.e. they are **carriers**.

Pedigrees

A pedigree is a 'family tree' showing the occurrence of a particular characteristic over two or more generations.

Pedigrees are shown by **pedigree charts** in which males are represented by squares, females by circles, and the characteristic is indicated by filling in the squares and/or circles. Related individuals are joined by lines.

From the phenotypes in the chart the genotypes, or *possible* genotypes, of the various individuals can be worked out.

Human pedigrees are particularly useful to **genetic counsellors**: they advise parents with a history of an inherited disease as to the chances of their children inheriting the disease.

The inheritance of blood groups

First be sure you understand blood groups (see page 66). A person's blood group (**A, B** or **O**) is caused by the presence in the cells of two out of three possible alleles $\mathbf{I^A}$, $\mathbf{I^B}$ and **i** (table 28.1). Alleles $\mathbf{I^A}$ and $\mathbf{I^B}$ show no dominance with respect to each other (i.e. they are **co-dominant**), but each is dominant to **i**.

The alleles are transmitted in a normal Mendelian manner. A person's blood group depends on which particular alleles he or she inherits from the parents (figure 28.3).

Table 28.1 *Blood groups and the alleles responsible for them.*

Groups (phenotypes)	Alleles (genotypes)
A	I^AI^A or I^Ai
B	I^BI^B or I^Bi
AB	I^AI^B
O	ii

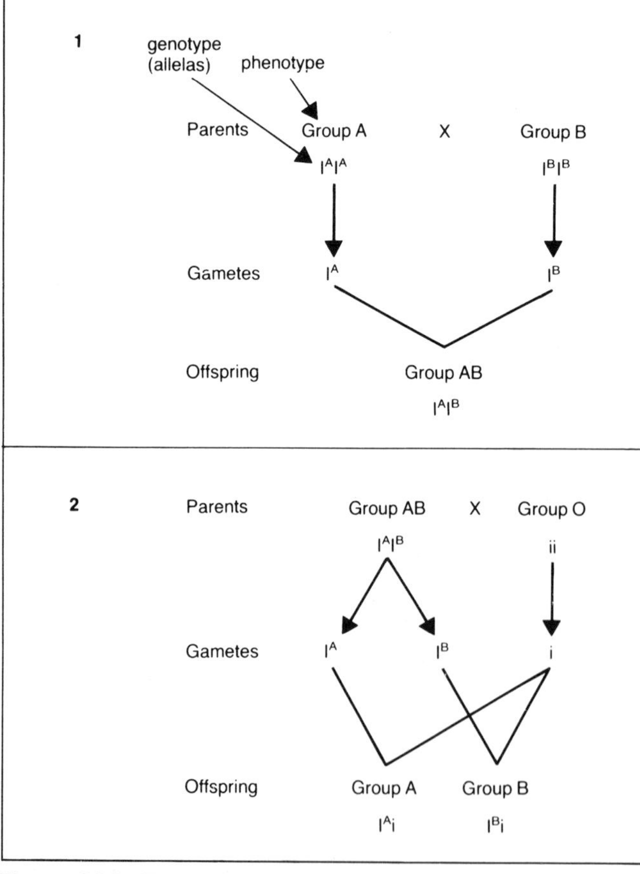

Figure 28.3 *Two examples of crosses to show how people inherit their blood groups from their parents.*

• Questions

1 (a) Look at the example of inheritance shown in figure 28.2.
 (i) A white-flowered plant can have only one possible genotype. What is its genotype and why can it have no other genotype?
 (ii) You can find out the genotype of an F2 red-flowered plant by crossing it with a white-flowered plant. Suggest one other way of finding out the genotype of a red-flowered plant.
 (iii) What would be the phenotypes of the F2 offspring if the allele for red flower was only partially dominant?

(b) Pure-breeding organisms, when crossed with one another, produce offspring all of which are like their parents. Which of the genes (pairs of alleles) in figure 28.1 would show pure-breeding when crossed with a genotypically identical individual?

2 The diagram below shows the offspring of crosses between pure bred Aberdeen Angus bulls, which are black, and pure bred Redpoll cows, which are red. The ratio of the colours of the offspring of the first generation is also shown. Coat colour is controlled by a single gene which has two forms (alleles): one for black and one for red coat colour.

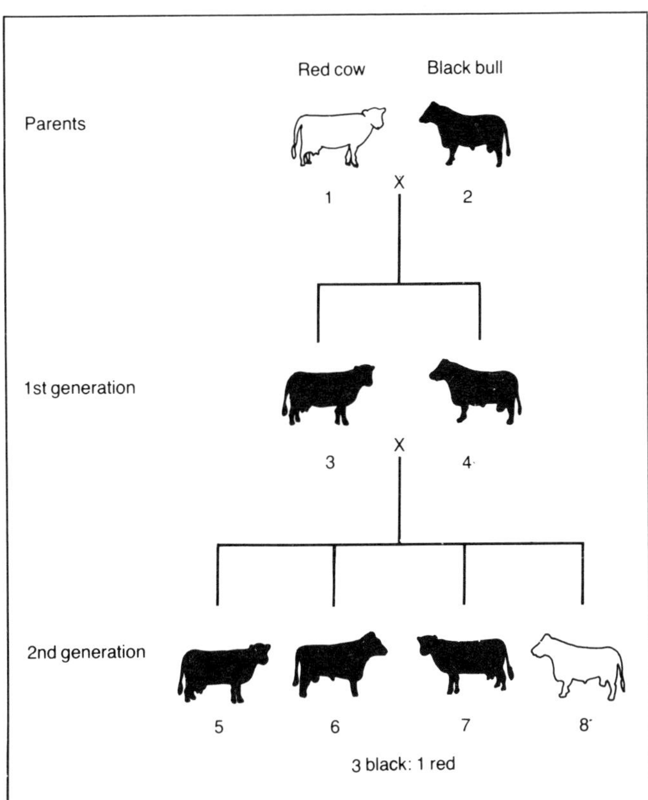

(a) What letters are suitable to represent the two forms (alleles) of the gene?
 (i) black coat colour (ii) red coat colour

(b) (i) Give the number of each animal in the diagram which is definitely homozygous for the gene for coat colour.
 (ii) Give the number of each animal in the diagram which is definitely heterozygous for the gene for coat colour.

(c) Explain why some of the animals in the diagram could be either homozygous or heterozygous for the gene for coat colour.

(LEAG)

3 A gardener has three plants of the same kind:
Plant A has red flowers
Plant B has red flowers
Plant C has yellow flowers
The gardener carried out the crosses shown below.

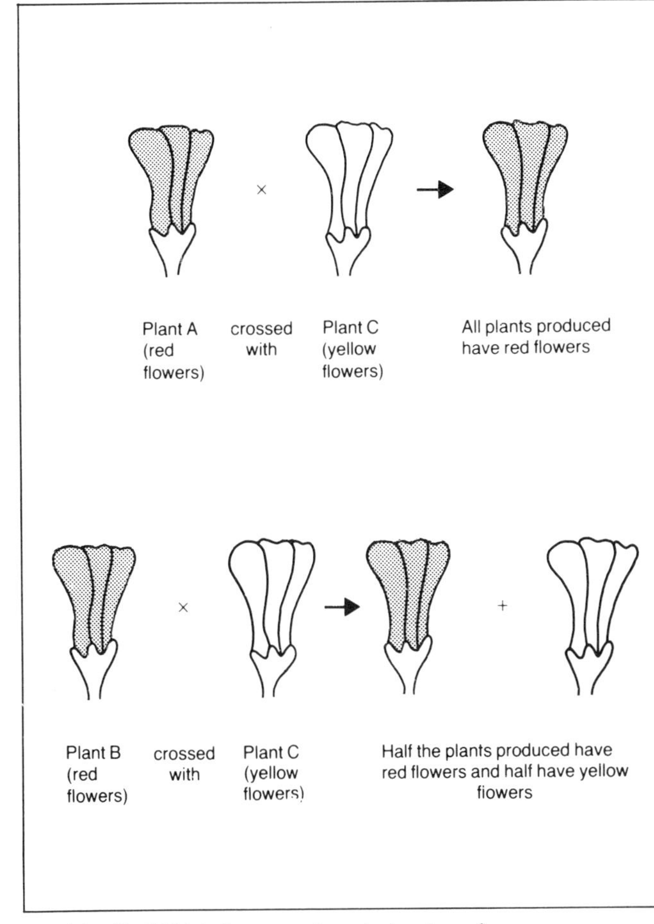

(a) (i) Which flower colour is dominant?
 (ii) Explain the reason for your answer to (i).

(b) Explain what the gardener would do
 (i) to get plants which all have red flowers,
 (ii) to get plants which all have yellow flowers.

(NEA)

4 Albinism is an inherited condition in which pigment fails to be produced in the skin. The diagram below shows a pedigree for albinism in a human family. Squares are males, circles are females; filled-in squares and circles are albinos, open squares and circles are phenotypically normal.

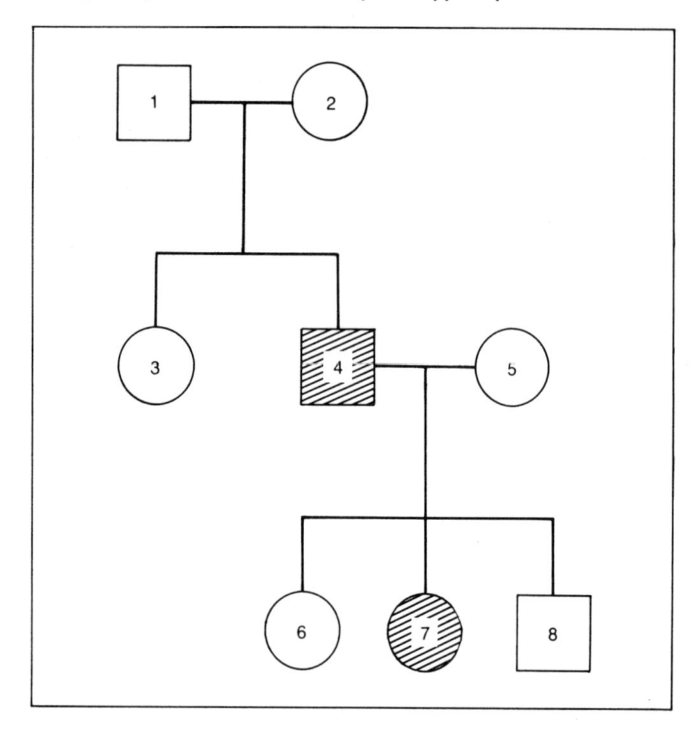

(a) What is meant by the phrase 'phenotypically normal'?

(b) Is albinism caused by a dominant or recessive allele? Give a reason for your answer.

(c) Using the appropriate letters, state the genotypes, or possible genotypes, of individuals 1 to 8.

(d) If individual 8 marries an albino, what is the chance that their first child will be an albino?

5 John belongs to blood group O, but his father belongs to blood group A and his mother to blood group B.

(a) Make a diagram to show how John inherited his blood group from his parents.

(b) What are the possible blood groups that John's brothers and sisters might belong to? Make a diagram to illustrate how they inherited their blood groups.

29

Variation and evolution

What is variation?

Variation is the differences which exist between individuals belonging to the same species.
There are two types of variation:

1. **Continuous variation**: smooth gradation between individuals, e.g. human height, hair colour, intelligence.
2. **Discontinuous variation**: sharp distinction between individuals with no in-betweens, e.g. different strains of fruit fly (*Drosophila*), human blood groups.

How can you show whether variation is continuous or discontinuous?

Measure the feature in a large number of individuals (sample) chosen randomly. Plot the measurements as a histogram (bar chart). If the feature shows continuous variation, it should give a **normal distribution curve** (figure 29.1).

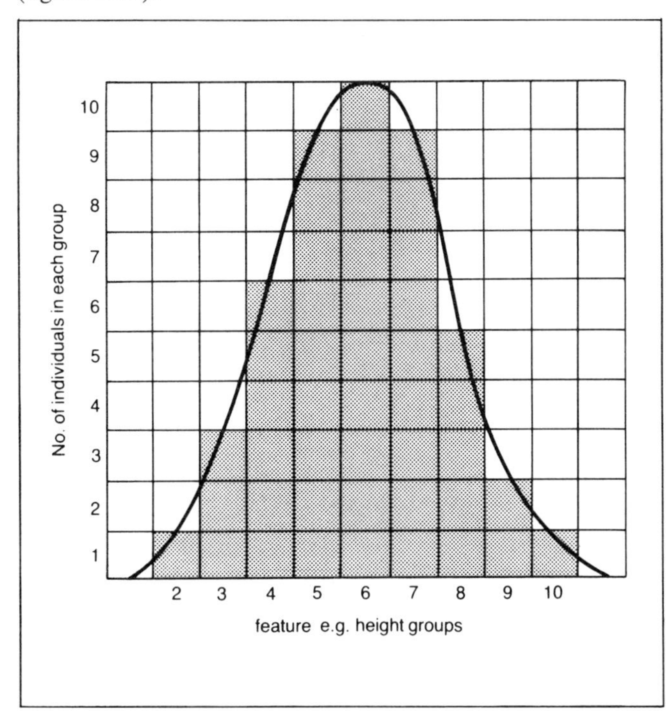

Figure 29.1 *Bar chart showing how a feature such as height varies in a population of individuals. In the case of height, each bar represents the number of individuals falling into a particular group of heights e.g. 120–125 cm etc. The curve formed by joining the tops of the bars is a normal distribution curve.*

What causes continuous variation?

Continuous variation is caused by reshuffling of genes which occurs when parents produce offspring. Gene-reshuffling is caused by:

1. **Free assortment of genes** during gamete-formation. This is due to the random manner in which homologous chromosomes come together and subsequently separate during meiosis.
2. **Crossing over of genes** during gamete-formation. This occurs when homologous chromosomes come together and become intertwined during meiosis.
3. **Fertilisation**, which is random. Individual gametes are genetically different and fertilisation brings different sets of genes together.

As a result of the above, every individual has a unique set of genes. The only exceptions are identical twins and offspring formed by asexual reproduction: they are genetically identical.

What causes discontinuous variation?

Discontinuous variation is caused by **mutation**. A mutation is a sudden change in the genetic constitution of an organism. Mutations occur during the formation of eggs and sperms.

Different types of mutation

There are two types of mutation:

1. **Chromosome mutation**: This is a change in the number or size of chromosomes in the cells of the organism. It is caused by abnormal behaviour of chromosomes during meiosis. *Example*: Down's syndrome caused by the presence of a single extra chromosome.
2. **Gene mutation**: This is a change in the order of bases in the DNA within a gene. It usually occurs in the gametes, so all the cells of the offspring may be affected. *Examples*: sickle-cell anaemia, caused by a change in *one* pair of bases in the gene which codes for haemoglobin.

What causes mutations?

Mutations occur spontaneously all the time. The *rate* is increased by various environmental factors, e.g. radioactivity, ultra-violet light and cyclamates. (Cyclamates are artificial sweeteners, now banned).

Harmful and useful mutations

Most mutations are harmful. Some, however, may be useful. A useful mutation gives an individual a beneficial characteristic which helps it to survive.

An example of a useful mutation is the dark form of the peppered moth. In industrial areas this moth is camouflaged against sooted tree-trunks. However, the white form is conspicuous, making it more liable to be eaten by birds.

Natural selection

Natural selection is the process by which individuals which are well adapted to their environment survive, whereas poorly adapted individuals die. This is referred to as **survival of the fittest**.

Individuals are constantly 'fighting' to survive. This is called the **struggle for existence.**

The struggle for existence arises largely from **competition** (see page 23). One of the first people to recognise this was Thomas Malthus: in 1798 he pointed out that the human population always grows faster than the food supply, thus creating 'famine, pestilence and war'.

Examples of natural selection

Here are some examples of natural selection:

- **The peppered moth**: the dark form flourishes in industrial areas, the white form in non-industrial areas.
- **Bacteria**: some types of disease-causing bacteria have become resistant to penicillin.
- **Sickle-cell anaemia**: the gene for sickle cell anaemia is surprisingly common in malarious areas because individuals with the mild form of the disease (sickle-cell trait) have a slight immunity to malaria.

Natural selection depends on variation

Natural selection will only occur if individuals of a species differ from one another. In other words they must show variation (see page 133). Moreover, the variation must be *genetic* so that beneficial characteristics can be passed from parents to offspring. Mutation provides the best kind of variation for natural selection.

Darwin's theory of evolution

According to Charles Darwin's theory (also put forward by Alfred Wallace, a contemporary of Darwin), natural selection over millions of years can lead to a species *changing* (i.e. **evolving**) so that eventually it becomes a new species.

Evolution by natural selection depends on the fact that every now and again a useful mutation occurs in a population which makes the organism better adapted to its environment. Again, the black form of the peppered moth provides an example.

Darwin's theory of evolution is summarised in figure 29.2.

Artificial selection

In artificial selection a human chooses (i.e. selects) individual animals or plants with desirable characteristics and allows them to reproduce (breed). Other individuals are prevented from reproducing.

There are two types of artificial selection:

1. **In-breeding**: crossing closely related individuals. If continued over many generations, in-breeding may result in physical and/or mental defects. This is because harmful genes accumulate in the offspring. *Example:* highly bred pedigree dogs.
2. **Out-breeding**: crossing unrelated individuals. This usually results in the offspring being strong and robust. *Example:* mongrel dogs.

What use is artificial selection?

Artificial selection has enabled humans to produce more efficient farm animals and crop plants. *Examples*: breeds of cattle with higher milk yield; varieties of wheat with higher yield of grain; varieties of potato which are resistant to potato blight fungus.

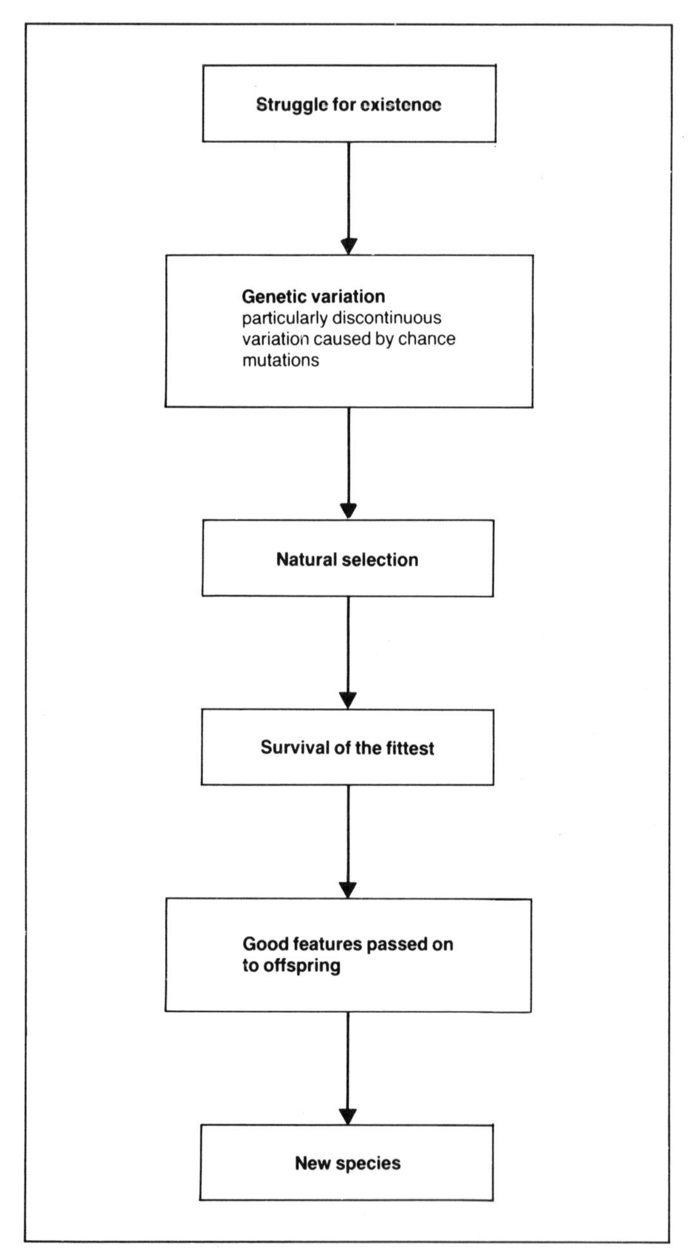

Figure 29.2 *The chain of events in Darwin's theory of evolution.*

Questions

1 Explain the difference between the following pairs of terms:

(a) Continuous and discontinuous variation.

(b) Chromosome mutation and gene mutation.

(c) Natural selection and artificial selection.

(d) In-breeding and out-breeding.

2 *The 'struggle for existence' arises largely from competition.*

(a) Explain *with an example* why competition should create a 'struggle for existence'.

(b) Suggest *one* other process in biology, apart from competition, which may contribute to the 'struggle for existence'.

3 *Mutation provides the best kind of variation for natural selection.*

(a) Give *one* example of natural selection.

(b) Why does mutation provide the best kind of variation for natural selection? What does 'best' mean in this context?

4 The heights of the 30 pupils in a biology class were measured and the results are shown in the table below.

Pupil	Height (cm)	Pupil	Height (cm)	Pupil	Height (cm)
1	150	11	152	21	141
2	151	12	140	22	144
3	148	13	145	23	147
4	151	14	146	24	154
5	146	15	149	25	150
6	150	16	150	26	152
7	152	17	153	27	155
8	144	18	154	28	156
9	147	19	160	29	158
10	151	20	139	30	149

(a) (i) Copy and complete the table below to show how many pupils there are at the various heights.

Height (cm)	138–139	140–141	142–143	144–145	146–147	148–149	150–151	152–153	154–155	156–157	158–159	160–161
Number of pupils												

(ii) Make a bar graph of the results on graph paper.

(b) Name the kind of variation shown by your bar graph.

(c) Comment on any interesting or unusual features of the graph.

5 The wildebeest were migrating across the great plains of the Serengeti: thousands of them, some large, some small, some strong, some weak. Lions followed them, intent on catching the slowest ones. Water was scarce and only the most persistent wildebeest could take advantage of the occasional water holes.

(a) Which phrases in this passage illustrate
(i) variation,
(ii) competition,
(iii) survival of the fittest?

(b) Explain how the passage illustrates Darwin's theory of evolution.

Specimen GCSE Questions

What sort of questions will I get in the examination? The answer to this depends on which examination group you use. To see the sort of questions set by your particular group you must look at specimen papers published by the group, or at papers which have been set in the past.

Despite small differences between the groups, certain generalisations can be made about the sort of questions which you are likely to get. The following notes should serve as a guide whatever group you are entered for. The examples are taken from all the groups.

You are usually given a certain number of lines on which to write your answers, and you are told how many marks each question carries.

Some questions are mainly or entirely factual. They usually require very short answers which are written in a space or box. Here is an example from WJEC:

1 The diagram below shows certain bones and muscles in the arm.

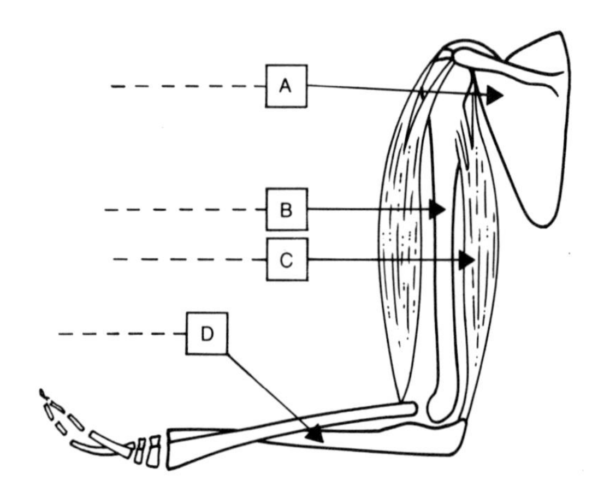

(a) Name the structures labelled A to D, using names chosen from the list below the diagram.

1. Ligament 2. Shoulder Blade 3. Biceps
4. Tendon 5. Ulna 6. Triceps 7. Humerus
8. Radius **(4 × ½ = 2 marks)**

(b) Describe how the muscles of the arm are used to raise it. **(1 mark)**

(c) Name the structure which attaches muscle to bone. **(1 mark)**

You may get a simple comprehension exercise. A comprehension exercise consists of a short piece of scientific writing followed by questions. The questions are designed to see if you understand the passage and can get information from it. Here is an example from WJEC:

2 Read the following extract from a newspaper article and answer the questions which follow:

LADYBIRDS TO SAVE EUCALYPTS

An Australian ladybird has been successfully introduced into New Zealand to <u>biologically control</u> the eucalyptus tortoise beetle, which has a hard exoskeleton, by eating its eggs.

The adult and larval stages of the beetle eat eucalypt leaves. This results in heavy and repeated leaf loss which may cause the death of the trees.

In Australia the tortoise beetle is controlled by <u>parasites</u> and predators and other beetles compete with it for food. In New Zealand there are no such predators and the use of <u>insecticides</u> has proved an unsatisfactory method of control.

In August, overwintering adult beetles emerge from under loose bark of eucalypts, feed and then mate. Eggs are laid on the underside of young leaves. The larvae which hatch feed on the new shoots. About November, mature larvae drop to the ground and pupate in the soil. Adults emerge and the cycle is repeated.

(a) Using organisms from the article above complete a food chain consisting of a producer, a herbivore and a carnivore. **(3 marks)**

(b) Suggest a suitable definition for **each** of the following terms which are underlined in the passage:
biological control **(1 mark)**
parasite **(1 mark)**
insecticide **(1 mark)**

(c) Suggest a possible reason why the use of insecticides has been ineffective in controlling the eucalyptus beetle. **(1 mark)**

(d) The tortoise beetle is a pest in New Zealand but not in Australia. Suggest a reason for this. **(1 mark)**

(e) State a precaution which would have to be taken prior to the release of the Australian ladybirds into the eucalypt trees of New Zealand. **(1 mark)**.

Some questions involve applying what you know about one topic to a different topic. In the following example, which comes from SEG, your knowledge of growth, transpiration and variation has to be applied to plant cuttings.

Certain parts of this question require slightly longer answers than the previous questions. The number of marks should give you an idea of how much you should write.

3 In late summer young stems of willow were cut into 30 cm lengths. The lower 10 cm of each stem cutting were stripped of leaves and buds and then buried firmly out of doors in coarse gritty soil, as shown in the diagram.

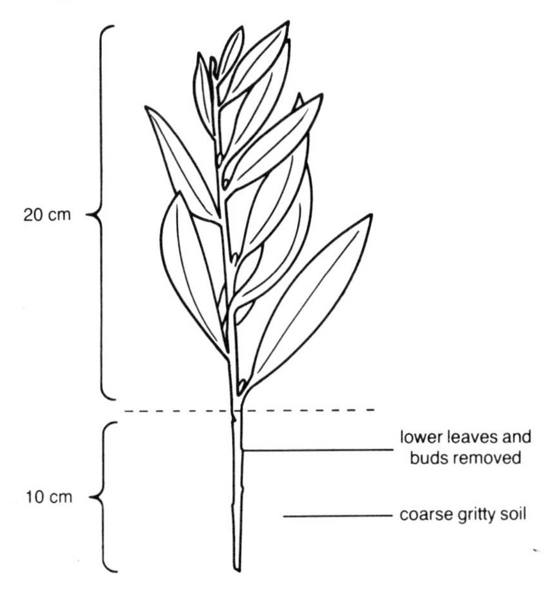

After a few weeks all the leaves dropped off. After a few months roots formed on the cut stem ends. The following spring 70% of the stem cuttings grew into new plants.

(a) (i) Suggest **one** reason for removing the leaves and buds from the lower 10 cm of stem. **(1 mark)**
(ii) Suggest **one** reason for using a coarse gritty soil instead of a fine soil. **(1 mark)**
(iii) Suggest **one** reason for burying each stem firmly (by pressing down the soil around it). **(1 mark)**

(b) (i) Suggest **one** way in which loss of all the leaves after a few weeks might have **helped** the rooting of the stems. **(2 marks)**
(ii) Suggest **one** way in which loss of all the leaves after a few weeks might have **hindered** the rooting of the stems. **(2 marks)**

(c) (i) Describe an improvement on this method of growing stem cuttings. **(1 mark)**
(ii) Explain how your improvement would work. **(1 mark)**

(d) A year after the cuttings were put in the soil one of the new plants was 1.2 m tall with six side branches and another was 0.5 m tall with eleven side branches.
(i) Describe an experiment you could carry out during the following year to help decide whether the differences between these two plants were caused by inheritance or had some other cause. **(4 marks)**
(ii) What results of your experiment would suggest that the differences were inherited? **(1 mark)**
(iii) What results of your experiment would suggest that the differences had some other cause? **(1 mark)**

There are bound to be questions which involve sorting out and interpreting data. Often the data are numerical, and graphs or charts may be involved. The following example comes from LEAG:

4 The diagram below shows a section through a wood where two samples of animals, sample A and sample B, were collected.

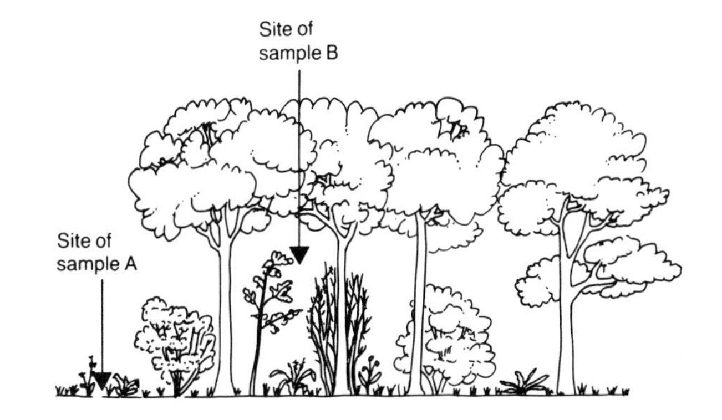

The table below shows the animals collected in each sample. The same method was used to take samples at each site.

Animal	*Number of animals	
	Sample A	**Sample B**
Snails	40	3
Mites	150	30
Spiders	10	40
True worms	10	0
Centipedes	5	1
Insects — Ants	30	5
Springtails	140	65
Aphids	70	100
Midges	110	20
Beetles	50	10

*(Numbers simplified from actual data)

(a) (i) Which animal was present in the largest number at site A? **(1 mark)**
(ii) Which animal was present in the largest number in the combined samples, A and B? **(1 mark)**

(b) Complete the circle below to form a pie-chart of the insects at site A. The circle has been divided into 20 equal parts. The sector for the ants has been completed on the pie chart to help you. **(3 marks)**

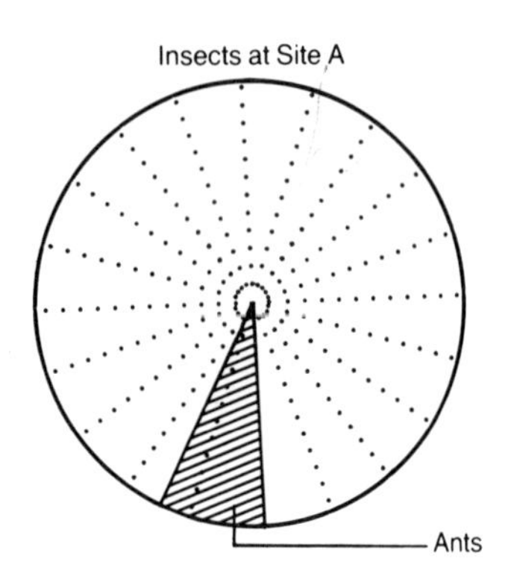

(c) From the numbers given in the table, which animal is likely to be a secondary consumer? **(1 mark)**

(d) Suggest **two** reasons why there are more snails in sample A than in sample B. **(2 marks)**

You may be asked to measure something and carry out simple mathematical calculations, as in the following question from WJEC. Note that in this question the only thing you need to know is the meaning of the term incubation period. It reminds us that GCSE is about understanding and doing, not just knowing.

5 A temperature chart of a patient suffering from a virus caused disease is shown below.

Day	Body temperature, °C
0	37
4	37
6	37
8	40.5
10	39.5
12	38
14	39
16	38.8
18	37.5
20	37
22	37
24	37

(a) How long did the incubation period last? **(1 mark)**

(b) What was the maximum increase in body temperature above normal? **(1 mark)**

(c) How long did the fever last? **(1 mark)**

Some questions may contain technical terms and facts which you have never met before. DON'T PANIC. The examiners are not trying to trick you. They want to see if you can work out the answer to the question by using your common sense and intelligence. Take this MEG question, for example:

6 The diagram below shows the volumes in cm^3 of the human lungs at various stages of inspiration and expiration.

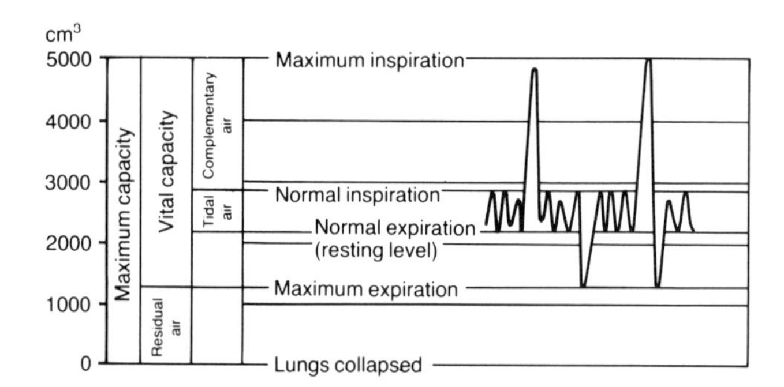

(a) What is meant by *complementary air*? **(1 mark)**

(b) What is meant by *residual air*? **(1 mark)**

(c) When are the lungs said to contain their *normal capacity*? **(1 mark)**

The examiners don't expect you to know the meaning of the terms complementary air and residual air. They want you to derive their meaning from the graph. From the graph you can see that complementary air is the volume of air which a person can breathe in following a normal inspiration: residual air is the volume of air which remains in the lungs after a normal expiration. What you do need to know are the meanings of the words inspiration and expiration.

Some questions involve writing a short account or explanation. Be as brief and concise as possible, and answer only what the question asks for. Candidates often waste time, and may even lose marks, because they write about things that are not required. You will be given a certain number of lines on which to write your answer, so you will know how much to write. Sometimes the examiners provide a list of words which you should include in your answer. Such is the case in this WJEC question:

7 Write a short account of the functions of the placenta and of how the embryo is protected while it is growing. In your account, use the words in the following list: Amniotic fluid, umbilical cord, glucose, vitamins, urea, carbon dioxide. **(6 marks)**

But you won't always be given guidance as to how to answer these kinds of questions. It's up to you to decide what you should write. The examiners will have decided beforehand the points which they want you to mention, and they will give you a mark for each one. So it is vital to cover all the relevant points, however briefly. Here are two examples of such questions from NEA.

8 Describe how man has polluted the environment. (Do NOT use smoke as an example.) **(8 marks)**

9 Each autumn, the trees in the deciduous forests shed their leaves. Explain how the elements contained in the carbohydrate cellulose in the leaves are made available for the growth of trees in subsequent years. **(5 marks)**

You may be asked to write an essay where you are not given a set number of lines to write. In this case you have to decide not only what to write, but also how much to write. You must plan your answer carefully. A certain number of marks will be given for each relevant point which you mention. Before you start writing jot down the points which you know you ought to cover. When you write the essay be sure to stick to the relevant points. Include diagrams whenever necessary.

In any kind of essay question:

ANSWER THE QUESTION, THE WHOLE QUESTION AND NOTHING BUT THE QUESTION

Answer the question which the examiner has set, not the question which you would have set if you'd been the examiner!

Don't do what a certain divinity candidate did when asked to compare Saul and David. He wrote: 'Who am I to compare two such great men? Here instead is a list of the kings of Israel'.

You will probably be given a choice between several different essays. The following question from LEAG was one of two alternatives.

10 Explain how urea passes from the liver, where it is made, to the bladder. **(14 marks)**

Some essay questions may be extremely broad in scope, like the following from MEG. You can't possibly cover all the ground in a question of this kind. What you must do is to select the most important ideas and express them clearly in a logical order. In this sort of question it is particularly important to plan your essay and make a list of relevant points before you start writing. Remember that marks will be given for each relevant point which you mention.

11 Write a clear account of the mechanisms by which growth and development occurs in plants and animals. **(15 marks)**

At the other extreme from essay questions are multiple choice questions. In this case all you have to do is to make a mark on a piece of paper—in the right place! In each question you are given four alternative answers. You have to choose the most correct answer. Sometimes you can spot the correct answer straight away. If you can't, you can usually arrive at it by eliminating the ones that are obviously wrong.

Not all examination groups use multiple choice questions, but if your group does you must make sure you get plenty of practice at doing them. They are more difficult than you may realise and it's very easy to make silly mistakes. The following examples will help you to see what's involved. They are all from MEG. The first five are factual, the remainder involve higher skills such as interpreting data.

12 Humans are classified as **mammals** because they
A have a vertebral column.
B produce live young.
C are warm blooded.
D have mammary glands.

13 Which one of the following properties makes soya bean particularly valuable as a meat substitute?
A It has a high mineral salt content.
B It has a high protein content.
C It has a high fat content.
D It has a high fibre content.

14 The diagram below represents the carbon cycle.

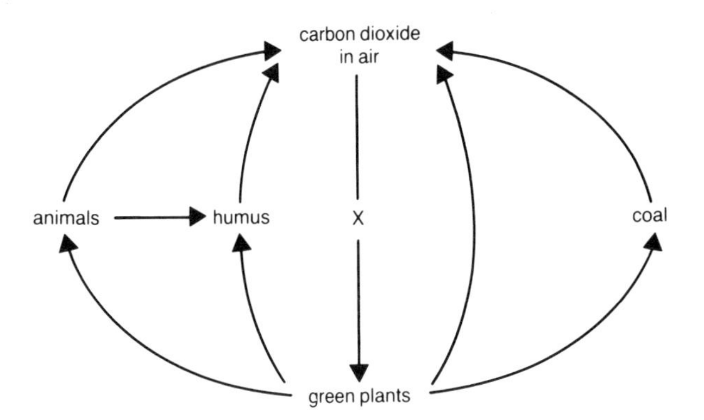

What does **X** represent?
A photosynthesis
B plant respiration
C combustion
D heterotrophic nutrition

15 Which of the following is the sequence of structures through which a mammalian sperm passes between leaving the sperm duct and fusing with an ovum?
A vagina, urethra, uterus, oviduct
B vagina, oviduct, uterus, ovary
C urethra, vagina, uterus, oviduct
D ureter, urethra, vagina, uterus

16 A division of one cell by mitosis will produce:
A two nuclei, each of which has the same number of chromosomes as the parent nucleus.
B four nuclei, each of which has the same number of chromosomes as the parent nucleus.
C two nuclei, each of which has a chromosome number half that of the parent nucleus.
D four nuclei, each of which has a chromosome number half that of the parent nucleus.

17 The diagram below shows a section through a hinge joint.

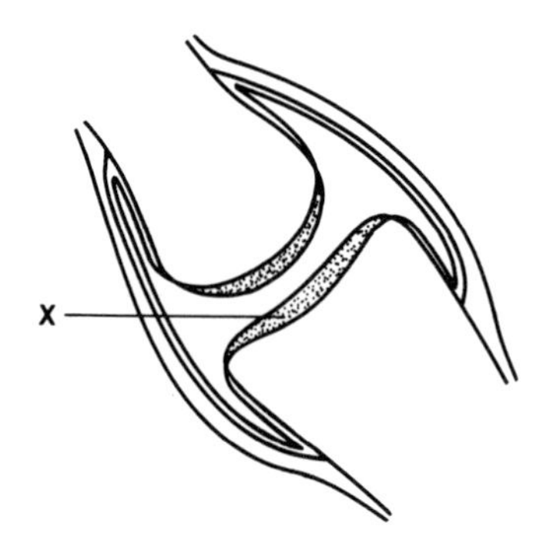

What is the structure labelled **X**?
A bone
B cartilage
C synovial membrane
D synovial fluid

18 The graph below shows the rate at which bubbles were produced by pond weed at different temperatures, all other factors, including light intensity, being kept constant.

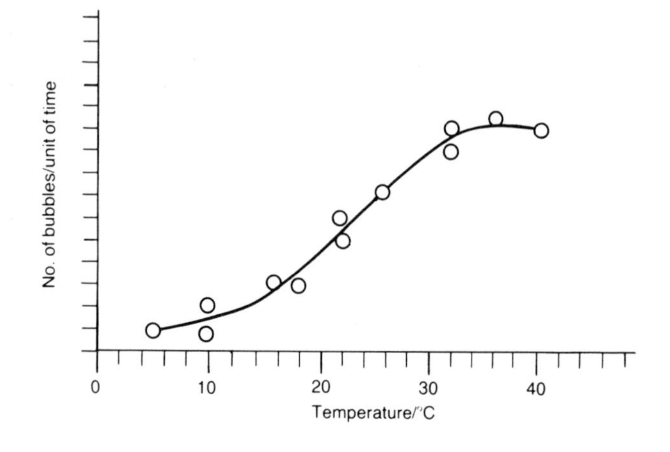

What conclusion can one draw from these results?
A Green plants produce oxygen when illuminated.
B Heat increases the rate of production of gas by green plants at all temperatures.
C Light intensity affects the rate of photosynthesis.
D The rate of photosynthesis increases with increasing temperature up to a limiting value.

19 A culture of yeast cells was started in a nutrient medium. Three days later a culture of the protozoan *Paramecium* was added. The numbers of yeast cells and *Paramecium* were recorded and are shown in the graph below. The units on the two vertical scales are different.

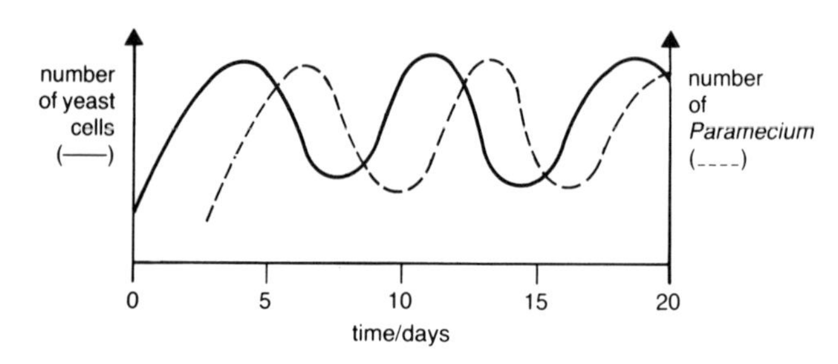

Why are the two curves so similar in shape?
A *Paramecium* feeds on yeast cells.
B Yeast feeds on dead *Paramecium* cells.
C *Paramecium* and yeast have the same life span.
D *Paramecium* and yeast compete for the nutrient medium.

20 Question 20 refers to the diagram below which shows an experiment designed to compare two soil samples **X** and **Y**.

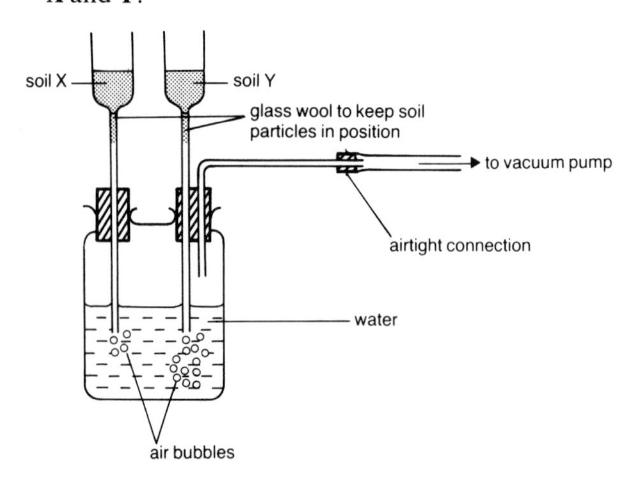

How do the size of the particles and the size of the air spaces in soil **X** compare with those in soil **Y**?

	size of particles	*size of air space*
A	smaller in **X** than **Y**	smaller in **X** than **Y**
B	smaller in **X** than **Y**	larger in **X** than **Y**
C	larger in **X** than **Y**	larger in **X** than **Y**
D	larger in **X** than **Y**	smaller in **X** than **Y**

21 A germinating bean seedling with a straight plumule and radicle was pinned to a vertically-held cork sheet so that the plumule and radicle were horizontal, as shown by the seedling in the diagram below, and maintained in a humid atmosphere.

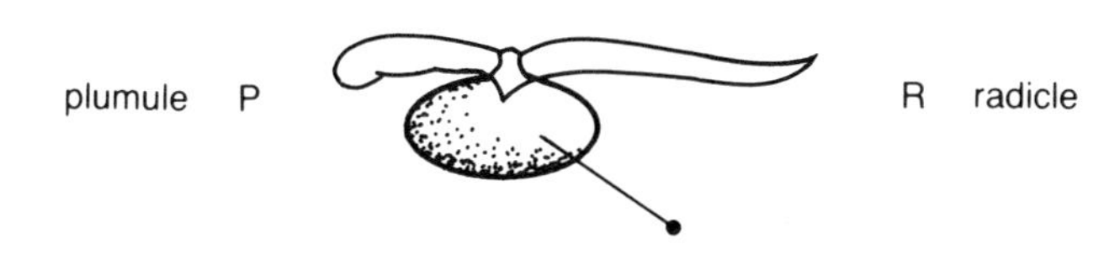

Which one of the following represents the seedling after 48 hours?

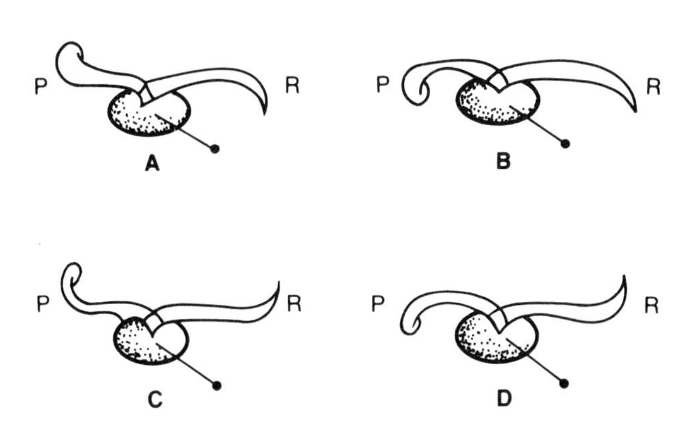

22 In the family tree shown below, squares □ represent males and circles ○ represent females. Healthy individuals are indicated by white symbols. Individuals who show a genetically controlled defect are indicated by black symbols.

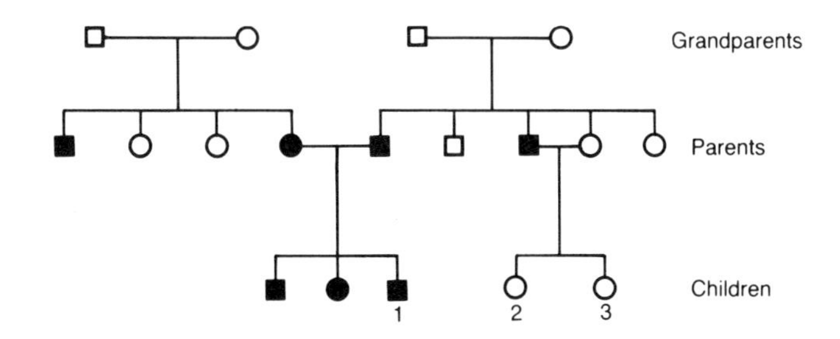

If you assume that the children labelled 1 and 3 in the diagram, grew up, married each other and produced a family, what is the chance that their first child will show the defect?

A 100%
B 75%
C 50%
D 25%

23 Apparatus was set up as shown in the diagram below.

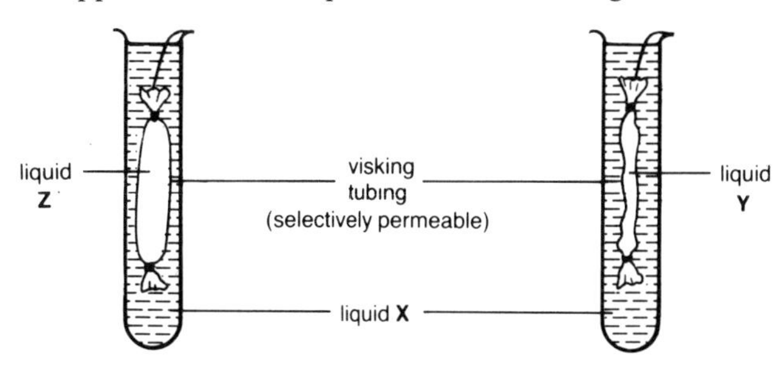

After some time had elapsed, the visking tubing containing liquid **Y** had collapsed while the tubing containing liquid **Z** was firm and hard. Which one of the following could be a correct description of the liquids at the start of the experiment?

	Liquid **Z**	*Liquid* **X**	*Liquid* **Y**
A	25 % sucrose solution	10% sucrose solution	water
B	water	25% sucrose solution	25% sucrose solution
C	10% sucrose solution	water	25% sucrose solution
D	water	10% sucrose solution	25% sucrose solution

Rather similar to multiple choice are matching pair questions. Each group of questions has a set of responses labelled A, B, C and D. In each case you have to select the best response and match it to the question. Each letter may be used once, more than once, or not at all. Here are two examples from NEA.

24 A Saprophytism
B Competition
C Parasitism
D Predation

From the above list of biological terms choose the one which describes
(a) yeast growing on bruised, windfall apples.
(b) birds eating worms.
(c) a fungus growing in the leaves of a living potato plant.
(d) decay of dead leaves in the soil.

25 The table shows the energy content and composition of four foods found in our diet.

Food	**Kilo-joules per 100g**	**Composition per 100 g**					
		Protein	**Fat**	**Carbo-hydrate**	**Vitamin C**	**Vitamin D**	**Iron**
A	3800	0.4 g	86 g	0	0	40 mg	0
B	130	1.2 g	0	8 g	220 mg	0	0
C	1150	8.8 g	1.5 g	60 g	0	0	0
D	400	2.0 g	0.1 g	25 g	10 mg	0	6 mg

From the table above select the food which would
(a) help to prevent anaemia.
(b) provide most energy.
(c) be best for growth.
(d) be best to prevent rickets.

Answers to topic questions

Chapter 1

1 (a) (i) Hand lens.
(ii) None.
(iii) Light microscope.
(iv) Light microscope.
(v) Electron microscope.

(b) Half the natural size.

2 (a) (i) Holozoism.
(ii) Parasitism.
(iii) Photosynthesis.
(iv) Parasitism.
(v) Saprotrophism.

(b) Vultures, though usually regarded as holozoic, are really saprotrophs because they feed on dead remains. *Euglena* can feed by photosynthesis in the light, or saprotrophically in the dark.

3 (a) The effect of light/dark on the growth of cress seedlings.

(b) (i) **Either** A if the effect of light is being investigated; **or** B if the effect of the dark is being investigated.
(ii) To serve as a standard with which to compare the results of the experiment.

(c) (i) Presence of cardboard box might make conditions more humid for the plant in A. Presence of lamp might make conditions hotter for the plant in B.
(**Note**: in a controlled experiment of this kind, all the conditions must be the same for both plants except for the condition which is being investigated.)
(ii) Cover B with a cardboard box and set a heatless light source into the top of it.

Chapter 2

1 (a) A species is a very small group of closely related organisms, whereas a phylum is a much larger group containing many different species.

(b) Some organisms have more than one common name, and some organisms have the same common name as other organisms.

(c) A small organism is less likely to be damaged by a paint brush than by forceps which could easily squash it.

(d) Plankton is made up of microscopic organisms which would pass through a coarse mesh.

(e) Size is a very variable feature of organisms and is of no value in identification, which is what keys are intended for.

2 Drawing 1 D
Drawing 2 B
Drawing 3 E
Drawing 4 C
Drawing 5 A
Drawing 6 F

(**Note**: from your knowledge of the animal kingdom can you say what each animal is?)

3 (a) Left or right hind leg may be chosen. The drawing should have a simple outline. Use a ruler to ensure that the length and width of the leg are twice those of the original. (**Note**: measuring things accurately is one of the skills required in GCSE.)

(b) A has no wings, B has wings.
A has claws, B has no claws.
A has no stripes on its abdomen, B has stripes on its abdomen.
A has hairs on its abdomen, B has no hairs on its abdomen.
A has straight feelers, B has bent feelers.
A has much smaller eyes than B.

(**Note**: other more subtle differences may be mentioned, but why bother when the above are so obvious?)

(c) Hard cuticle (exoskeleton). Jointed limbs.

(d) Hooked claws at ends of legs. Hairs projecting from abdomen.

4 (a) (i) Flowers, fruits. (**Note**: 'seeds' wrong because they are possessed by gymnosperms as well.)
(ii) Feathers, egg with hard shell, beak.
(iii) Gills, scales. (**Note**: certain other animals have gills (e.g. crabs) and scales (e.g. reptiles), but gills and scales are so characteristic of fish that they are acceptable answers here.)
(iv) Hair, milk-production, mammary glands.
(v) Single celled.

(b) (i) Fungus lacks chlorophyll whereas alga has chlorophyll.
(ii) Insect has three pairs of legs whereas arachnid has four pairs of legs.
(iii) Reptile has dry, scaly skin whereas an amphibian has moist, non-scaly skin.
(iv) Vertebrate has a backbone, invertebrate has no backbone.
(v) Virus lacks cell structure and can only be seen in the electron microscope, whereas bacterium has cell structure and can be seen under a light

microscope. (**Note**: the cell structure of bacteria is simpler than that of higher organisms and there is no proper nucleus.)

5 (a) D (**Note**: it is actually bladderwrack, a brown seaweed.)

(b) B (**Note**: the structure projecting from the top is the spore capsule.)

(c) C

(d) A (**Note**: the mushroom is the spore-producing body of a mycelium which is in the soil.)

Chapter 3

1 (a) Precise answers depend on habitat chosen. The following are guidelines.

1 Light: light meter (of the type photographers use).
2 Temperature: thermometer. (**Note**: maximum–minimum thermometer gives temperature range.)
3 Moisture/humidity: estimate how long it takes for dry cobalt chloride or thiocyanate paper to change from blue to pink.
4 Rainfall: collect rain in measuring cylinder over a known period of time. (**Note**: this is the principle of a rain gauge.)
5 Wind: direction can be found by means of a wind vane; speed measured by means of an anemometer which is like a mini windmill.
6 Waterflow: time how long it takes for a floating object (e.g. ping pong ball) to be carried a certain distance.
7 Pressure: various types of pressure gauge may be used.
8 pH: universal indicator paper (colour change gives pH).
9 Murkiness (of water): lower a white disc into the water and note how far it can be let down before it becomes invisible.
10 Salinity (of water or soil): chemical analysis.

(**Note**: the methods listed above are only meaningful if they are carried out in two different habitats and the results compared.)

(b) The earthworm may be taken to illustrate the sort of answers which this question requires.

1 Light: earthworms move away from light which therefore keeps them underground during daylight hours.
2 Temperature: low temperatures may freeze soil water, so burrowing becomes impossible; high temperatures may dry soil (see below).
3 Earthworms need moist soil since they lose water rapidly through permeable skin in dry conditions; earthworms come up onto surface of soil and copulate only in moist conditions.
4 Rainfall: affects moisture content of soil (see above); excess may cause soil erosion.
5 Wind: no direct effect but excessive wind may dry soil and/or cause erosion.
6 Waterflow: may be affected by excessive rainfall (see above).
7 Pressure: not relevant, except that heavy objects on surface might squash worms.
8 pH: most species of earthworm move away from acid soil.
9 Not relevant.
10 Earthworms do not like salty soil; salt causes them to lose water by osmosis.

2 There are all sorts of possibilities. Here are some rather obvious examples. You can probably think of others.

1 Tadpoles are food for water beetles.
2 Thrushes are predators of snails.
3 Head lice are parasites of humans.
4 Antelopes compete with wildebeest for grass.
5 Sea anemone protects small fish which lives amongst its tentacles.
6 Trees shelter woodlice from drought.
7 Humans are host for threadworms.
8 Bees pollinate plants.
9 Foxes disperse hooked fruits of burdock ('burs') which cling to fur.
10 Trees provide attachment for climbing plants, e.g. ivy.

3 (a) Conditions more constant in water than on land. For example, temperature of water in sea/lake/river fluctuates less than air temperature.

(b) Flowering plant which produces fruits and seeds: geranium
Herbaceous plant with perennating organ: iris (perennating organ is a rhizome).

(c) Pine tree. Survives winter by having needle-like leaves with thick cuticle, which reduces water loss when soil water is frozen.

4 (a) (i) Marram grass (grows on sand dunes).
(ii) Roses.
(iii) Heather (grows on moors).

(b) Increases humus content, so prevents leaching, and helps to keep the soil moist and warm.

(c) Waterlogged soil has reduced oxygen content, decomposers respire anaerobically producing lactic acid.

5 (a) Light (and/or heat) from lamp drives small animals downwards through funnel into beaker.

(b) Light (and/or heat) will not penetrate far enough to reach all the animals.

(c) The animals remain alive in water; they would be killed by ethanol.

(d) Mites, insect larvae (e.g. wireworms), ants.

(e)

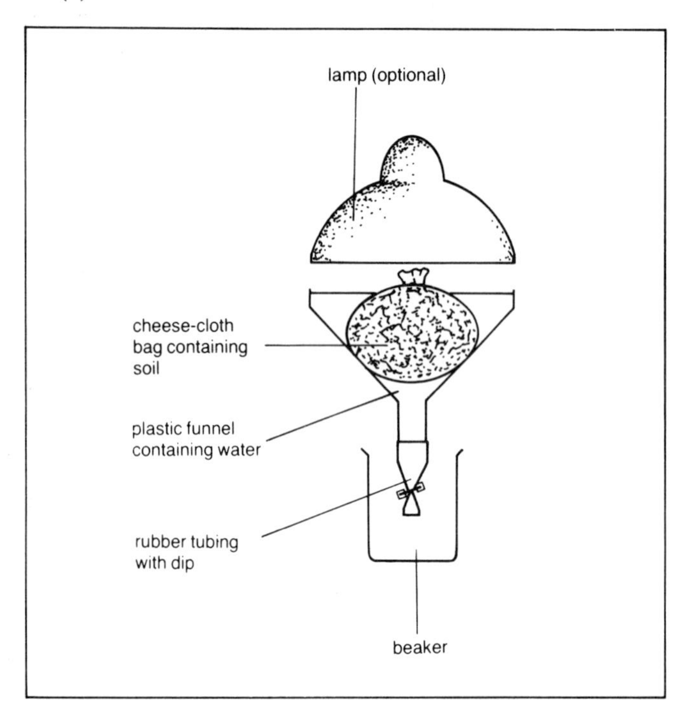

6 Various choices could be justified. Here are the most obvious ones:

(a) 1 (and perhaps 3)

(b) 4 and 5

(c) 3 and 7

(d) 8

(e) 3

7 (a) (i) More domestic fires/industrial burning of coal, oil, etc., towards city centre.
(ii) More factories/power stations/steel works/ smelting towards city centre. (**Note**: most fuels contain sulphur which, when burned, give off sulphur dioxide.)

(b) Women might be less susceptible to bronchitis than men, or might smoke less.

(c) The greater the air pollution, the lower the percentage lichen cover.

(d) Use of smokeless fuels in city, extraction of sulphur dioxide from factory smoke.

Chapter 4

1 Habitat is the place where the organism lives.
Environment is the conditions which exist in the habitat.
Ecological niche is the position of the organism in the food chain/web, i.e. what it feeds on and what feeds on it.

2 (a) Increase: if B increases there will be more B (food) to support C.

(b) Decrease: if D increases, more B will be eaten so there will be fewer B to support C.

(c) Decrease: if E increases, more C will be eaten.

(d) Increase: if F decreases, fewer C will be eaten.

(e) Increase: if G increases, more D will be eaten so there will be fewer D eating B; this will result in more B to support C.

(**Note**: this is an easy question so long as you don't make careless slips!)

3 (a) A food chain is a sequence of organisms in which each organism provides food for the next organism. A food web is a series of interconnected food chains.

(b) In a food chain each consumer has only one source of food, so if one of the consumer species is destroyed all the consumers further on in the chain will die. In a food web each consumer may have several different sources of food, so if one of the consumer species is destroyed the others can still survive.

(c) Energy is lost when food passes from one step to the next in a food chain or web.

(d) More humans can be supported by a given mass of crops than by livestock which has eaten that mass of crops.

(e) (i) An organism is usually larger than the organism it feeds on.
(ii) Exceptions include parasites (invariably smaller than their hosts), and aphids feeding on the leaves of a plant.
(**Note**: predators are usually larger than their prey, but there are exceptions to this too, e.g. a lion may tackle an animal larger than itself.)

4 (a) Sunlight.

(b) (i) Fungi/earthworm/insect/mite.
(ii) Insect/mite/ground beetle/mole.
(iii) Ground beetle/mole.
(iv) Fungi.
(v) Mole.

(c) (i) Leaf litter → earthworms → moles
(ii) Leaf litter → earthworms → ground beetles → moles
Leaf litter → insects and mites → ground beetles → moles
(iii) Leaf litter → fungi → insects and mites → ground beetles → moles

(d) Leaf litter. Its role is to provide food (which the leaves made by photosynthesis) for the rest of the community.

5 (a) Peat is formed from dead plants when oxygen is absent. The decomposers respire anaerobically with the formation of lactic acid. As lactic acid builds up it kills the decomposers, preventing decay from being completed.

(b) The peat is buried and hardens into coal.

(c) When coal is burned (combustion) the carbon is released as carbon dioxide gas which goes into the atmosphere.

(d) Because we use up coal far more quickly than it is formed.

6 (a) Respiration adds carbon dioxide to the atmosphere; photosynthesis removes it.

(b) (i) Protein → ammonia
(ii) Ammonia → nitrite
Nitrite → nitrate
(iii) Nitrogen → nitrate/protein

(c) (i) Harmful. Nitrites cannot be used by plants (in fact they poison them.)
(ii) Nitrate → nitrogen

Chapter 5

1 (a) (i) The number of individual plants per unit area.
(ii) The percentage proportion of the ground occupied by the plants.

(b) (i) One way is to throw the quadrat over your shoulder so that you cannot see where it is likely to land.
(ii) *Either*
Include the plant if more than half of it is inside the frame.
Or
Include those plants which are touching or overlapping two sides of the frame (say the top and left-hand sides) and exclude those which are touching the other two sides even if most of the plant is within the frame.

(**Note**: the trouble with the first method is that some plants may be half inside the quadrat and half out. For this reason the second method is better. The important thing is to adopt the same procedure every time you take a count. In other words be consistent.)

2 (a) (i) It is best to present your results as a table:

Distance from trunk (m)	1	2	3	4	5	6	7	8	9	10
Number of daisies per m^2	1	4	9	9	11	12	15	16	14	17

Method: draw ten one-centimetre squares (equivalent to ten one-metre quadrats) in a straight line from the tree trunk outwards; count the number of dots (daisies) in each square to give the number of daisies per m^2.

(**Note**: your figures may differ slightly from those above, depending on exactly where you have drawn your squares.)

(ii)

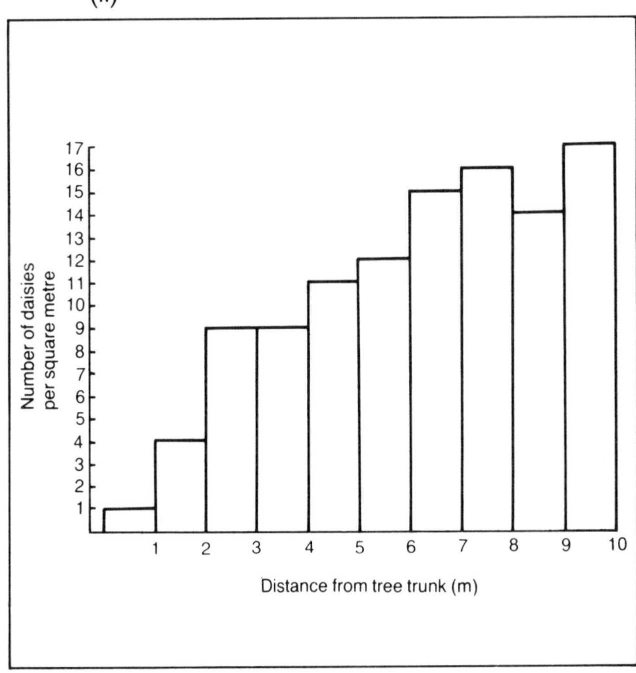

A bar chart is more appropriate than a line graph and just as useful for detecting a pattern and drawing conclusions. A line graph assumes that the number of daisies changes smoothly, which is not necessarily the case.

(b) (i) Lack of light caused by shading from canopy of tree. (**Note**: other reasons are possible, e.g. shortage of water and/or mineral salts due to competition with tree roots; the important thing is that your suggestion should be sensible and testable by an experiment — don't make it difficult to answer.)
(ii) First use a light meter to make sure that the light intensity beneath the tree canopy really is lower than that elsewhere. At the beginning of the next season set up a canvas shade, equivalent to the canopy of the tree, over part of the lawn where you know daisies normally grow abundantly and compare the daisy population with that in an adjacent unshaded part of the lawn. Make sure that all other conditions (availability of water, mineral salts, etc.) are the same in the two parts of the lawn.

(A number of variations on this experiment are possible. For example, you could investigate the effect of the lower light intensity on daisy seedlings grown in seed boxes indoors. Be sure to include any necessary controls.)

3 (a) At this stage there are relatively few reproducing individuals.

(b) Four of the following:
disease (e.g. myxomatosis)
shortage of food (plant leaves, etc.)
shortage of suitable ground for burrows
failure to reproduce successfully
fighting between individuals

4 (a) Continent A: 2.65%. Continent B: 0.59%. (**Note**: the percentage growth rate is the difference between the birth rate and the death rate.)

(b) Better birth control (family planning).

(c) Poorer living conditions, poorer medical care, more disease.

(d) Continent A: Asia, Africa, South America.
Continent B: Europe, North America, Australia.

5 (a)

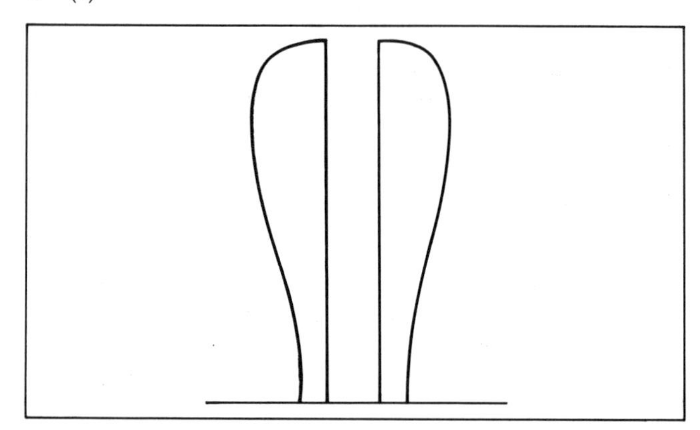

(b) Narrow base because more efficient/widespread birth control results in fewer babies being born. Broad top because advances in medicine (e.g cure for cancer) results in old people living longer.

(c) (i) Fewer schools and teachers will be needed because there will be fewer pupils.
(ii) Health services will have to look after an increasing number of old people.

Chapter 6

1 (a) (i) Larva (caterpillar) of cabbage white butterfly.
(ii) Wireworm (larva of beetle), leather jacket (larva of daddy longlegs), millipede eat roots of crop plants.
(iii) Mosquito (transmits malaria and yellow fever).
(iv) Greenfly (aphid) (transmits plant virus diseases).
(v) Thistles, poppies and grass (various species) may compete with crop plant (e.g. wheat) for soil nutrients.

(b) Some pests are parasites and cause disease, e.g. blood fluke.
Some plant pests clog up rivers and reservoirs, e.g. water hyacinth.
Some pests destroy buildings, e.g. dry rot fungus, woodworm (larva of beetle), termites.
Some pests make food go bad, e.g. flies, fungi, bacteria.

2 (a) Type C

(b) Type D

(c) Type A

(d) Type B

3 (a) A small dose of dead or inactivated germs (vaccine) of a particular disease is put into a person's body by injection, scratching the skin or by mouth. In the case of poliomyelitis the vaccine is taken by mouth.

(b) Once in the bloodstream, the vaccine causes the body to produce antibodies against the polio virus, thereby protecting the person against polio in the future.

(c) (i) $\frac{5314 - 395}{5314} \times 100 = 92.6\,\%$

(ii) Improvements in personal and public hygiene.
Introduction of vaccination against the disease.

(d) 1963. In that year the number of cases fell to a very small number.

4 (a) Southern part, where the mean winter temperature is relatively high.

(b) (i) released; (ii) wind; (iii) north-easterly; (iv) cereals.

(c) By treating cereal crops with a fungicide which kills the rust fungus.
By developing new strains (varieties) of cereal which are resistant to rust.

(d) Rust disease can ruin cereal crops, particularly wheat and maize, resulting in food shortage and/or lowered income from export of cereals.

(**Note**: this question involves mainly common sense; little knowledge is necessary except for a general understanding of fungi.)

Chapter 7

1 (a) A: cytoplasm
B: cell membrane
C: nucleus

(b) D: mitochondrion
E: glycogen granule/food storage granule
(**Note**: mitochondrion is larger than glycogen granule.)

(c) Epithelial tissue.

(d) (i) It would have been scraped from the inner surface of the person's cheek with a sterilised spatula.
(ii) The cheek tissue on the spatula would have been placed in the centre of the slide, a drop of water or stain added, and then a coverslip lowered carefully onto the drop.

(e) The scale bar in the photograph is 1 cm long. The magnification of the cell in the photograph is the number of times 5 μm would fit into this distance. The answer is 2000. (**Note**: this is greater than the magnification achieved by the microscope; it includes the degree to which the photograph, which was taken down the microscope, was enlarged.)

2 (a) Cell wall, starch grains, chloroplasts, vacuole.

(b) Glycogen granules.

(c) Root cells, cells of storage (perennating) organ, e.g. bulb, potato tuber.

(d) Endoplasmic reticulum, ribosomes.

(e) Mitochondria.

3 (a) *Genes:* chemical structures (DNA/deoxyribonucleic acid) which determine an individual's characteristics and by which the characteristics are transmitted (handed on) to the next generation.

Characteristics: the features of the individual, e.g. colour of eyes, colour of hair, etc.

Controls: allows certain chemical substances to enter or leave the cell, but not others.

Cellulose: the chemical substance (carbohydrate/polysaccharide) of which plant cell walls are made.

Chlorophyll: a chemical substance/green pigment which plays an essential part in photosynthesis (it traps/harvests the necessary light energy).

Photosynthesis: the method of nutrition/feeding used by plants and some single-celled organisms in which carbon dioxide and water are built up into complex substances, using energy from sunlight.

Carbohydrate: an organic substance which contains carbon, hydrogen and oxygen in certain proportions.

Energy: the ability/capability to do work.

Synthesis: the process by which a chemical substance is made/manufactured.

Proteins: organic substances which contain carbon, hydrogen, oxygen and nitrogen in certain proportions. Sometimes sulphur is present too.

(b) (i) Nucleus: controls the various processes which go on inside the cell.
(ii) Cytoplasm: carries out various functions (chemical reactions) as directed by the nucleus.

(**Note**: these statements are very general; for details of how the nucleus controls the cytoplasm see Chapter 27.)

4 (a) A tissue is a group of cells which performs a particular function.
An organ is a distinct part of the body, composed of several different tissues, which carries out one or more specific functions.

(b) (i) Epithelial, connective, blood, nerve and muscle tissues.
(ii) Epithelial, blood and nerve tissues.
(iii) Epidermal, packing, vascular and strengthening tissues. (**Note**: green stems also contain photosynthetic tissue.)
(iv) Epidermal, packing, vascular and strengthening tissues.

5 (a) (i) The various structures are mainly arranged symmetrically on either side of a line drawn down the middle of the body.
(ii) The various structures are arranged in a circle round a central point like the spokes of a wheel.

(b) The digestive system. This consists of a single tube which is asymmetrically arranged for much of its length.

(c) Bilaterally symmetrical: earthworm, insect (e.g. locust), etc.
Radially symmetrical: *Hydra*, sea anemone, jellyfish.

(d) Radially symmetrical: sepals, petals and stamens in radically symmetrical flower e.g. buttercup; vascular bundles in stems and roots of flowering plants.
Bilaterally symmetrical: two-winged fruit of sycamore; sepals, petals and stamens in bilaterally symmetrical flower, e.g. sweet pea.

Chapter 8

1 (a) (i) Cube A

Surface area $= 2 \times 2 \times 6$ (the area of one surface $\times$ the total number of surfaces) $= 24\ \text{cm}^2$

Volume $= 2 \times 2 \times 2 = 8\ \text{cm}^3$

$$\text{Surface-volume ratio} = \frac{24}{8} = 3$$

(ii) Cube B

Surface area $= 4 \times 4 \times 6 = 96\ \text{cm}^2$

Volume $= 4 \times 4 \times 4 = 64\ \text{cm}^3$

$$\text{Surface-volume ratio} = \frac{96}{64} = 1.5$$

(b) Cube A because, being smaller, it has a larger surface–volume ratio than cube B.

(c) (i) Cube B because it has a smaller surface–volume ratio; its surface area is small relative to its

volume, so gas exchange across the surface cannot supply it with enough oxygen to meet its needs.
(ii) Lungs and gills.

2 (a) A membrane which will let water molecules through but not larger particles (in this case sugar molecules).
(b) (i) Water molecules pass by osmosis from the beaker into the bag, thus causing the fluid to rise in the capillary tube.
(ii) Sugar (or salt).
(iii) If sufficient sugar is added, the solution in the beaker will have the same concentration as the solution in the bag and osmosis will stop.
(c) Peel a potato tuber, cut it in half and remove the central tissue so as to make a potato cup. Half fill the cup with a strong/concentrated sugar solution. Place the cup in a dish of distilled water so that the surface of the water is at about the same level as the sugar solution inside the cup. Water should pass through the selectively permeable potato tissue into the cup, causing the sugar solution to rise.

3 (a) (i) Swells/expands.
(ii) Turgid.
(iii) Shrinks.
(iv) Plasmolysed.
(b) The cell wall which, being made of tough cellulose, does not break.
(c) (i) It would burst.
(ii) There is no cell wall; the cell is lined only by the delicate cell membrane which breaks easily.
(d) Turgid cells, tightly packed, maintain the shape of the plant and help to keep it erect.

4 (a) (i) Tubes B and D
(ii) Tube A
(iii) Tube C
(b) Tube A: the leaf has photosynthesised in the light, using up more carbon dioxide than it has produced from its respiration.

Tube B: the animals have been respiring, producing carbon dioxide.

Tube C: the carbon dioxide produced by the animals' respiration has been used by the leaf for photosynthesis.

Tube D: because it is dark the leaf has not been photosynthesising and so has not been using up carbon dioxide; however, it has been respiring, producing carbon dioxide.
(c) The carbon dioxide concentration would increase. (**Note**: the increase would be even greater than in tubes B and D because the leaf and the woodlice would both be producing carbon dioxide from respiration.)
(d) The oxygen produced by photosynthesis is used for respiration, and the carbon dioxide produced by respiration is used for photosynthesis.

Chapter 9

1 (a) Starch is insoluble and consists of numerous glucose molecules packed close together. Glucose, when broken down in respiration, gives energy, so starch is a concentrated store of energy.
(b) It makes the cuticle waterproof.
(c) Cellulose. Consists of glucose molecules linked together in a specific way.
(d) It insulates them against heat loss. (**Note**: this is a better answer than 'it keeps them warm' because it is more precise.)
(e) Proteins differ in the amino acids they contain and the order in which they are linked together in protein molecules. The proteins we eat are broken down (digested) in the gut into their constituent amino acids. Inside our cells the amino acids are linked together in a different order to give the proteins characteristic of the human.

2 (a) High temperatures (above about 45°C) destroy (denature) the protein enzymes so that they no longer work/act. (**Note**: the enzymes in biological washing powders remove stains by digesting them; if the enzymes are denatured their digestive action will not take place.)
(b) The protease partially digests the protein in the meat, softening it.
(c) Cellulase breaks down cellulose into sugars. As cellulose is responsible for the toughness of a cabbage, treatment with cellulase will soften it.
(d) Amylase breaks down starch into the disaccharide sugar maltose. Maltose is sweet and runny.

3 (a) (i) At 20°C: 1.7 mg product per minute.
At 30°C: 3.3 mg product per minute.
(ii) $\times$ 2 approximately.
(b) Raising the temperature increases the speed of movement of the reactant (substrate) molecules which therefore collide with the enzyme molecules more often.
(c) Above 40°C the enzyme, being a protein, is denatured (destroyed) and stops catalysing the reaction.
(d) An acid (e.g. hydrochloric acid) or a poison (e.g cyanide). The acid would stop the reaction because the enzymes which control metabolic reactions do not work/act at low pH. The poison would stop the reaction by inhibiting the enzyme molecules.

4 (a) (i) Require energy: reactions 1, 2 and 3.
(ii) Release energy: reaction 4.

(b) Reaction 1 takes place in plants for storing glucose/energy.

Reaction 2 takes place in animals for storing glucose/energy.

Reaction 3 takes place in plants for making cell walls.

Reaction 4 takes place in nearly all living cells for releasing energy.

(c) In an animal's gut during digestion.
In a potato tuber when it starts to grow/sends out a shoot.

(d) In chloroplast-containing cells of plants in the light when photosynthesis takes place.

Chapter 10

1 (a) To prevent carbon dioxide leaving the test tube or entering it from the surrounding air.

(b) (i) A second test tube, identical to the first one but not containing any animals or plants.
(ii) A control is necessary because otherwise you could not be sure that any carbon dioxide detected by the indicator had been given out by the organisms.

(c) Hydrogencarbonate indicator solution is more sensitive than lime water and is therefore more suitable for small organisms which give out relatively little carbon dioxide.

(d) In the light the leaf would photosynthesise and therefore use up/absorb carbon dioxide.

2 (a) A second set-up, identical to the first one but not containing any animals.

(b) Absorption of oxygen by the small animals should cause the coloured water to rise in the capillary tube. If carbon dioxide given out by the organisms was not absorbed it would accumulate in the test tube and interfere with the movement of the coloured water in the capillary tube, invalidating the results.

(c) Set up a respirometer, containing the small animals, in a water bath at room temperature. Measure how far the coloured water rises in the capillary tube in a certain time. Repeat the experiment in a water bath at 10°C and 30°C. In each case express the rate of respiration as the distance travelled by the coloured water per unit time.

3 (a) Adenosine triphosphate.

(b) Cells do not obtain energy from glucose direct. They obtain energy from ATP. The function of glucose in cells is to provide energy for making ATP.

(c) All traces of glucose and ATP were removed from the muscle fibres. This was necessary because if they were present, adding them to the muscle would be pointless.

(d) Sending messages (impulses) through nerves, transporting substances, keeping warm (temperature regulation), growth, cell division, moving molecules and/or ions against a concentration gradient.

4 (a)

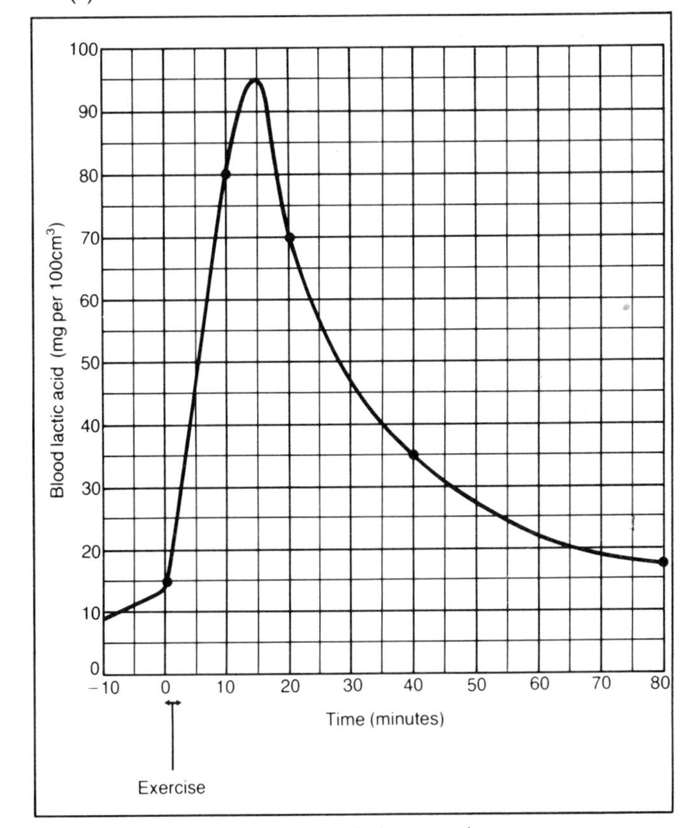

(b) (i) Muscles, particularly leg muscles.
(ii) Oxygen cannot be delivered to the muscles quickly enough to keep pace with their needs, so they respire anaerobically producing lactic acid.

(c) (i) It takes time for the lactic acid, produced by the muscles, to get into the general bloodstream.
(ii) The lactic acid is broken down into carbon dioxide and water.

(d) Worm in stagnant mud: little or no oxygen in its environment.
Tapeworm in the gut: no oxygen in its environment.
Whale/seal: runs out of oxygen during prolonged dive.

Chapter 11

1 (a) Carbohydrate is our main energy food.

(b) Citrus fruits contain a lot of vitamin C (ascorbic acid) which prevents scurvy.

(c) Cod liver oil contains a lot of vitamin D (calciferol) which prevents rickets.

(d) Margarine contains mainly unsaturated, as opposed to saturated, fat. Saturated fat is associated with heart disease.

(e) Milk protein contains more essential amino acids than wheat protein.

2 (a) Plant food (particularly vegetables) contains a relatively small amount of nutrients per unit mass, so more must be eaten to ensure that the person receives enough of each necessary nutrient.

(b) A diet containing lots of different plant foods.

(c) Certain vitamins (e.g. vitamin D) and essential amino acids.

(d) The diet would not need to be so bulky, and the plant components would not need to be so varied.

3 (a) Starch occurs as starch grains in the cells of the tuber.

(b) (i) It provides energy.
(ii) Being a storage organ, starch is particularly concentrated in a potato.

(c) (i) Add a drop of dilute iodine solution: blue-black colour indicates starch.
(ii) Crush up a small piece of potato, place it in a test tube and add water, shake well, add Benedict's/Fehling's solution and heat to boiling: green or brown precipitate indicates reducing sugar.
(iii) Extract juice from the potato, add the juice drop by drop to a drop of blue DCPIP solution: disappearance of the blue colour indicates vitamin C.

(d) (i) Keeping the cells turgid.
(ii) Maintaining the correct concentration of the blood and tissue fluids.

4 (a) (i)

Age (years)	**Energy** (kJ/kg)
1	550
11	308
18	258
25	180

(ii)

Age (years)	**Protein** (g/kg)
1	2.5
11	2.06
18	1.6
25	0.87

(iii) Expressing the energy and protein requirements in this way makes it possible to compare the requirements of the cells/tissues at different ages.

(b) (i) Growth, which requires energy, is taking place rapidly in the one-year-old whereas in the 25-year-old growth has stopped.
(ii) Protein is needed in the one-year-old for the formation of new cells and tissues which occurs during growth. This function is no longer required in the 25-year-old.

(c) (i) Replacement or repair of worn out cells and tissues.
(ii) Milk, eggs, meat.

(d) His diet was likely to become unbalanced; in particular he was likely to run short of energy foods.

Chapter 12

1 (a) (i) Pepsin, trypsin and peptidases.
(ii) Amylase, maltase.
(iii) Sucrase.
(iv) None.
(v) Lipase. (**Note**: bile salts are wrong because they are not enzymes.)

(b) Bile salts have no chemical effect on fat; they only change it physically.

(c) Set up two test tubes, one containing fat and lipase only, the other containing fat, lipase and bile salts. The amounts of fat and lipase should be the same in both test tubes, as should all conditions such as temperature. Estimate how long it takes for the fat to disappear in each test tube.

2 (a) Pepsin and hydrochloric acid. (Water would also be correct!)

(b) Two possible hypotheses:
(1) Messages (impulses) might travel in nerves from taste receptors in the mouth (and/or smell receptors in the nose) to the stomach wall, causing the glands to produce (secrete) gastric juice.
(2) A hormone (chemical messenger) might travel in the bloodstream from the mouth to the stomach wall, causing the glands to produce gastric juice.

(In fact the first hypothesis is correct. However, the question asks you to suggest an explanation. The important thing is not that your explanation is correct but that it is sensible and based on a sound understanding of biological principles.)

(c) (i) 5 minutes.
(ii) The delay is the time taken for the nervous/hormonal reflex to take place.

(d) The food should be clean (sterilised) because there is no acid to kill germs, and fluid or semi-fluid because there are no muscle contractions and enzymes to turn the food into chyme. Also, little meals should be taken frequently (rather than large meals occasionally) because, without the stomach, there is nowhere to store the food.

3 (a) As controls.

(b) To obtain evidence that the substance in saliva which breaks down starch to sugar really is an enzyme. (A property of enzymes is that they are destroyed by excessive heat.)

(c) The test tube should not be heated with a naked flame because the contents are likely to shoot out of the test tube when they boil. Instead the test tube should be heated by being placed in a boiling water bath or beaker of boiling water.

4 (a) (i) 4 hours
(ii) 20 hours

(b) In Pat's graph the pH falls sharply at meal times. In Ann's graph the pH falls not only at meal times but at other times as well, probably when snacks or sweets are eaten.

(c) Stop eating between meals; clean teeth more often.

(d) (i) Without oxygen.
(ii) In anaerobic conditions the bacteria produce lactic acid. The lactic acid eats into the enamel, so that the bacteria get into the dentine and pulp cavity.

5 (a) (i) Incisors.
(ii) Cutting/gnawing: they have sharp, chisel-like ends.

(b) (i) Cheek teeth/molars.
(ii) Crushing/grinding: they have broad tops, with sharp ridges like a file.

(c) Canines.

(d) The gap provides space for storing food before it is swallowed.

Chapter 13

1

Structure	Function
Blood capillaries	Warm
Cilia	Clean
Mucus glands	Clean, moisten
Sensory cells	Test

2 (a) Nose (nasal cavity) or mouth (mouth cavity), pharynx (throat), glottis, larynx (voice box), trachea (windpipe), bronchus, bronchiole, alveolus.

(b) The force is a negative pressure/suction. It is brought about by expansion of the chest (thorax) achieved by flattening of the diaphragm and upward and outward movement of the ribs.

3 (a) 500 cm^3

(b) 4375 cm^3 (4.375 litres)

(c) $\frac{500}{3400} \times 100 = 14.7\%$

(d) It leaves a large reserve capacity which can be brought into use when necessary, e.g. during exercise.

(e) (i) The breathing movements become deeper. (Inspiration and expiration are both more prolonged.)
(ii) The frequency of the breathing movements might become greater. (More breaths per unit time.)
(iii) During exercise the muscles need more oxygen; deeper and/or more frequent breaths help to get oxygen to the muscles more quickly.

4 (a) (1) Healthy lung has wide bronchiole, diseased lung has constricted (narrow) bronchiole. (2) Healthy lung has deep folds (partitions) between alveoli, diseased lung has shallow folds (no partitions) between alveoli.

(b) The narrow bronchiole means that less air will pass through it. The shallow folds (lack of partitions) means that there will be less surface area for gas exchange.

(c) The depth and/or frequency of breathing would be expected to increase so as to make up for the defects in the lungs and ensure that enough oxygen gets to the tissues.

(d) Cigarette smoke, factory smoke, smog.

Chapter 14

1 If haemoglobin combines with carbon monoxide it cannot combine with oxygen. So oxygen stops being sent to the cells, which therefore cannot produce energy with the result that the cells die.

2 (a) (1) The legs are below the heart so blood is moving against gravity. (2) The blood has been through the capillaries and is therefore under reduced pressure in the veins.

(b) (1) The contractions of the heart push the blood from behind. (2) The relaxations of the heart pull (suck) the blood towards the heart. (3) Breathing in (inspiration/inhalation) helps to suck the blood towards the heart. (4) The veins are wide and have stretchable walls which minimises resistance to the flow of blood. (5) The veins have valves which prevent the blood slipping back.

3 (a) The cell is a biconcave disc and is therefore thinner in the centre than further out.

(b) The function of the white blood cells is to destroy germs (bacteria and viruses), so the number increases when you have a disease.

(c) Contact of the platelets with the surface of the glass rod triggers the blood-clotting mechanism.

(d) The muscle in the wall of the left ventricle is thicker than the muscle in the wall of the right ventricle: the left ventricle has to pump blood to almost all parts of the body whereas the right ventricle has to pump blood only to the lungs.

4 (a)

	Test panels		
	anti-A	**anti-B**	**Neither andi-A nor anti-B**
Blood group A	√	×	×
Blood group B	×	√	×
Blood group AB	√	√	×
Blood group O	×	×	×

(b) (i) Sarah: group AB
Tom: group O
(ii) Sarah's

(c) It shows what a negative result looks like and provides a standard for comparison.

(d) (i) The lancet (needle) for pricking the finger or thumb must be sterilised. The lancet (needle) must be thrown away afterwards and never used again. Finger or thumb must be sterilised immediately before and after the blood has been obtained.
(ii) Safety precautions prevent the finger or thumb going septic, and they prevent diseases such as viral hepatitis and AIDS spreading from one person to another.

5 (a) A pulmonary artery (right)
B anterior vena cava (great vein)
C right atrium (auricle)
D right ventricle
E posterior vena cava (great vein)
F aorta
G left atrium (auricle)
H pulmonary vein (left)
I left ventricle
J aorta

(b) B/E, C, D, A, H, G, I, F, J

(c) I

(d) It will prevent the left ventricle contracting or at least severely reduce its ability to contract.
(**Note**: the coronary arteries supply the heart muscle with oxygen; a clot in one of these arteries will prevent oxygen reaching the part of the heart wall served by that particular artery.)

(e) (1) Smoking, particularly if the smoke is inhaled. (2) Eating too much food which is rich in fat/saturated fat/cholesterol.

6 (a) 0.8 seconds

(b) 0.6 seconds. The valve is closed when the ventricle pressure falls below the aortic pressure, so the length of this period was measured on the graph.

(c) 8–9 kilopascals. The aortic pressure is highest when the ventricle has contracted and blood has been forced into the aorta; the aortic pressure is lowest when the ventricle is relaxing and the blood has moved on along the aorta.

(d) The muscle in the wall of the ventricle contracts. So the wall moves inwards against the blood inside/volume of ventricle is reduced. The aortic valve is closed, so pressure increases in the ventricle.

Chapter 15

1 (a) The effect of lack of nitrogen on the growth of broad bean seedlings.

(b) Tap water contains some mineral elements/ions, whereas distilled water does not.

(c) As a salt, e.g. potassium nitrate.

(d) To serve as a control with which the growth of the group A seedlings could be compared.

(e) To compare the growth of plants in two different conditions you may need to have a large sample of plants in each group and average any measurements that you make. For your results to be valid the samples should be as large as possible.

2 (a) From fastest to slowest: inorganic fertiliser, compost, green manure. (**Note**: inorganic fertiliser is already in a form which is available to plants; compost is less readily available because it has not yet been broken down completely into inorganic substances; green manure is the least readily available because the breakdown process (decay) has not yet started.)

(b) Farmyard manure improves the texture of the soil. It enables soil particles to stick together in small clumps, helps to aerate the soil and releases useful chemicals slowly over a long period.

(c) They contain bacteria in their roots (root nodules) which fix atmospheric nitrogen (convert atmospheric nitrogen into nitrates).

(d) The bacteria/microbes which bring about decay are aerobic and will break the compost down only when oxygen is present. (**Note**: if the compost heap is too tightly packed and oxygen cannot get into it, decay cannot take place and silage will be formed.)

3 (a) N nitrogen, P phosphorus, K potassium.

(b) They are major elements needed by plants/crops for successful growth and development; they are used for fertilising the soil.

Chapter 16

1 The increase in mass of the tree was greater than the decrease in mass of the soil. Therefore the tree had gained mass from a source other than the soil. We now know that it had taken in carbon dioxide from the air and built it up into complex carbon compounds. Some of these compounds had become incorporated into the structure of the tree.

2 (a) 1: light, because increasing the light intensity increases the rate of photosynthesis.
2: carbon dioxide, because increasing the light intensity has no effect on the rate of photosynthesis unless the concentration of carbon dioxide is raised.
3: light, because increasing the light intensity increases the rate of photosynthesis.

(b) Carbon dioxide or temperature.

(c) (i) Middle part of the day.
(ii) Night.
(iii) Dawn and dusk.
(iv) Dawn and dusk.
(v) Middle part of the day.

(d) When the weather is exceptionally cold.

3 Place a short length of the pondweed in a beaker of water. Illuminate it to ensure that it gives off a steady stream of bubbles in the light. Then illuminate it with different coloured lights (i.e. wavelengths) in turn (red, orange, yellow, etc.). Find out in which colours the pondweed gives off bubbles.

This method assumes that the giving off of bubbles is an indication that photosynthesis is taking place.

4 (a) (i) 410 p.p.m.
(ii) 0800 and 1900 hours.

(b) (i) It stays constant at 410 p.p.m. except in the last 15 minutes or so when it begins to fall.
(ii) It falls from just below 410 p.p.m to 275 p.p.m.

(c) Carbon dioxide is used by the grass for photosynthesis during daylight hours. As the light gets more intense, the rate of photosynthesis increases, so more and more carbon dioxide gets used up.

Chapter 17

1 (a) The upper side of the leaf is exposed to the sun, and water would evaporate quickly from it. To prevent this the number of stomata on the upper side is reduced.

(b) The water content is lowered/reduced.

(c) Soil water content, atmospheric pressure. (**Note**: low atmospheric pressure increases the rate of transpiration; for this reason plants living at high altitudes have adaptations for reducing water loss.)

(d) Summer.

(e) To reduce transpiration (water loss) from the side of the leaf which is exposed to the sun.

2 (a) (i) Reduces the surface area across which transpiration (water loss) may take place.
(ii) The hairs trap humid air and hold it close to the leaf, thus reducing the rate of transpiration.
(iii) The pits trap humid air and prevent it being blown away from the leaf; this reduces the rate of transpiration.

(b) (i) Cactus.
(ii) The spines prevent herbivores eating the stems.

3 (a) (i) Heartwood: towards the centre of the trunk. Sapwood: towards the outside of the trunk.
(ii) Heartwood is denser, harder and drier than sapwood.

(b) (i) Allow it to dry so that it does not warp later, and treat it against attack by fungi and insects.
(ii) It should be strong and springy.

(c) (i) Conifers grow more quickly than oak trees.
(ii) (1) Good drainage from sloping land; (2) sloping land unsuitable for other crops because of the difficulty in harvesting.

4 (a) Transpiration is the evaporation of water from above-ground parts of a plant.

(b) (i) As the morning progresses the sun's heat increases, raising the temperature and speeding up transpiration.
(ii) As the afternoon progresses the sun's heat decreases, lowering the temperature and slowing down transpiration.

(c) The stem and/or leaves may be drooping. More water has been lost from the leaves than has been taken up by the roots, so the cells have lost their turgidity (turgor) and the plant has wilted.

(d) Measurements of transpiration were not made often enough, so there are insufficient points on the graph. The data would be improved by obtaining measurements at 0930 and 1130. As it is, we cannot be certain about the rate of transpiration at these times.

5 (a) The curves for the absorption of bromine ions and the rate of respiration follow each other closely/are very similar.

(b) Two experiments are possible:
(1) Find the effect of increasing the temperature on the absorption of bromine ions and the rate of respiration. If temperature has the same effect on both processes, it is likely that the two are connected.
(2) Treat the roots of the plant with a poison which prevents (inhibits) respiration, and see if this also prevents (inhibits) the absorption of bromine ions.

(c) Absorption of ions by roots takes place by active transport which requires energy from respiration. (Active transport is the movement of particles against a concentration gradient.)

(d) See table 15.1, page 71.

Chapter 18

1 (a) Liver.

(b) Islets of Langerhans in pancreas.

(c) Sweat glands in dermis of skin.

(d) Liver.

2 (a) The erector muscle would be shorter and fatter, having contracted, and the hair would be more vertical. With the hairs in a vertical position, air gets trapped between them. The layer of air helps to insulate the body against heat loss in cold weather.

(b) (i) The skin is wet because sweat has been secreted (produced) by the sweat glands and has passed along the sweat ducts onto the surface of the skin. The sweat evaporates, cooling the skin and the blood which flows through it. The cooled blood then circulates, helping to cool the rest of the body.
(ii) The face is hot because blood, warmed by the muscles during the exercise, flows close to the surface of the skin. This is achieved by dilation (widening) of the vessels serving the surface layers of the skin. Heat is lost from the blood by radiation and/or conduction. The cooled blood then circulates, helping to cool the rest of the body.

(c) The sebaceous glands secrete oil onto the surface of the skin. The oil keeps the skin moist and the hair supple, preventing drying. Dry skin tends to crack, allowing entry of germs (bacteria, etc.) which may cause infection. Also the oil may kill certain harmful germs.

3 (a) (i) A hepatic vein
B hepatic artery
C hepatic portal vein
D bile duct
E pancreatic duct
(ii) B (hepatic artery).

(b) (i) Insulin.
(ii) It is carried in the bloodstream.
(iii) It causes the liver to metabolise more sugar and/or store sugar as glycogen, thereby lowering the concentration of blood sugar.
(iv) Diabetes (diabetes melitus), excess sugar in the blood.

(**Note**: Glucagon would also be a correct answer to this question. It, too, is carried in the bloodstream to the liver. However, its effect is the reverse of that of insulin: it causes the liver to metabolise and/or store less sugar, thereby raising the concentration of blood sugar.)

4 (a) 1 cortex
2 medulla
3 ureter

(b) (i) $\frac{0.12}{1.2} \times 100 = 10\,\%$
(ii) 2 (cortex)
(iii) It is reabsorbed from the kidney tubules back into the blood.

(c) (i) No fluid lost by sweating; blood too dilute (watery).
(ii) Intake of water and salt must have been the same each day. The same total amount of salt would therefore have been lost each day via the kidney (osmo-regulation) and sweat, thus ensuring that the concentration of the blood and tissue fluid was kept constant (homeostasis).
(iii) It is always necessary to get rid of urea (nitrogenous waste) and other toxic substances.
(iv) Drink water and eat salt. (**Note**: drinking compensates for the loss of water in sweat; however, this dilutes the blood and tissue fluids so salt must be eaten to keep up the concentration.)

5 (a) Osmosis, the one-way net movement of water molecules across a partially permeable (selectively permeable) membrane.

(b) (i) Urea.
(ii) Diffusion.

(c) To increase the surface area over which osmosis and diffusion can occur.

(d) A kidney transplant is more economic because, once the operation has been carried out and the patient leaves hospital, there is little or no further expense. Treatment with an artificial kidney has to take place regularly, the machine is expensive and its continued maintenance means further expense. For the patient the kidney transplant is more satisfactory because it provides a permanent solution to his/her kidney problem, and it avoids the inconvenience and discomfort of having to be attached to a machine for hours on end at regular intervals.
(**Note**: the question does not ask for disadvantages of a transplant, but one obvious disadvantage is the possibility that the kidney transplant may be rejected by the body's immune system.)

Chapter 19

1 (a) D, E, B, C, A

(b) Receptor: a structure which, when stimulated, sends off impulses in a nerve/nerve fibre. Examples: eye, ear; touch, temperature, taste and smell receptors.

Effector: a structure which, when it receives nerve impulses, responds in some way. Examples: muscles, glands.

2 (a) A synapse is a connection between one nerve cell and the next, or between a nerve cell and a muscle.

(b) By means of a chemical transmitter which diffuses across the synapse and starts up an impulse on the other side.

(c) The chemical transmitter can only be produced on one side of the synapse, and it can only start up an impulse on the other side.

(d) The chemical transmitter is easily destroyed/inhibited/neutralised by other chemical substances, e.g. drugs.

3 (a) (i) A synapse
B cell body
C nucleus
D axon/axoplasm
E myelin sheath
(ii) A is located in the grey matter of the spinal cord. B is located in the dorsal root ganglion.
(iii) To protect (insulate) the axon/axoplasm; to speed up the rate at which impulses travel along the axon.

(b) Approximately one hundredth of a second (0.01 s). The impulses have to travel from the tip of the finger, along the nerve in the arm to the spinal cord, then back along the nerve in the arm to the muscle which bends the arm (biceps). The total distance is approximately 1 m. We are told that impulses travel in the nerve at 100 m/s, so the time taken from an impulse to travel through the reflex is

$$\frac{1}{100} = 0.01\,\text{s}$$

(**Note**: the exact answer will depend on the length of the arm and on where you assume that the nerve is connected to the muscle. The important thing is not so much getting the right answer as explaining your method.)

4 (a) The choice chamber should have a perforated floor. Water should be placed under the damp compartments, and a drying agent under the dry compartments. Suitable drying agents are anhydrous calcium chloride or silica gel. The choice chamber should have a lid so as to keep the damp/dry conditions inside.

(b) Woodlice prefer damp to dry conditions.
Woodlice prefer dark to light conditions.

(c) Repeat the experiment many times and/or use a larger number of woodlice.

Chapter 20

1 (a) Retina.

(b) Lens and cornea.

(c) Iris and ciliary body.

(d) Ciliary body.

(e) Optic nerve.

2 (a) At first the touch receptors in the skin send off impulses to the brain, but after a time they stop sending off impulses (sensory adaptation).

(b) The bright light has broken down the pigment (visual purple) in the rods. The rods will not start functioning in the dark room until the pigment has been regenerated (re-made).

(c) When you look straight at an object, the light rays are brought to a focus on the centre of the retina (fovea); this part of the retina contains densely packed cones and is responsible for precision vision.

(d) In dim light the cones do not function, only the rods. The cones give colour vision, the rods black-and-white, so in dim light colours cannot be seen.

3 (a) A change from dim light to brighter light.

(b) The eye/retina detects the increase in light intensity, impulses travel via the optic nerve to the brain and from the brain to the iris muscles. The circular muscle contracts and the radial muscle relaxes — this constricts the iris and the pupil gets smaller.

(c) The sclerotic (also called the sclera) protects the eyeball from damage.

(d) The choroid. It absorbs the light, preventing it being reflected within the eye, and it nourishes the retina.

4 (a) The graph tells us that cones are sparse towards the edge of the retina but plentiful in the centre, and that rods are plentiful towards the edge of the retina but absent (or sparse) at the centre.

(b) The blind spot (point of attachment of optic nerve).

(c) The numbers of rods and cones are not given on the vertical axis. As a result it is impossible to tell if there are any rods at the centre of the retina. The graph would be improved by giving the numbers and drawing the curves on a grid/graph paper.

Chapter 21

1 (a) Female sex hormone/oestrogen.

(b) Male sex hormone/androgen.

(c) Thyroxine or adrenaline.

(d) Adrenaline.

(e) Insulin.

2 (a) Hormone (chemical messenger).

(b) Bloodstream.

3 (a) (i) Amylase, lipase, trypsin (all enzymes).
(ii) Pancreatic duct and then the small intestine (duodenum).

(b) (i) Islets of Langerhans.
(ii) Insulin or glucagon.

(c) It would no longer be possible to control the concentration of blood sugar.

4 (a) (i) Adrenaline.
(ii) Adrenal glands, close to the kidneys.

(b) (i) Blood flow through the arteries increases, so oxygen reaches the muscles more quickly.
(ii) Blood flow through these arteries decreases, so blood can be diverted to the muscles which have a greater need for it.
(iii) Muscle/muscle fibres.

Chapter 22

1 (a) In pairs: shoulder blades, collar bones, ribs, humerus and any other arm or hand bones, pelvis, femur and any other leg or foot bones.
Singly: skull (cranium), breastbone (sternum), vertebrae.

(b) Between successive vertebrae and at joints. In both these situations the function of the cartilage is to form a springy cushion between the bones and prevent jarring.

2 Rounded head on right for smooth articulation at ball and socket joint.
Flat process (projection) on left for attachment of muscle/tendon.
Dense (solid) bone in shaft for strength.
Spongy bone (lattice) in head to take strains in different directions.
Cavity (hollow) in centre for lightness.

3 (a) (i) The inner lining of the capsule secretes (produces) synovial fluid. By surrounding (enveloping) the joint it prevents germs from entering and synovial fluid from leaking out.
(ii) The ligaments hold the two bones together.

(b) The ligaments are tough (strong) and elastic. Their toughness (strength) prevents bones being pulled apart at the joints (dislocated) even when subjected to strain. Their elasticity allows them to be stretched when the bones are moved by contraction of muscles. (**Note**: when stretched, considerable energy can be stored in ligaments and this plays an important part in locomotion, particularly jumping.)

(c) The tendon of the biceps muscle. (**Note**: it would be tempting to just write 'muscle'; however, the question asks you to be as precise as possible.)

4 (a) Antagonistic muscles are muscles which, when they contract, produce opposite effects. When one muscle contracts, its antagonist relaxes. When muscle A contracts, muscle D relaxes and the upper leg bone is pulled backwards; when muscle D contracts, muscle A relaxes and the upper leg bone is pulled forwards. When muscle B contracts, muscle E relaxes and the lower leg bone is pulled forwards; when muscle E contracts, muscle B relaxes and the lower leg bone is pulled backwards. When muscle C contracts, muscle F relaxes and the foot is pulled backwards; when muscle F contracts, muscle C relaxes and the foot is pulled forwards.

(b) Muscle C has to propel the body forward over the ground. It therefore has to take much more strain than muscle F which simply has to return the foot to its original position after it has been pulled backwards by muscle C.

5 (a) (i) The body becomes long and thin.
(ii) The body becomes short and fat.

(b) (i) The skeleton is the fluid in the body cavity.
(ii) The muscles contract against the fluid skeleton, changing the shape of the worm. First the front end becomes long and thin, forcing itself into the soil. Then the front end becomes short and fat, anchoring the front end in the soil. Then the rest of the worm becomes short and fat and the back end is drawn up towards the front end. The process is then repeated.

Chapter 23

1 (a) (i) A gamete is a cell which cannot develop any further until/unless it fuses (unites) with another gamete. Eggs and sperms (spermatozoa) are gametes.
(ii) Fertilisation is the fusion of a sperm with an egg.

(b) Gametes may be released at a particular time of the year, e.g. spring when the daily hours of light ('day-length') reaches a certain level, or when the moon is at a particular phase. The development (maturation) of one sex may influence the development of the other sex, so the gametes are released together.

(c) One sex may attract the other by releasing a chemical substance into the surrounding water.

(d) (1) The egg is much larger than the sperm. (2) The egg is a simple spherical shape whereas the sperm is more elaborate with head, middle piece and tail. (3)

The egg is surrounded by a jelly coat which is absent in the sperm. (4) The chromosomes in the egg nucleus are more spread out than the chromosomes in the sperm nucleus. (5) The egg cannot move itself whereas the sperm can swim by waving its tail from side to side.

2 (a) In spring/summer there is plenty of food (water fleas, etc.) for the newly formed offspring resulting from asexual reproduction. In autumn/winter there is little or no food available; the zygospore is resistant to adverse/wintry conditions and remains dormant until the following spring.

(b) (i) Several offspring may be formed from a single parent.
It does not necessitate two partners.
The chancy nature of fertilisation is avoided.
Offspring are produced relatively quickly.

(ii) It produces genetic variety in the offspring.
Harmful genes in the parents will not necessarily be handed on to the offspring.
It helps the species to survive the winter.

(c) Reason why it should be regarded as a method of reproduction: two or more individuals can be formed from one original.

Reason why it should not be regarded as a method of reproduction: the *Hydra* does not break up into pieces of its own accord; it is therefore not a natural process.

3 (a) They can be blown about/dispersed by the wind.

(b) The stalk raises the spore capsule well above the ground. The stalk is flexible and allows the capsule to be shaken by the wind.

(c) In dry conditions the spores are less likely to stick together and can be dispersed over a wider area.

(d) Before they can germinate the spores must absorb water so as to burst open.

(e) Splitting has the effect of jerking the capsule, with the result that the spores are thrown out.

Chapter 24

1 (a) Testes.

(b) Ovaries.

(c) Urethra.

(d) Funnel of oviduct/Fallopian tube.

(e) Uterus.

2 (a) In the male the pituitary gland produces (secretes) gonad-stimulating hormones which stimulate the testes to produce (secrete) male sex hormones (androgens). The male sex hormones stimulate the testes to start producing sperms and cause secondary sexual characteristics to develop, such as growth of body hair and breaking of the voice.

(b) In the female the pituitary gland produces (secretes) gonad-stimulating hormones which stimulate the ovaries to produce (secrete) female sex hormones (oestrogens). The female sex hormones stimulate the ovaries to start producing eggs and cause secondary sexual characteristics to develop such as growth of the breasts and laying down of fat in the thighs.

3 (a) Condom, cap, male and female sterilisation.

(b) Oral contraceptive, injectable contraceptive.

(c) IUD.

(d) Spermicide, oral contraceptive, injectable contraceptive.

(e) Male and female sterilisation.

4 (a) A umbilical cord
B placenta

(b) C carries mother's blood from placenta.
D carries mother's blood to placenta.
E carries foetal blood to placenta.
F carries foetal blood from placenta.

(c) Oxygen, glucose, fats, amino acids, vitamins, salts, antibodies.

(d) Carbon dioxide passes from the foetal capillaries to the maternal blood space, from which it is taken in the mother's bloodstream to her lungs and breathed out/exhaled/expired.

Urea (nitrogenous waste) passes from the foetal capillaries to the maternal blood space from which it is taken in the mother's bloodstream to her kidneys and excreted in her urine.

(e) The villi increase the surface area across which exchange of chemicals can take place between foetal and maternal blood.

5 (a) Sperm cannot reach the egg to fertilise it, and even if a sperm did fertilise the egg, the fertilised egg/embryo would be unable to pass down the Fallopian tube to the uterus and continue its development.

(b) (i) Usually only one egg matures per month and this would be too difficult to find, so hormone treatment increases the chance of finding eggs. It also means that more eggs can be fertilised, which increases the chance of one of them developing successfully.

(ii) The eggs are near the surface of the ovary where they are easy to get at/remove. Also they are ready to be fertilised. The doctor would do this in the middle of the cycle, i.e. mid-way between one menstrual period and the next (12th to 14th day).

(iii) The culture solution should contain soluble food substances (glucose, amino acids, etc.) and the right concentration of salts. It should be aerated so that it contains oxygen, and kept in sterile conditions at normal body temperature (37°C).
(iv) Fusion of the sperm with the egg to form a zygote/fertilised egg.
(v) It takes three days for the lining of the uterus to be ready for receiving the embryo, and for the embryo to be sufficiently developed for implantation. (The embryo will not implant until it consists of a ball of cells.) Moreover, waiting three days gives the doctors time to see if fertilisation has been successful and if the embryo is developing satisfactorily.
(vi) The cervix is the natural opening and leads straight into the uterus; no surgery is required and no damage is done to the body wall or uterus. However, the eggs have to be obtained through the body wall because they cannot be reached by the natural opening.

(c) The term 'test-tube baby' is not a good one: the eggs/embryos are placed not in a test tube but in a flat dish in which they can be observed under a microscope; also the dish contains a very young embryo (ball of cells), not a baby. The only merit of the term is that it does imply the idea of something being done in a laboratory.

(**Note**: this question does not require much knowledge as most of the relevant information is given in the passage; however, it does require carefully written answers which cover all the relevant points — see advice on answering examination questions on pages 136–41).

Chapter 25

1 (a) Male
(b) Neither
(c) Neither
(d) Neither
(e) Female

2 (a) (i) Colour attracts insects and/or other pollinating agents.
(ii) Nectar serves as a bait for insects.
(iii) Spikes enable the pollen to cling to the hairs on insect's body.

(b) (i) Stamens shaken by wind/air currents, scattering pollen.
(ii) Feathery structure has large surface area for catching pollen.
(iii) Small size makes pollen grains light so that they are more easily carried by wind and air currents.

3 (a) Coloured and/or shiny so as to be easily seen (conspicuous).

(b) Contains air, making it capable of floating (buoyant) like a boat.

(c) Has extensions/wings/hairs which enable it to be carried by the wind like a glider or parachute.

4 (a) (i) Ovule.
(ii) The point of attachment of the stalk which connected the ovule with the wall of the ovary.

(b) (i)

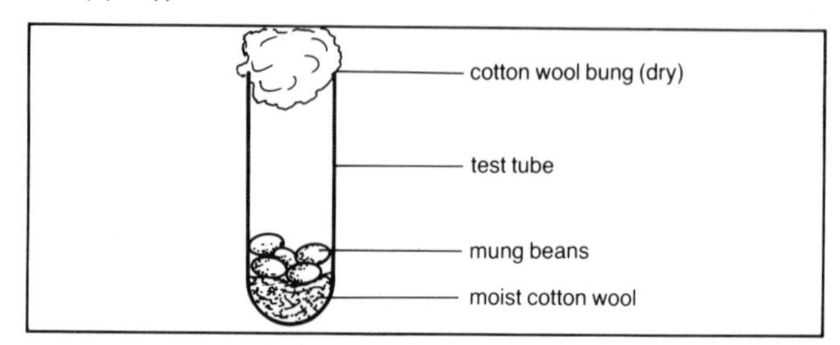

(ii) Soak the seeds. Place a minimum of five seeds in each test tube. Place each test tube in a constant temperature water bath at a different temperature (e.g. 5, 10, 15, 20, 25, 30, 35, 40, 45°C). All other conditions must be the same in all the test tubes. Measure the time it takes for the first seed to germinate or for 50% of the seeds to germinate, or count the number of seeds which have germinated after an appropriate time (say 5 days). Take the splitting of the seed coat or emergence of the radicle as an indication that germination has taken place.
(iii) Difficulties might include keeping the temperatures constant, checking the progress of the experiment sufficiently often, and telling exactly when germination has taken place.

5 (a)

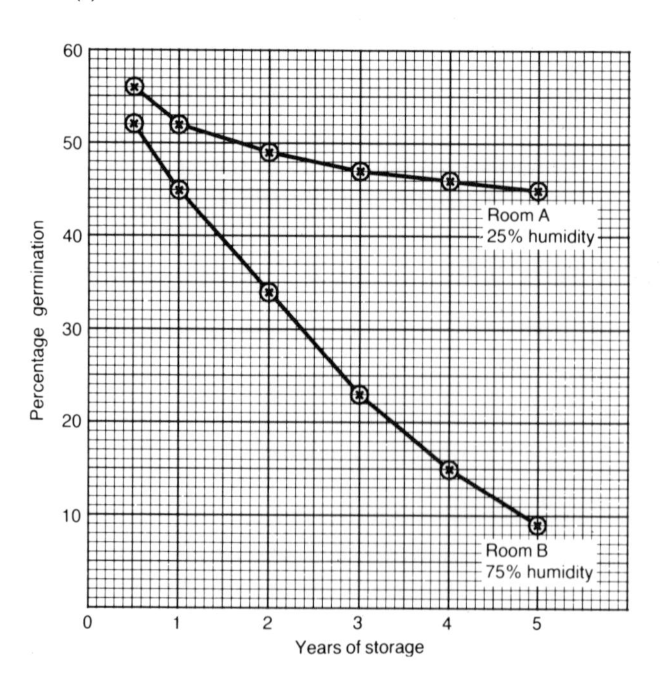

(b) (i) 19 %
(ii) 1.6 years.

(c) Room A had a lower humidity and was therefore dryer; seeds survive for longer in dry conditions.

(d) It would be best to pack the seeds in containers containing a drying agent, thereby reducing the moisture content/humidity in the containers to a minimum.

Chapter 26

1 (a) Increase in dry mass is due solely to the addition of new organic material and does not include the absorption (or loss) of water.

(b) There is little or no oxygen in waterlogged soil and so the roots cannot respire aerobically and obtain sufficient energy for good growth.

(c) The tip of the shoot produces auxin which prevents the lateral buds from developing into side branches. Cutting off the apex removes the source of auxin, and so side branches develop.

(d) Growth is always slowed down on the side of the shoot to which auxin sinks under gravity. This means that, whatever way up the seed is planted, the shoot will always grow upwards.

(e) Absorption of water by the seed enables the tissue inside to expand, splitting (rupturing) the seed coat and allowing the root (radicle) to emerge. Water is also needed for movement of food reserves and for growth of the shoot and root.

2 (a) Swallowing (intake of water) by the insect.

(b) New material (chemical substances), derived from the insect's food, is added to the existing tissues.

(c) (i) The graph/line to the immediate right of each arrow would be horizontal rather than sloping.
(ii) After the insect has expanded its body by swallowing air, the new cuticle hardens. Once the new cuticle has hardened further expansion is impossible, though the body mass can still increase by the addition of internal structures.

(d) The graph/line would be a smooth curve rather than a series of steps.

3 (a) Approximately 8 cm, 14 cm and 9 cm.

(b) Better diet.

(c) The growth rate is particularly high in adolescence and the food requirement is correspondingly high. More food would have been available in 1958 than in 1874.

(d) Genetics (inheritance), more healthy living, exercise.

4 (a) Growth takes place towards the tip.

(b) 87–88 %.

(c) Growth takes place by cell division at the tip of the root, followed by cell expansion just behind the tip.

(d) Cotyledons.

(e) (i) The root anchors the seedling before the shoot starts pushing upwards through the soil.
(ii) Being bent back, the delicate tip of the shoot is protected from being scratched/damaged by the soil particles as it grows upwards.

5 (a) Yes.

(b) A shoot should be covered with a transparent cap and lit from one side. This is necessary because failure to bend towards light might be due to contact of the cap with the shoot.

(c) Repeat the experiment with the tip of the shoot exposed to the light and the part just behind the tip covered with a foil sleeve. If the hypothesis is correct the shoot should bend towards the light.

(d) The light causes auxin, produced by the tip, to accumulate (gather) on the darker side of the shoot, with the result that the shoot grows faster on that side.

Chapter 27

1 (a) A gene is a structure within a chromosome which determines an individual's characteristics and by which those characteristics are handed on to the offspring.

(b) A chromosome is a thread-like body which occurs in the nucleus of a cell and which contains (carries) genes.

(c) Homologous chromosomes look exactly alike and contain genes which control the same characteristics.

2 (a) In the formation of gametes (eggs and sperms).

(b) Meiosis ensures that the number of chromosomes remains the same (i.e. 46) in each generation, in other words that the number is the same in the offspring as in the parents. (**Note**: if this was not so, the number of chromosomes would double in every generation — what an idea!)

3 (a) Cell B. You can tell from the fact that there are two pairs of chromosomes, i.e. each chromosome has an identical partner.

(b) (i) Cell A.
(ii) Haploid means that the cell contains only half the full number of chromosomes, i.e. only one of each homologous pair.

(c) Cell B. A gamete would be haploid, like cell A; cell B, however, is diploid.

(d) Chromosomes hold (carry) the genes in an orderly/organised manner within the cell, transmit/convey the genes from parents to offspring, and bring genes together in different combinations thus producing genetic variety.

4 (a) 46

(b) 23

(c) One of the pairs of homologous chromosomes (the bottom right hand pair) differ in size. The larger one is the X chromosome, the smaller one is the Y chromosome. Males possess an X and Y chromosome, females possess two X chromosomes.

(d) One of these chromosomes (there is no way of telling which one) came from the mother, the other one from the father.

5 (a) DNA, deoxyribonucleic acid, is the chemical substance of which genes are made. It occurs in the chromosomes inside the nucleus.

(b) Genetic engineering is the process, carried out by humans, in which an organism's genes (genetic constitution/genetic make-up) are altered so that its characteristics are changed.

(c) A human.

(d) A bacterium.

(e) Multiply (divide repeatedly).

(f) Insulin.

Chapter 28

1 (a) (i) The genotype of a white-flowered plant is rr. It can have no other genotype because r is recessive and a white-flowered plant cannot possess an R allele.
(ii) By self-pollinating it. If the red-flowered plant has the genotype RR all the offspring will be red-flowered. If its genotype is Rr, some of the offspring will be red-flowered and some will be white-flowered. (Note the reason: it is like the second generation of offspring in figure 2.)
(iii) Pink.

(b) BB and dd. (**Note:** only homozygous individuals show pure-breeding/breed true.)

2 (a) (i) Black coat colour: B.
(ii) Red coat colour: b.

(b) (i) 1, 2 and 8.
(ii) 3 and 4.

(c) Black coat colour is dominant, red coat colour is recessive. Black-coated individuals all have the same colour (phenotype) but their genotype may be BB (homozygous) or Bb (heterozygous) because only one B allele needs to present for the black colour to be produced.

3 (a) (i) Red.
(ii) The cross between plants A and C (red flower × yellow flower) produced only red-flowered offspring.

(b) (i) Either self-pollinate plant A or cross it with another pure-breeding red-flowered plant (i.e. one which gives only red flowers when crossed with C.)
(i) Either self-pollinate plant C or cross any two yellow-flowered plants.

4 (a) Individuals who are not albinos.

(b) Recessive. Albinism appears in the offspring of parents neither of whom are albinos (both are phenotypically normal).

(c) Let **a** represent the allele for albinism and let **A** represent the normal allele.
1 **Aa**
2 **Aa**
3 **AA** or **Aa**
4 **aa**
5 **Aa**
6 **Aa**
7 **aa**
8 **Aa**

(d) One half (50%). (Be sure you understand the reason. Half the sperms produced by individual 8 will carry the **A** allele, and half will carry the **a** allele. All the eggs produced by the albino will carry the **a** allele. So, when mating occurs, there will be an equal chance of an egg being fertilised by an **A** sperm or an **a** sperm.)

5 (a)

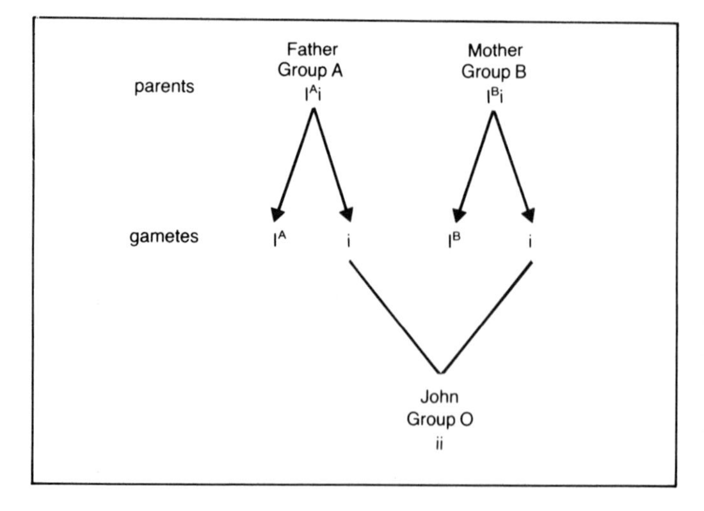

(b) John's brothers and sisters might belong to groups A, B, AB or O.

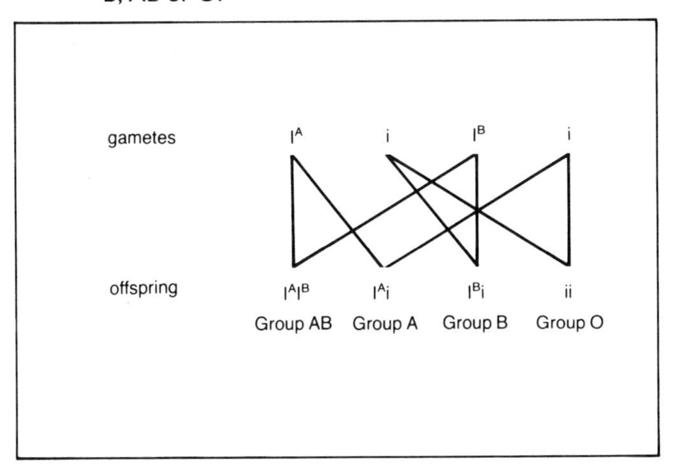

Chapter 29

1 (a) Continuous variation shows a smooth gradation between individuals. Discontinuous variation shows a sharp distinction between individuals.

(b) Chromosome mutation is a change in the number or size of chromosomes in a cell. Gene mutation is a change in the chemical structure of a gene i.e. the order of bases in the DNA within a gene.

(c) Natural selection is the process by which individuals which are well adapted to their environment survive, whereas poorly adapted individuals die. Artificial selection is the process by which individuals with desirable characteristics are chosen by humans and allowed to breed, whereas other individuals are prevented from breeding.

(d) In-breeding is crossing (mating between) closely related individuals. Out-breeding is crossing (mating between) unrelated individuals.

2 (a) Competition arises when organisms share/use/require the same commodity and there is not enough to satisfy the needs of all the individuals. For example, there may be insufficient plants in a certain area to produce food for all the rabbits that live there. As a result, the rabbits fight/bully/chase each other away so as to get sufficient food for survival.

(b) Predation, susceptibility to disease.

3 (a) Any of the three examples described on page 134 will do. Other examples: rats have become resistant to the rat poison warfarin and mosquitoes to DDT, and certain grasses have become tolerant to heavy metals and can grow successfully on the spoil from, e.g. lead mines.

(b) Mutations are persistent: once a mutation has occurred it may remain in a population indefinitely. 'Best' in this context means discontinuous variation of the kind which carries advantages in the natural environment.

4 (a) (i)

Height (cm)	138–139	140–141	142–143	144–145	146–147	148–149	150–151	152–153	154–155	156–157	158–159	160–161
Number of pupils	1	2	0	3	4	3	7	4	3	1	1	1

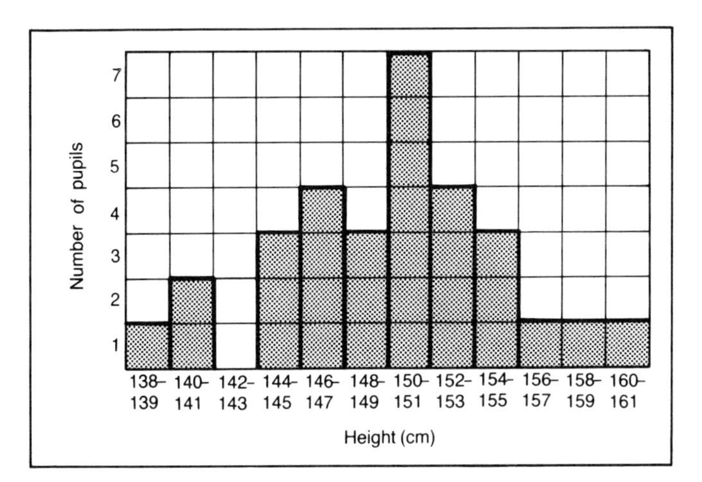

(b) Continuous variation.

(c) The graph is almost a normal distribution but (1) it is slightly asymmetrical, the highest point being slightly to the right of centre; (2) there are no students in the 142–143 cm height group; (3) there are only three students in the 148–149 cm height group.

5 (a) (i) 'some large, some small, some strong, some weak'.

(ii) 'only the most persistent wildebeest could take advantage of the occasional water holes'.

(iii) 'intent on catching the slowest ones' and 'only the most persistent could take advantage of the occasional water holes'.

(b) Natural selection weeds out the weakest/slowest/poorly adapted wildebeest which are caught/eaten by the lions. The others survive and hand on beneficial genes to their offspring. In this way, over many generations, wildebeest as a species become better adapted to the environment/improve/may change to a new species.

Answers to Specimen GCSE Questions

Note. These answers are the author's own suggestions, not those of the examining groups. They are only suggested answers: other correct versions may be possible.

1 (a) A shoulder blade/shoulder girdle/pectoral girdle
B humerus
C triceps
D ulna

(b) The biceps contracts and the triceps relaxes. Because of the way the muscles are attached to the bones, the lower arm (radius and ulna) is raised.

(c) Tendon.

2 (a) Producer: Eucalyptus
Herbivore: tortoise beetle
Carnivore: ladybird

(b) Biological control: the introduction of one organism to control (reduce the numbers of) another organism.
Parasite: an organism which feeds on, and thereby harms, another living organism.
Insecticide: a chemical substance used to kill insects.

(c) The hard exoskeleton (cuticle) protects the beetle from the insecticide.

(d) In Australia other predators keep its numbers under control.

(e) Make sure they feed only on the Eucalyptus beetle.

3 (a) (i) Removal of the leaves and buds may stimulate root growth, reduce transpiration during the period when there are no roots to absorb water from the soil. Also they might decay, and the decay might spread to the stem.
(ii) There would be more air (and therefore oxygen) in coarse gritty soil, and larger spaces for the new roots to grow through.
(iii) This would fix the cutting firmly in the pot and prevent it rocking about, and it would provide closer contact between the developing roots and the water in the soil.

(b) (i) There would be no water loss (transpiration) from leaves and less resistance to wind.
(ii) There would be no food from photosynthesis to nourish the growing roots, and microorganisms might enter the stem and cause disease or decay.

(c) (i) Dip the cut end of the stem in rooting powder before placing it in the soil. Cover the cutting with polythene and/or spray with a fine mist of water. Spray with fungicide.
(ii) Rooting powder contains growth substance/auxin which stimulates root growth. Polythene cover and/or mist spray increases humidity, thereby reducing water loss (transpiration) and promotes root growth. Fungicide kills fungi, preventing disease/decay.

(d) (i) Take an equal number of cuttings from each plant. Grow them under the same conditions of light, temperature and humidity, etc. Measure the heights and numbers of side branches and work out the average for each group of plants.
(ii) A difference in height and number of side branches between the two groups of cuttings, similar to the differences between the two parent plants.
(iii) No difference in height and number of side branches between the two groups of cuttings.

4 (a) (i) Mites.
(ii) Springtails.

(b)

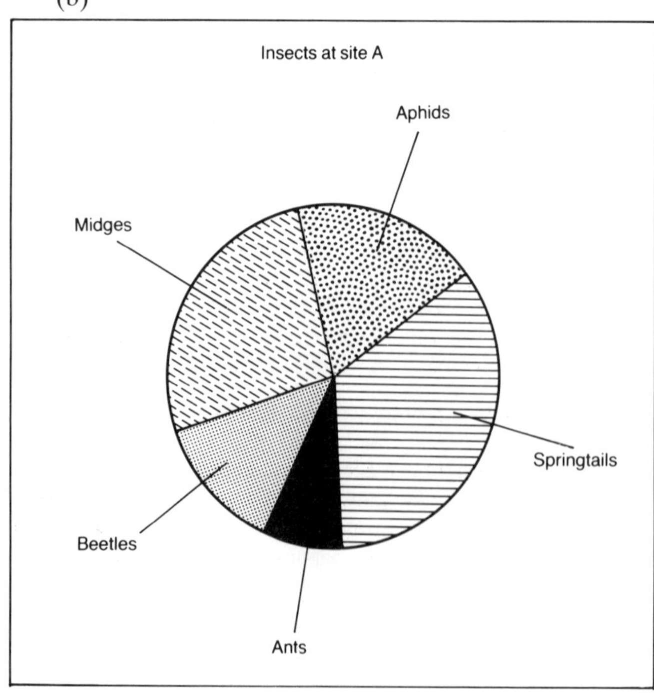

(c) Centipede. (Note the reason: they are the smallest number.)

(d) Site A has more food for snails, more shelter, fewer predators.

5 (a) Between 6 and 8 days.

(b) 3.5°C.

(c) 10 days.

6 (a) Complementary air is the maximum volume of air which can be inspired (breathed in) after a normal respiration.

(b) Residual air is the volume of air left in the lungs after a maximum expiration (i.e. after as much air as possible has been breathed out).

(c) The lungs contain their normal capacity when they contain air at the resting level, i.e. after a normal expiration.

7 The function of the placenta is to exchange materials between the mother's blood and the embryo's blood. Glucose and vitamins are two of the foods that pass from the mother, via the umbilical cord, to the embryo. Urea and carbon dioxide are waste materials that pass from the embryo to the mother, again via the umbilical cord. The embryo is protected by the thick wall of the uterus and the amniotic fluid. (**Note**: one mark is given for each of the words in the list, correctly used.)

8 Your answer should include a definition of pollution, where it comes from (i.e. its source), what part of the environment it gets into (e.g. air, water, soil) and what its effects are. There is no need to give a long list of pollutants (like textbooks do!), just several well chosen examples to illustrate the principles. Something like this:

Pollution is any process carried out by humans which adds a harmful agent (pollutant) to the environment. The main types of pollution are chemical, thermal and radioactive, and they may affect the air, water (lakes, river, sea) or land, particularly the soil.

An example of chemical pollution is the insecticide DDT. This may build up in food chains and poison top carnivores such as birds of prey. Chemical pollution also occurs when fertilisers run off the land into lakes and rivers. The extra nutrient (nitrates, etc.) speed up the growth of plants and this leads to an increase in the number of animals. When the plants and animals die they decompose. The decomposers use up all the oxygen, making it impossible for animals and plants to live there at all. This is called eutrophication.

Radioactive pollution may result from an accident at a nuclear power station or as a result of the testing of nuclear bombs. A high level of radioactivity in the environment can cause an increase in the number of cases of cancer, particularly cancer of the blood.

(**Note**: Other examples could be given. However, be careful that you don't spend so long on one example that you don't have time or space to cover all three types of pollution.)

9 The answer to this question involves giving a sequence of events, starting with cellulose in dead leaves and finishing up with new growth. Marks are given for each step in the sequence, so you must be sure to cover all the steps:

The elements in the carbohydrate cellulose are carbon (C), hydrogen (H) and oxygen (O). When leaves decay/decompose they are digested by bacteria and fungi in the soil. Digestion of the cellulose in the leaves produces soluble glucose which contains C, H and O. The glucose is absorbed by the bacteria and fungi and used in respiration, producing carbon dioxide gas (CO_2) and water (H_2O). The water is released into the soil and can be absorbed by the roots of trees. The carbon dioxide passes into the atmosphere/air and can subsequently be used by the leaves for photosynthesis. In this process carbon dioxide and water are built up into sugars and other substances which provide energy and materials for growth of the trees.

10 The following points should be included:

Urea passes from the liver cells into the blood in which it is dissolved. It then passes to the kidneys in the circulation: hepatic vein, vena cava, heart, pulmonary circulation to lungs and back to heart, aorta, renal artery, kidney. In the kidney the urea passes in the bloodstream to the glomerulus of nephron where it is filtered into capsule (mechanism should be explained briefly). Urea, in solution, then passes down tubule (which should be described briefly) and some may be added by secretion/active transport from adjacent capillaries. Urea then enters collecting duct which leads into pelvis, then down ureter to bladder for storage.

(Credit given for a good diagram with relevant labelling and information.)

11 The framework of the answer should be the three processes by which growth and development in general take place, namely cell division, cell expansion and differentiation. Each of these processes should then be explained, with diagrams where necessary.

A brief summary of how these processes differ in animals and plants should be included. For example, cell division occurs all over the body in a developing animal, whereas in plants it is restricted to certain regions (meristems). In animals growth stops after a certain age, whereas in plants it continues throughout life. In plants cell expansion is achieved by developing a large vacuole inside the cell which takes in water by osmosis; this does not occur in animals whose cells tend to remain small and in which increase in bulk is achieved more by cell division than by cell expansion.

In both animals and plants, cells differentiate by acquiring particular shapes (e.g. nerve cell in animal, sieve tube in plant), but in plants additional specialisation is achieved by the cell wall which may, for example, become lignified.

12 D

13 B

14 A

15 C

16 A

17 B

18 D

19 A

20 A

21 A

22 C

23 A

24 (a) A
(b) D
(c) C
(d) A

25 (a) D
(b) A
(c) C
(d) A

Index